RED NOVEMBER

ALSO BY W. CRAIG REED

DNA

Tarzan, My Father (with Johnny Weissmuller Jr. and William Reed)

RED
NOVEMBER

Inside the Secret U.S.-Soviet Submarine War

W. CRAIG REED

WILLIAM MORROW *An Imprint of* HarperCollins*Publishers*

HarperCollins books may be purchased for educational, business, or sales promotional use. For information please write: Special Markets Department, HarperCollins Publishers, 10 East 53rd Street, New York, NY 10022.

FIRST EDITION

Designed by Jamie Lynn Kerner

Library of Congress Cataloging-in-Publication Data

Reed, W. Craig.
    Red November: inside the secret U.S.-Soviet submarine war / W. Craig Reed.—1st ed.
        p.   cm.
    Includes bibliographical references.
    ISBN 978-0-06-180676-6
    1. United States—Military relations—Soviet Union. 2. Soviet Union—Military relations—United States. 3. Cold War. 4. Submarine warfare—United States—History—20th century. 5. Submarine warfare—Soviet Union—History. 6. Submariners—United States—Biography. I. Title.
    E183.8.S65R435 2010
    359.9'30973—dc22

                                                                2009053269

10 11 12 13 14   WBC/RRD   10 9 8 7 6 5 4 3 2 1

This book is dedicated to my father, Lieutenant William J. Reed, Retired, who helped devise and deploy the top-secret Boresight program, and to the underwater sailors and civilians who sacrificed so much. The following pages are a tribute to the commitment, courage, and constant vigilance of those who sacrificed so much to ensure that our world did not end by way of fire and fallout.

INTRODUCTION

In wartime, truth is so precious that she should always be attended by a bodyguard of lies.
—Winston Churchill

In March 2009, beyond the frosted windows of an arthritic building in downtown Saint Petersburg, Russia, a callous wind forced its will upon millions of helpless snowflakes. Inside the hotel ballroom, hundreds of Russians ignored the weather as they laughed, danced, hugged, and drank. Vodka flowed. Music blared and platters of food beckoned. I stood at my table as a husky man with playful eyes and Santa cheeks approached. He beamed and introduced himself as Sergei. He said he once served as the commander of a Soviet submarine and told me that NATO code-named his class of boat the "Victor III." He asked if I recognized this name. I smiled and said that my submarine, the USS *Drum,* had once come too close to such a boat near Vladivostok.

Eyes wide, Sergei took two steps backward. "K-324?"

"Yes," I said. "K-324."

Sergei reached his stubby arms around my shoulders and gave me a bear hug. In my ear he said, "You should be dead."

I nodded and said nothing.

Sergei pointed at a shiny pin on my lapel.

"U.S. Navy diver," I said.

His eyes lit up again as he tapped a similar emblem on his Russian Navy uniform. He unhooked his pin and attached it to my shirt. I did the same for him. Sergei then grabbed two glasses and filled each with a shot of vodka.

He handed one to me and in broken English quoted an old Russian proverb, "After a storm there is fair weather, after sorrow there is joy."

I clicked my glass against his and downed the burning liquid. Before me, and all around me, were former enemies. Submariners who once pointed the barrels of their guns at my head, fingers poised and aims steady. Now, with the passage of time, at an annual Russian event that honors submariners, we laughed and joked about our escapades from decades past.

In Russia, submariners are revered and respected, as their profession is considered dangerous, their sacrifice worthy of praise. For this select group of volunteers, camaraderie runs as deep as their vessels. None care about nationalities, creeds, or skin color. That night, dozens of former submariners treated me as a brother among brothers. Even though we were strangers whose governments once fought as enemies, we greeted one another with firm handshakes, warm hugs, and broad smiles. I felt honored and humbled.

After dinner, a small group of submariners walked to the dance floor. Side by side they raised their glasses and voices as they sang a Russian submariner's song. Though I didn't understand the words, I felt the meaning touch the deepest part of my soul. More and more submariners joined the throng as the voices reached a crescendo. Tears filled my eyes. Words can never do justice to the feelings that overcame me when I stood alongside my brothers and toasted all submariners, especially those lost at sea who now serve "on eternal patrol."

As I left the event, I wondered if those who consider themselves enemies today could do as we had done that night. Lay down their swords and find a common bond. I realized that until that day, there could be no fair winds, and many in the world were destined to remain captured by the storms of sorrow.

If one believes the Mayans, the world will end in the year 2012. Whether by global warming, menacing asteroids, or bioterrorism, we are always on the brink of annihilation. Skeptics voice their doubts, but for those of us who served during the forty-six-year Cold War, such fears are not without merit, for never did we come closer to nuclear self-destruction than in October 1962 and again in May 1968. Conflicts involving U.S. and Soviet submarines were common factors in both.

No discernible fanfare marked the final moments of a war that cost taxpayers $8 trillion and the lives of more than 100,000 Americans—almost 87,000 of those in the conflicts with Korea and Vietnam. There were no ticker tape parades, no blowing horns, and no mothers waving flags when the Cold War finally ended. The U.S. Senate voted against the Cold War Medal Act of 2007, which would have awarded official recognition to thousands of veterans who fought secret battles around the world. Now they must remain unsung heroes.

Some carried M-16s and trudged through rice paddies. Others listened with breathless anticipation to the secrets revealed in foreign tongues captured from cable taps 700 feet deep. Still others prayed to the gods of their faith as depth charges shattered the ocean and enemy torpedoes threatened to turn their vessels into twisted metal coffins. My father and I were among these few, and this history and personal narrative are long overdue.

Most submariners, Navy SEALs, divers, and "spook" intelligence operators, sworn to secrecy, are to this day reluctant to discuss their secret Cold War operations. Many, especially those who worked in compartments outside operational areas or did not have a "need to know," were unaware of the details surrounding the missions they undertook. A few, like me, recall every second of the more eventful assignments. For the first time ever, these veterans have come forward to tell their stories, perhaps to release the secrets held captive in their minds for decades by official mandate.

In 1998 Sherry Sontag and Christopher Drew's *Blind Man's Bluff* captured public attention by revealing many of the details about these clandestine and dangerous submarine missions. Most of us submariners agree that this book delivered an informative, interesting, and reasonably accurate accounting of Cold War espionage operations. However, few submariners or operators gave the authors information about their involvement in top-secret Holystone and Ivy Bells programs. Furthermore, none discussed two other top-priority submarine projects code-named Boresight and Bulls Eye. *Red November* is the first book to take readers deep inside all four of these programs and reveal firsthand details about the harrowing events that veterans have been reluctant to discuss.

While I acknowledge that some submariners, cryptanalysts, and government operatives argue that "insider" details about these missions, which many historians believe were instrumental in ending the Cold War, should remain untold, I believe that history is robbed by this posture. What if the world never knew about the Manhattan Project? What if governments never revealed undisclosed details about the Cuban Missile Crisis? What if these once top-secret historical events remained labeled classified forever? National security demands secrecy, but at some point technological advances and world events make this stance obsolete. Many of us who served frontline in the underwater Cold War signed gag orders to maintain our silence for decades. Our duty to one another also held our tongues until the passage of time could ensure we would not violate our oaths as submariners. Now, for many of us, our days of silent running are over.

AUTHOR'S NOTE

Names used, including those for persons, boats, and ships, as well as dates, titles, event details, and geographic locations noted herein, are, for the most part, accurate. A few exceptions occur where memories are incomplete or national security concerns take precedence. No individuals depicted are composite portraits or fictionalized, but some dialogue and details have been reconstructed or paraphrased and time frames compressed.

RED NOVEMBER

CHAPTER ONE

Red sky at night, sailors delight.
Red sky in morning, sailors take warning.
—OLD SAILING PROVERB

WITH ORDERS TO CONDUCT A TOP-SECRET espionage mission, the USS *Blenny* (SS-324) sped toward danger on the last day of April 1952. A bright sun warmed the black deck of the World War II–vintage submarine as she cruised past a dozen colorful sailboats off the coast of San Diego. A cold wave crashed across the bow of the boat and dotted Lieutenant Junior Grade Paul Trejo's lips with the taste of salt. Still a freshman to the fraternity of underwater warriors, Trejo stood on the bridge and admired the beauty of his diesel-powered sub as she cantered across the water with the smooth gate of a stallion. In the white churn of her wake, dolphins played, and when guided beneath the waves, she became a silent assassin worthy of respect.

Less than two miles from the submarine base at Ballast Point, the diving alarm sounded. Trejo slid down the ladder past the conning tower—the tiny space just above the control room, then descended one more deck into the control room. There, assuming the duties of diving officer, he joined a half-dozen sailors on watch and stared at the indicator lights on the "Christmas tree" panel. When the horizontal bars changed from red to green, verifying closure of all hull openings, he held up an open hand and called out an order: "Bleed air."

A petty officer nodded and spun open a valve to bleed high-pressure air into the boat. The needle on a manometer twitched and then inched up a few millimeters. Trejo closed his hand, and the petty officer shut the valve. Both men focused on the pressure indicator. A half minute later, satisfied that no air leaked from the boat through an unwanted hole, Trejo closed his hand. "Green board, pressure in the boat," he said.

Commander James S. Bryant called down from the conning tower, "Diving Officer, make your depth six-zero feet."

Trejo repeated the order to the bow planesman seated at the front of the control room. As the chief of the watch blew air from the *Blenny*'s ballast tanks to submerge the diesel-driven craft beneath the waves, Trejo visualized a plume of seawater shooting skyward toward a blue dawn. The boat submerged and made a turn toward paradise.

After a short stop in Hawaii, the *Blenny* departed for Japan on Monday, May 13, crossed the 180th meridian—the domain of the Golden Dragon—and arrived in Yokosuka eleven days later at 0800 on May 24, 1952. Bound for enemy waters, the *Blenny* sailed away from Yokosuka on May 29. Commander Bryant gathered his officers in the wardroom as excited banter and cigarette smoke filled the air. The skipper's blue eyes and movie star persona reminded Trejo of Cary Grant. Bryant's forehead wrinkled as he pointed at their assigned station on a map spread across the wardroom table. The officers' talk faded to silence.

The Soviet city of Vladivostok had a population of almost a half million Russians in the early fifties. Nestled near the Strait of Korea to the south and Petropavlovsk Naval Base to the north, "Vlad" served as a staging area for an extensive segment of the Red Bear's Eastern Fleet, including ballistic missile submarines. The Korean War was in full swing, and the Soviets were shipping weapons to the North Koreans, so the commander, United States Naval Forces, Far East (COMNAVFE), ordered U.S. attack submarines to patrol the La Pérouse Strait near Vlad to gather as much intelligence as possible. Reconnaissance patrols in this area occurred between early March and late October, as too much ice encircled the seaport throughout the long winter months.

Trejo recalled that the USS *Besugo* (SS-321) had attempted a patrol in the strait in December 1950, but dire weather conditions made reconnaissance all but impossible. U.S. subs now patrolled only during the warm window when the Russians sent ships to Korea and other countries and bombarded Vlad with incoming cargo vessels—or "megaton squirrels," as they called them—delivering "acorns" to the Soviet Northern Pacific Fleet.

American boats were ordered to observe and photograph the payloads of these vessels during peak traffic times to find out what the enemy might be building, planning, or thinking. Such missions, thought Trejo, were akin to flying a passenger plane unnoticed into Chicago's recently renamed O'Hare International Airport on Christmas Eve.

The more aggressive sub drivers—the skippers of these diesel-powered "smoke boats"—in search of intelligence rewards, often snuck past Soviet warships and tiptoed to within a mile of Vlad's coastline. While most of the world claimed a three-mile coastal sovereignty, the Soviets insisted on twelve. Either way, Trejo knew that a U.S. boat caught in the act of spying near Vlad faced but two choices: escape or die.

The *Blenny* arrived in her operating area on the second day of June after a four-day jaunt from Japan. She snorkeled throughout the night, and then near dawn, Commander Bryant maneuvered the boat closer to the harbor. In the cramped conning tower, now silent and rigged for battle stations, Trejo's temples pulsed. The stale air smelled of fear. Wiping away a bead of sweat, sitting on a bench in front of the torpedo data computer, he stared at the blinking lights on a panel not more than an arm's reach away. As the assistant TDC operator, and part of the fire control tracking party, his job entailed helping track—and potentially prosecute—enemy targets. The electromechanical TDC fed target tracking information into an attached torpedo programming system, which Trejo monitored with eagle eyes. The term *fire control* related to the firing of weapons, not the traditional snuffing of fires. If an unsunk target retaliated, however, then fire control could very well take on its literal meaning.

The conning tower sat just above the control room and served as the boat's nerve center. This small area normally held seven to ten men, including a sonarman, a radar operator, the TDC operator plus an assistant, the skipper, a helmsman, and a navigator stationed at the back near the navigation plot. After countless drills, the team hummed in unison like the pistons in a well-tuned engine.

Commander Bryant stepped toward the periscope well. "Up scope number one," he said.

The oil-covered mast slid upward with a hydraulic hiss. Commander Bryant slapped the handles on the attack periscope horizontal. Seawater dripped onto the deck, and Trejo watched the skipper plant the right side of his face against the rubber eyepiece. "Mast, funnel mast, clipper bow, king posts, transom stern, down scope."

Hydraulics whispered again as the mast slid downward. In his head, Trejo translated his skipper's jargon: the observed contact was a clipper-bowed cargo ship with two masts, a king post, and a transom-style stern.

"The big light is on," Bryant said, scratching at the stubble on his chin.

Trejo knew that his skipper referred to a bright searchlight on the southern tip of Sakhalin on Nishi-notoro. When the Soviets lit up the dawn with that light, they illuminated a surge in shipping traffic in the strait. They also unknowingly helped the *Blenny* capture some photographic intelligence—or PHOTINT—since everything navy needed a truncated moniker.

Bryant issued another order. "Ready the camera."

Chief Radioman Donald Byham, holding a thirty-five-millimeter Canon camera, stepped toward the periscope. Trejo imagined a handful of skinny Coke-bottle-glassed nerds back at the Naval Security Group headquarters in Fort George G. Meade, Maryland, poring over each photograph taken by the *Blenny* with a magnifying glass to see what the Red Bear might be up to this month. Thousands of snapshots of Soviet vessels delivered by dozens of U.S. submarines probably lay scattered across the desks of high-ranking officials at NSG. Were cargo ships delivering a new type of missile that could hit the White House from the other side of the Atlantic or parts for a new class of submarine that could run circles around U.S. boats? Were Soviet warships just conducting an exercise or preparing for a full-scale nuclear war? The navy needed to know, so much so that hundreds of lives were considered expendable in the search for that knowledge.

"Ready with the eyes?" Bryant asked, his arms dangling over the periscope handles.

"Good to go, Cap'n," Chief Byham said as he moved closer to the periscope.

The *Blenny* came with two periscopes—the number one scope, with a small diameter of less than two inches, and the four-inch-wide number two scope. The former was the wiser choice when stalking prey at close range, as the smaller diameter lowered the risk of detection. The larger scope served as the best PHOTINT platform, as the mast contained better optics.

The advanced optics in the number two scope were better but not perfect, so Chief Byham's "eyes" served as backup. The medium-build chief had worked as a talented commercial artist in civilian life and possessed uncanny drawing skills. Recalled to active duty at the outset of the Korean War, he also came equipped with a photographic memory. After a few short glances at a target through the periscope, he could re-create the images he saw as detailed hand sketches within minutes. God gave the human eye far greater acuity than a camera lens, so Byham could see subtle details hidden in the shadows that the Canon could not detect. These Byham drawings became part of the intelligence stash delivered to NSG for review. Chief Byham never got a dime for his artwork, but he did receive a letter of commendation.

"Up scope number two," Bryant said. "Raise the ESM mast."

Electronic surveillance measures, thought Trejo. Along with visual information, the NSG wanted recordings and measurements of wavelengths, frequencies, and pulse repetition rates emanating from enemy radar signals. They called it SIGINT. The ESM mast captured this type of data for subsequent perusal by the NSG. An alarm in the conning tower beeped when the ESM mast detected that Soviet radar might be "painting" one of *Blenny*'s masts. Too many beeps equated to "caught," which also meant they were screwed.

As the masts sped toward the surface, Bryant ordered a steady depth and buoyancy trim, as an inch too shallow could spell disaster.

Beep!

The hair on Trejo's neck bristled. The radar detection system in the ESM mast just got a hit from a nearby warship.

Beep! Beep!

Two more hits.

"Camera," Bryant said, stepping back from the scope. "Make it fast."

Chief Byham snapped the Canon onto the periscope's eyepiece and moved away. Commander Bryant hurried back to the scope, peered through the camera lens, and clicked off several shots. He spun the scope a few degrees and clicked off a couple more.

"Pull the camera," the skipper said, again stepping away.

Chief Byham removed the Canon.

"Eyes, you're up," Bryant said.

Chief Byham gave a quick nod and seated his face against the scope's eyepiece.

Beep! Beep! Beep!

"Five seconds," Bryant said.

Trejo glanced at his watch. Five seconds felt like fifty.

"I'm done," Chief Byham said.

"Down scope."

The sonarman, seated just aft of the periscope, turned his torso toward the captain. "Active pings in the water!"

Trejo's heart stopped. A Soviet warship must have gotten a good hit on the scope with a radar beam. The bad guys just caught *Blenny* spying, so they were compelled to pummel the ocean with active sonar. Depth charges and torpedoes might be next.

Through the open hatch, Bryant called down to the control room. "Diving officer, twenty degree down bubble, make your depth 300 feet!"

The diving officer echoed the command as the boat now angled toward the bottom. At 300 feet, Bryant glanced at the bathythermograph. The BT, fed by a small device installed on the boat's hull, displayed pressure and temperature related to depth. As colder water layers tend to reflect sonar beams upward, the skipper wanted to stay under one. To ensure accurate and timely readings, the boat dove to test depth every morning, then inched back toward the surface, all the while taking temperature and density readings to feed the BT and find the layers. Today, the best acoustic thermal layer started at 400 feet. Bryant called down to control and issued a new order. "Diving Officer, take us down to 450 feet."

Trejo's heart started again and raced to full throttle. Even though the *Blenny*'s newer "thick skin" design made her a deeper diving boat than her "thin skin" predecessors, her test depth topped out at 412 feet. Too much beyond that could result in the proverbial crushed beer can effect.

The boat groaned as she descended beyond 300 feet and the ocean's grip tightened.

350.

375.

412.

Steady at 450 feet.

Trejo could now hear Soviet fifty-hertz sonar pings through the hull. He moved his eyes upward, as if he could see more than just pipes and cables. The pings grew louder. The boat crawled along at three knots, her twin screws generating no more noise than a slow-speed fan. Trejo held his breath and prayed that Soviet sonar beams would not penetrate through the acoustic layer.

Then the explosions started.

Faint at first, they grew louder until the boat shook with each clap. Trejo hoped the Soviets were using warning depth charges, which were "light" versions of the subkilling kind, but he did not know for sure.

Click, whang! Another explosion.

The flooding alarm sounded, and men ran to damage-control stations. Leo Chaffin, the *Blenny*'s executive officer, darted out of the control room. Part of his duties included leading the damage control team. Minutes later he made a call to the skipper over a sound-powered phone. Bryant uttered a few words, nodded a couple times, then hung up the phone.

Facing the team in the conning tower, Bryant said, "The XO reports that we have a leak around the sound dome shaft in the forward torpedo room. He's sealed off the area from the forward battery and is pressurizing to thirty psi, but it won't be enough. We've got to take her deeper until they get the leak fixed."

Bryant leaned over and called down to the control room. "Diving Officer, make your depth 500 feet."

The diving officer's voice cracked as he confirmed the order.

The boat moaned in defiance to the added pressure. Trejo's mouth went dry. He understood the strategy but didn't like it. More depth equals more pressure, and more pressure makes things smaller, like pipes and shafts. This can sometimes make leaks easier to fix. *Sometimes.* Still, 500 feet could be 88 feet deeper than dead.

Seawater leaked from a half-dozen pipes in the overhead as the boat descended, showering men in the conning tower. Sonar pings bounced off the *Blenny*'s thick skin, followed by the staccato clap of more depth charges. The volume of both increased as the Soviets continued to close.

A dozen long minutes passed before the sound-powered phone rang in the conning tower. Commander Bryant answered, listened, nodded, and hung up. Again he addressed the tracking party: "XO says the leak is fixed. Skelly just earned your respect and a navy commendation." Bryant bent down and called through the lower hatch, "Diving Officer, make your depth 450 feet."

The diving officer echoed the order, and Trejo let his shoulders relax. He later learned that Auxiliaryman First Class Skelly, being a skinny kid, volunteered to wiggle his way into the well, head down, to make repairs. Using two main engine semicircle bearing shells fitted together, he wrapped them around the shaft packing to stop the leak. Hell of a jury rig, but it worked.

After twelve hours of hiding under a thermal layer, with pings and depth charge smacks filling the ocean around the *Blenny,* Trejo's lungs ached. His chest tightened as he struggled to pull in a breath. *Carbon dioxide,* the Achilles' heel of diesel submarines and the odorless killer of sailors. The most any smoke boat could stay underwater running on batteries before the air became stale and sailors risked CO_2 poisoning was three to four days. They were then forced to come shallow and run the diesel engines to push out the old air and pull in the new.

Trejo looked around the control room. Other men, bent over and coughing, also struggled to find enough oxygen to survive. Trejo glanced at his watch. The second hand ticked away, counting down the last few hours of his life. If the Soviets held the *Blenny* down on the bottom much longer, he'd spend the rest of eternity in a cylinder twelve feet longer than a football field.

His eyes blurry and his head dizzy, Trejo wondered what it might be like to serve on a boat that could stay down for months at a time versus only a few days. Although the world's first nuclear-powered submarine had not yet put to sea, the Department of the Navy announced on December 12, 1951, that her name would be the USS *Nautilus.* Within hours after hearing the news, Trejo joined his fellow crewmates in denigrating Admiral Hyman G. Rickover, the "father" of the nuclear submarine navy, by chanting the mantra "Diesel boats forever!" Now, with his body growing weak from lack of air, he promised God that he would never again malign Rickover or nuke-driven subs.

An hour later, with a third of the crew incapacitated by CO_2 poisoning, the sonar operator announced that the Soviet pings and explosions were finally subsiding. Bryant ordered all ahead two thirds and a thirty-degree turn toward freedom. The boat sprinted away and surfaced an hour later. Hatches opened, and the diesel engines pulled fresh air into the boat. Trejo took in a deep breath and smiled. He was still alive, at least until their next SpecOp.

When Paul Trejo returned to San Diego, he and his crewmates received a Navy Expeditionary Medal. The medal is awarded only to those "of the Navy and Marine Corps who shall have actually landed on foreign territory and engaged in operations against armed opposition." Submariners who chased the Red Bear added a star to their ribbon for each SpecOp mission they completed. Trejo wore his ribbon with pride and earned several stars over the next few years, painfully aware that the diesel-powered limitations of his smoke boat placed his life on a knife's edge each time he went to sea.

THE U.S. INTELLIGENCE COMMUNITY CAUGHT A glimpse of the potential for submarine espionage SpecOps during World War II, when, in preparation for beach landings, they sent a dozen diesel boats to island coastal waters to pop up antennas and listen to Japanese radio traffic. When the Cold War began in 1946, they remembered these wartime undercover missions and decided to call again upon their spies of the deep for "special operations." While submarines were ideal as stealth platforms to undertake these clandestine trips into danger, their need to snorkel every few days became a serious limitation when operations required extended endurance.

To solve this problem, the navy turned to Hyman G. Rickover, who put into motion the wheels that rolled the navy toward nuclear power. The Polish-born Rickover immigrated to the United States with his family and graduated from the U.S. Naval Academy in 1922. He received his appointment as the director of the Naval Reactors Branch in 1949, which led to his supervisory role in the planning and construction of the first ship submersible nuclear (SSN). Electric Boat Corporation laid the keel to the USS *Nautilus* on June 14, 1952. While the solution to the shortcomings of diesel power appeared well in hand, the navy now needed a new generation of nuclear-trained sailors and officers to man their growing armada. One of these officers started off as a seaman on a diesel boat.

When Gardner Brown boarded his first submarine in 1946, the World War II–vintage craft, with her long, thin frame squatting amid a swirl of shimmering oil and dock debris, reeked of diesel fumes. The boat's topside watch beckoned him aboard, whereupon his nose wrinkled even more at the smell inside—something akin to a gas station garage manned by sweaty rednecks. Below decks, another machinist's mate guided Brown on a tour of the boat, from the torpedo room in the bow, past the control room, mess decks, and berthing spaces, through the hot engine room, where large diesels stood at the ready, and finally to the aft torpedo room, where green MK-14 torpedoes waited patiently for something to kill.

Brown's tour guide explained that the USS *Cubera* (SS-347) was a relative of the *Balao*-class diesel submarine and gained her name from a large fish of the snapper family found in the West Indies. Commissioned on December 19, 1945, the *Cubera* never saw wartime action. She sailed to Key West in March 1946, where her crew assisted with the testing of a new top-secret submarine detection system. A few months later, she reported to the Philadelphia Naval Shipyard for an extensive GUPPY modernization.

GUPPY stands for Greater Underwater Propulsion Power. This $2 million overhaul installed a new battery system and technology based on pilfered German XXI submarine designs that returned an investment of faster, deeper, longer, and the ability to snorkel. Just like her World War II U-boat counterparts, the snorkel modification allowed the *Cubera* to pull in air while submerged so the diesel engines could recharge the batteries, or "fill the can," and refresh the crew's lungs.

After the *Cubera* completed her GUPPY upgrade, she received orders to head into harm's way. Gardner Brown closed the hatch above his head. As the *Cubera* prepared to dive, he thought about their destination. Prior to leaving port, nobody told him they were headed to the Black Sea. No one said they'd be conducting one of the first SpecOps of the Cold War, and not a soul talked about their odds of returning home alive.

Brown already knew a lot about death. After attending Governor Dummer Academy in Byfield, Massachusetts, he joined the navy in 1944. He passed the V-5 and V-12 program examinations to enter Dartmouth College in August 1944. The Naval Reserve Officers Training Corps established the V-12 in 1942 to recruit officers for the war effort. Assigned to a marine unit, Brown boarded a ship headed to a remote Pacific island. On February 19, 1945, as part of the Fourth Marine Division, Twenty-fifth Regiment, Third Battalion, I Company, Brown landed at Blue Beach Two and watched his companions die on the sands of Iwo Jima. The Japanese fought a fierce battle that lasted forty-five days, aided by fortified bunkers, hidden artillery, and eleven miles of underground tunnels. The battle of Iwo Jima cost the lives of almost 21,000 Japanese and over 6,800 U.S. Marines and is forever frozen in time by the iconic raising of the American flag atop the island's peak.

Brown did not join the ranks of the fallen and transferred to the submarine navy in January 1946. A little over a year later, he sped toward a new enemy—the Japanese red circle replaced by a Soviet red star. He described the boat's skipper, Commander George W. Grider, as "the perfect example of brains and brass ones." It turns out Grider would need both.

Commander Grider earned his submarine qualifications aboard the USS *Skipjack* (SS-184)—one of the most accomplished submarines of the war. He served as executive officer on the USS *Pollack* (SS-180) and as CO of the USS *Flasher* (SS-249) before transferring to the *Cubera*. Later in life, Grider became a U.S. congressman, but for now, he was a leader of men in an underwater world.

After crossing the Atlantic, Grider drove the *Cubera* to within a few miles of Sevastopol. Long considered the "jewel of the Crimea," this glittering white city housed a predominantly Russian population and the Soviet Black Sea Fleet. The Soviets had erected a steel gate across the entrance to the Strait of Dardanelles to prevent errant underwater visitors from sneaking into the harbor, but Grider viewed this as nothing more than a speed bump.

Equipped with a surveillance antenna and a Russian-speaking "spook" intelligence officer nicknamed Grab One, the *Cubera* waited near the harbor's entrance. Commander Grider raised the number one attack periscope and swung the cylinder back and forth. For several hours he studied the navigation light planted on the starboard side of Tenedos Island, at the northern end of the strait. When the light finally came on, signaling the approach of a warship, he swung the scope back toward the entrance. He reported to the crew in the small conning tower that he'd spotted a Soviet aircraft carrier and intended to follow her into the harbor.

Employing the "brass ones" that earned him the crew's respect, Grider nudged his boat to within several yards off the stern of the carrier. As the steel gates parted to admit the Soviet warship, the *Cubera* followed her in. Once inside, now surrounded by dozens of enemy ships, Grider raised a surveillance antenna, and Grab One went to work. The intelligence spook captured, recorded, and analyzed Soviet transmissions for two days. He then convinced Grider to "get a little closer." Forgoing caution, Grider complied.

Not more than a few minutes after raising the periscope for another look, the Soviet navy stirred, a subtle rumbling at first, soon followed by an all-out barrage. Grider dove the boat and tried to evade, but the Soviets slammed the ocean with active sonar pings. With the steel gate shut tight across the exit to the harbor, the *Cubera* had nowhere to run. Out of options, Grider searched for a place to hide.

Bottom-sounding sonar detected something strange on the ocean floor 300 feet down. Grider took the boat deeper. Near the bottom, he peered through the periscope into the murky water. Although the Black Sea usually provided clear viewing this time of year, not much daylight found its way down to this depth. Grider couldn't believe what his eyes told him, so he asked the chief of the boat, Gaines "Whirly" Smith, to have a look. Whirly gazed through the periscope and said, "I'll be damned, we're in the middle of Main Street."

The gods had smiled upon the *Cubera* by allowing her to stumble across an ancient city flooded by time. The tall earthen buildings afforded the perfect hiding place and just enough cover to deflect Soviet sonar beams.

Smiles quickly faded when, hours later, the boat's battery power and air supply dwindled. Grider knew they'd have to surface in less than sixteen hours or die on Main Street.

Twelve hours later, Gardner Brown experienced the curse of the diesel sub firsthand. Carbon dioxide replaced what little air remained.

Brown's head spun, and his lungs felt like dried prunes. With less than two hours remaining before the battery ran dry, the gods intervened once more. They sent a Russian destroyer through the gate. The *Cubera* snuck into her wake and swam out of the whale's belly. Having dodged death yet again, Brown wondered if he might have been better off staying in the Marine Corps.

Intelligence experts were ecstatic. Grab One had grabbed plenty, and the navy wanted more. With the bar now set ultra-high by Commander Grider, every submarine driver needed to risk as much or more to earn a "brass ones" title—and, perhaps, another stripe on his sleeve. This mission and others like it set the stage for a deadly high-seas contest between the United States and the Soviet Union that raged on for another forty-plus years. They also propelled the navy's quest for stealth platforms with greater endurance and range, which could only be accomplished with atomic-powered engines.

The invention of nuclear propulsion fanned the flames of the underwater Cold War when the atom stepped onto the stage with the commissioning of the USS *Nautilus* (SSN-571) on January 21, 1954. The navy's first nuclear-powered submarine signaled the end of diesel-driven boats and changed forever the life of under-ocean sailors. Now submariners could stay down for months versus days and could do so without needing to run noisy diesel engines. The USS *Seawolf* (SSN-575) followed in the *Nautilus*'s nuclear wake seven months later, and like a redheaded stepchild, it received little of the notoriety bestowed upon her older cousin.

Unlike most other diesel boat sailors, after his near-death experience in the Black Sea aboard the USS *Cubera,* Gardner Brown readily accepted an offer to "go nuke." He became one of seventeen submariners handpicked for the nuclear power program in the fall of 1953. Brown and his classmates spent five days a week completing intensive academic courses at Union College and two more days at a nuclear prototype putting into practice what they'd learned. They studied both pressurized water reactor technology—used on the *Nautilus*—and alkaline metal sodium technology used by the *Seawolf*. When the instructors discovered that learning sodium complexities required a higher IQ, they divided the class in half. Those with a little more brainpower, like Brown, wound up in the *Seawolf*-bound class.

Seawolf's namesake, a solitary fish with gnarled teeth and savage tusks, underscored the vessel's gritty demeanor and set the tone for her turbulent future. Fashioned from a vintage fleet diesel boat, *Seawolf* employed a superheated steam power plant versus the more traditional saturated steam reactor, which reduced machinery space size by almost half. Although more advanced and quieter than the *Nautilus*, *Seawolf*'s propulsion system carried additional risks and earned her the nickname "Blue Haze" when sodium coolant leaked from the reactor in the shipyard.

Such accidents fueled Admiral Hyman Rickover's consternation and fanatical focus on safety. Rickover ruled his atomic roost like a straw boss on a pyramid. Most navy subordinates and Electric Boat shipyard contractors feared his controlling management style, which Rickover exploited to further his safe sub agenda. While some hated the man, Gardner Brown became a follower and a friend. Like Rickover, Brown knew firsthand that from a power plant engineer's point of view, a diesel boat, which runs on "dead dinosaur juice," was to a nuclear submarine as a flint rifle was to a submachine gun. Nukes cost more and take far longer to build due to power plant complexities and safety concerns. More complexities equal more problems, and in *Seawolf*'s formative years, there were many. Brown recalls having to leap dozens of metallurgical hurdles caused by ultra-high reactor temperatures and flux densities while striving to build this boat.

Brown's cousin, Gene Centre, who also worked on the *Seawolf*'s reactor as a project manager for Bettis Atomic Power Laboratory, heard rumors about flaws in the A1 and A3 reactors being built by the Soviets. He also heard that the Russians considered their sailors expendable and so did not place a high value on radiation shielding and pump seals. Centre figured that the rumors might be more propaganda than truth, but nonetheless he concurred with Rickover that safety and reliability were paramount.

Armed with that priority, Dennis B. Boykin III, Electric Boat's power plant manager, insisted on keeping the *Seawolf* in the yards an extra year to develop a rod drive mechanism with special seals that prevented coolant leaks. Gene Centre and others helped engineer this, along with additional "over spec" components to ensure that, years later when the *Seawolf* replaced her sodium engine with a water-cooled type, she came out of the garage with the "heart of a '57 Chevy and the soul of a Mack truck." Although Centre never envisioned such, his dedication to ensuring that the *Seawolf* could be pushed well past her red line saved the lives of 190 men trapped on the ocean floor a few miles off the coast of Russia more than two decades later.

Seawolf finally received her commission on March 30, 1957. Lieutenant James Earl "Jimmy" Carter, who'd one day be the only U.S. president qualified in submarines, had received a billet as her engineering officer but resigned his commission after his father died in 1953. Without Jimmy Carter on board, Commander R. B. Laning grabbed the reins as the boat's skipper and galloped the navy's second nuclear racehorse around the track. Brown and his crewmates pushed her hard for the next several months during rigorous sea trials. Although her heart was willing, the *Seawolf* groaned in defiance at the high-speed runs, tight turns, deep dives, and steep angles ordered by Laning.

Brown was not anxious to endure any further SpecOps after his Black Sea experience on the *Cubera* under Commander "Brass Ones" Grider, but fate dictated otherwise. He went with the *Seawolf* in 1958 on her first special operations run into the Barents Sea, where, he says, "We stayed underwater forever, but at least we never had to worry about running out of air."

The *Nautilus* made history that same year by sliding under the ice floes and paving the first underwater trail to the North Pole. She remained underwater during the entire transit and hit speeds of more than twenty-three knots. In contrast, diesel submarines could push no more than ten knots submerged and needed to snorkel every few days to stay alive.

Nuclear power enabled submarines to accomplish their missions with greater safety and efficiency by solving the problems of endurance and speed while submerged. A large hurdle remained, however. Current submarine sonar systems could hear no further than a few miles away, making them all but deaf, dumb, and blind. That made them vulnerable to detection and destruction. By the early 1950s, American engineers had failed at every attempt to solve this monumental problem.

CHAPTER TWO

For thou cast me into the deep,
Into the heart of the seas,
And the floods surrounded me,
All Your billows and Your waves passed over me.

—JONAH 2:3

IN THE FALL OF 1953, DR. Donald Ross, an underwater engineer from Pennsylvania State University's Ordnance Research Laboratory, walked through the large glass doors of the Bell Telephone Laboratories building in Whippany, New Jersey. Roland Mueser, one of Ross's former classmates from Penn State and now a Bell Labs employee, met him in the lobby. Mueser beamed broadly, whisked a tousle of hair away from his forehead, and urged Ross to the front desk. Ross signed in with the guard, and Mueser pointed toward an elevator. As they hurried through the capacious foyer, Mueser told Ross that he'd been working for Captain Joseph Kelly on an exciting new project called Jezebel. He explained how they'd installed underwater low-frequency sonar arrays to help detect snorkeling submarines from over one hundred miles away. Ross raised one of his thick brown eyebrows and let loose a whistle.

The elevator reached an upper floor, and Mueser led Ross through a maze of corridors and hallways, all the while talking about how Project Jezebel received a name change to SOSUS (Sound Surveillance System) after the initial tests proved successful. Six SOSUS listening stations were now deployed in the North Atlantic basin, and nine more were authorized under Project Colossus—three in the Atlantic and six in the Pacific. These new arrays, to be installed in 1954, incorporated advanced upward-looking sonar capabilities.

Ross let out another whistle.

Turning a corner, Mueser said, "I suppose you're wondering why you're here."

Ross shrugged. "I figured you'll tell me soon or later."

"I will," Mueser said, "and then you'll pee your pants."

Walking across plush carpet, trying to keep his curiosity at bay, Ross wondered how Mueser convinced Bell Labs to hire him. He was a submarine hydroacoustic expert, which only peripherally related to something like SOSUS. Ross received his Ph.D. from Harvard in 1942—the first to graduate from that Ivy League university in only three years. He launched his career at the Harvard Underwater Sound Laboratory in January 1945, moving later that year to the Ordnance Research Laboratory at Penn State, where he met Mueser.

Ross's work at Harvard near the end of World War II contributed to improvements in propeller designs for the twin-screw GUPPY-upgraded diesel submarines. He'd also been assigned to the propeller team to produce quieter props for submarines. Still trying to connect his work to the reasons why he'd been invited to Bell Labs, Ross followed Mueser around the facility. His propeller experience provided few intersections with something like Project Jezebel, so why was he here? As Mueser strode past an office in the building, Ross peeked inside. Engineers in white shirts and dark ties scribbled on blackboards and wrestled with reams of desk paper. The place smelled almost antiseptic, not too unlike the waiting room in a medical clinic.

Mueser opened a lab door and motioned Ross inside. Three slide rule–clutching engineers, pens stuffed into pocket protectors, turned from a chalk-covered blackboard. Mueser introduced the trio as Larry Churchill, Herman Straub, and Rich Carlson.

Mueser found a chair, leaned back, and said, "What I'm about to tell you boys is way above top secret."

Herman Straub, a former World War II submarine skipper, let out a grunt. "Does that mean if I tell my wife you're going to kill me?"

"No," Mueser said, "we'll kill your wife."

Straub smiled. "For free?"

After a shared laugh, Mueser's face turned serious. "The navy has a problem they need us to solve."

"Such as?" Ross said.

"The Soviets are building a lot more submarines, and their ASW (antisubmarine warfare) forces have become much more aggressive. Our boats can't hear these guys if they're more than a few miles away."

"So what's the answer?" Straub asked.

"I brought Dr. Ross here to help us build a new long-range passive submarine sonar system," Mueser said.

"Passive?" Straub wondered. "As in no active pings?"

"As in strapping on headphones and listening," Mueser said.

"What do you mean by long range?" Ross asked.

"I mean like a hundred nautical miles," Mueser said.

"How the hell do we do that?" Straub asked.

"Using LOFAR technology developed for SOSUS," Mueser said.

Although he didn't pee his pants, a lightbulb turned on in Ross's head. Now he knew why Mueser encouraged him to join Bell Labs: they needed his expertise in submarine hydroacoustics to revolutionize how submarines hear.

The team went to work on an electronic "breadboard" passive sonar suite using a modified version of LOFAR (Low Frequency Array). The effort proved more difficult than expected, resulting in chain-smoking late-night sessions in the lab, followed by a battery of simulation tests. All of them failed. After more than a month, Ross was about to concede defeat when a crazy idea popped into his head.

He marched into Mueser's office and said, "Standard LOFAR won't work."

"Why not?" Mueser asked.

"Subs are too short and noisy to be good platforms for low-frequency sonar."

"Wonderful," Mueser said dryly. "So now what do we do?"

"We use a higher frequency."

"We tried that during the war," Mueser said. "Didn't work. We needed to get close enough to smell a fart before we could hear anything."

"I know," Ross said, "but I think I can solve that problem with a new approach to demodulation."

Mueser grinned. "I told 'em you were a genius."

Demodulation, in short, is the science of extracting information from a carrier wave. Ordinary radios do this by snatching signals from the air and decoding the AM or FM frequencies into something we can hear at 1210 on our dial. Simply put, Ross conceived the idea of using a form of demodulation to "decode" the various frequencies received by a passive sonar system on a submarine. Once completed, the breadboard design, which consisted of a small circuit board filled with electronic components and dangling wires, passed all the simulation tests. Now they needed to prove it could work in the real world.

The team traveled to Submarine Development Group Two in New London, Connecticut, and boarded the USS *Cavala* (SS-244). His heart thumping, Ross peered through the open hatch of the dark submarine. The pungent smell of diesel fumes filled his nose as he climbed down the ladder. Standing on the tile in the narrow passageway, claustrophobia prompted him to suck in his gut and narrow his shoulders. He followed an officer toward the control room, wondering how anyone could stand to live in such a tight space for weeks on end.

The *Cavala*'s skipper, Lieutenant Commander Bill Banks, greeted the team in the control room. Ross unveiled the sonar suite breadboard.

"What's that?" Banks asked.

"Your new passive sonar system," Ross said.

"No shit," Banks said. "Does it work?"

"That's what we're here to find out."

"So it's a beta breadboard."

"More like an alpha," Ross said.

Banks lowered his eyes, shook his head, and walked away.

The *Cavala* pulled out of port, and the team set about connecting the breadboard to the sub's passive sonar system. After running a couple of functional tests, Ross sat in front of the sonar console in the *Cavala*'s conning tower and slipped on a pair of headphones. Closing his eyes, he listened. Nothing. He adjusted some settings and listened again. *Something?*

Banks peered over Ross' shoulder. "Can't get it to work?"

Ross removed the headphones and handed them to the skipper. Banks pulled them over his ears.

A few seconds later, Banks's eyes widened. "I'll be damned. I can hear him!"

The "him" Banks referred to was a snorkeling *Balao*-class submarine more than one hundred nautical miles away. Over the next few weeks, the team confirmed detection of other submarines and classified a carrier task group operating 150 nautical miles distant. They code-named their invention DEMON, short for demodulation. This new design extended the hearing range of previous sonar systems by more than thirty times, and Ross soon found himself spending more time cruising underwater than almost any other civilian engineer. He and teammate Herman Straub spent ten days on the USS *Nautilus* installing a new DE-MON sonar suite, measuring radiated noise and gathering spectra. Their analysis and technical advice helped the crew reduce the boat's noise signature, making her less detectable by SOSUS or shipborne passive sonar arrays.

Following his *Nautilus* experience, Ross produced the first report on noise emanation from nuclear submarines in August 1957. The Naval Scientific and Technical Intelligence Center (NAVSTIC), impressed by the report, asked Ross to analyze a top-secret DEMON recording of a new Soviet submarine. The sounds emanating from the sub were strange—definitely not the piston slapping one might expect from a diesel engine or the hushed quiet of a battery-driven propeller. After a day of looking over the LOFARgrams made from the recordings, Ross confirmed that the acoustic signature came from a nuclear power plant. Jaws dropped at NAVSTIC.

The next day, the U.S. Navy announced that the Soviets had just deployed their first nuclear submarine. NATO bestowed the code name November, which happened to coincide with the month when the keel was laid. American submariners nicknamed her Red November, but the Soviets simply called their new Project 627 submarine the K-3. They designed this twin-reactor attack boat to hinder U.S. shipping lanes in the event of a war, but the crew of *Red November* secretly trained for another top-secret mission: sneaking close enough to New York harbor to fire a twenty-seven-meter-long nuclear torpedo at the Statue of Liberty.

Faster, deeper diving, and more heavily armed than the USS *Nauti-*

lus, the cigar-shaped K-3 put the fear of God into most of the navy brass. That paranoia prompted an underwater race that cost trillions of dollars and pitted the United States against the Soviet Union for another three decades.

The introduction of K-3, which employed a cylindrical hull better suited for underwater speed, motivated the U.S. Navy to accelerate a radical new design for its own fast-attack nuclear subs. The USS *Skipjack* (SSN-585) entered the fray on May 26, 1958, to counter the *Red November* threat. Sporting improved SW5 pressurized water reactors, *Skipjack*-class submarines could go from zero to thirty-plus knots in a few dozen heartbeats. A teardrop-shaped hull and single in-line propeller improved underwater speed and agility and sent a clear signal to the Soviets that they were once again behind in the game.

Unable to match American ingenuity, the Soviets responded with numbers. By 1959, they were the proud owners of 260 long-range offensive submarines. This sobering development forced the United States to concede that its nemesis planned to shift the mission of these vessels from coastal defenders to front-line strike weapons. Several *Zulu* and *Golf*-class boats could now hurl 3,200-ton nuclear warheads at the United States from over a thousand miles away. Despite the monumental breakthroughs in underwater sound technology used in ocean-mounted SOSUS and sub-installed DEMON sonar systems, the only distant subs these inventions could reliably detect were noisy snorkeling diesel boats. Finding an underwater craft running near silent on battery power, or semiquiet on nuclear power—well before she launched her nuclear payload at the United States—seemed as insurmountable as landing men on the moon.

CHAPTER THREE

*Man is a great blunderer going about in the woods,
and there is no other except the bear that makes so much
noise.*

—MARY AUSTIN

HAVING WORKED HIS WAY UP FROM a seaman to a communications technician chief (CTC) in the navy, William J. Reed received orders to report to the Naval Security Group at Fort George G. Meade, Maryland, for a top-secret assignment briefing. He was unaware that his new orders would place him at ground zero for one of the most important discoveries of the Cold War. As his son—at the time only three years old—I had no way of knowing that our move to a small Turkish village, and our adoption of a wounded baby bear, would become the impetus for that discovery.

Istanbul is situated on the Bosporus, which cuts across the Anatolian Peninsula and exits into the Black Sea, dividing Turkey into a European and an Asian composite. Constantine renamed the city Nova Roma in A.D. 330, but everyone started calling the eastern capital Constantinople. In 1930 Turkish authorities officially named the city Istanbul, which translates loosely as "downtown."

Turkey is an important part of NATO and has always welcomed U.S. military and technical assistance. In 1959 the United States maintained several strategic stations throughout that country. One of these, where my dad now worked, was a radio listening station near Karamür-

sel. The antenna array at this facility was operated by Air Force personnel and "borrowed" by the navy for high frequency direction finding
(HFDF)—a way of pinpointing the location of a ship or submarine by
determining the direction to its radio transmissions. Operators called
these HFDF stations "Huff Duffs." A small detachment of NSG personnel, including my dad, operated the HFDF equipment as part of The
United States Logistics Group, or TUSLOG Detachment 28. Most of
the families assigned to serve at this base lived about fifteen miles away
in Yalova, where at night the lights of Istanbul twinkled like Santa's village from across the Sea of Marmara, but Dad insisted on living some-
place a bit more tranquil.

After a week of searching, my father, a six-foot-two navy chief with
Dick Tracy features, found us an apartment in Değirmendere, a small
town six miles from the base. The only thing modern here was the automobile garage on the highway at the village entrance that was built
of sticks and stones. The entry road came with an overhanging of oak
and elm trees and led to the main square, a dirt clearing dominated by
majestic hundred-year-old oaks flanked on both sides by shops. At the
open-air bakery, we watched the baker, a master artist, pulling brown
loaves from an ancient kiln. My mouth watered as the rich aroma wafted
through the open door and pulled in hungry shoppers.

My dad often went hunting with a fellow navy chief and Halidere,
our Turkish neighbor. Now and then Dad brought back a couple pigs,
but Halidere refused to eat any of that "filthy pork." A week later, Halidere knocked on our door and gave my mom a little squealing animal
of some kind. We were clueless as to what this tiny thing might be, so
Mom called my dad at the base. He came home a couple hours later
and scratched his head. He went next door and talked to Halidere.

Dad discovered that our neighbor trained bears for a living. Halidere taught them how to stand up on their hind legs and dance around
in the streets while people giggled and threw money at their paws. He
stole two bears from a cave when the momma bear went hunting for
food. While running away, he dropped one, and the poor thing hurt its
leg. He figured the wounded animal could never dance, so he gave it to
my mom. Dad offered him some American cigarettes as a thank-you.
Halidere loved American cigarettes. Dad said he loved American women,

too, and told my mom to never get her fanny too close to Halidere's hands. Being three, I didn't know why he told her that, but it made me laugh anyway.

The baby bear weighed only sixteen ounces, and, except for his paws, he looked like an overgrown rat. No fur at all. I thought all bears looked like little cuddly teddy bears, but not this one. Dad named him Ayi Bey, which translates roughly as "Sir Bear." Not knowing what else to do, Mom put him on a human baby's schedule and diet of baby formula. He woke up every few hours crying for food, crying to be held, or just crying. Mom started feeding Ayi Bey with an eyedropper, but that didn't last long. He wanted more. She borrowed a baby bottle from an American family, and Dad poked more holes in the nipple so Ayi Bey could get enough formula mixed with Pablum.

Ayi Bey stayed furless for about a month. I wanted to call him Fuzzy Wuzzy, like the bear with no hair, but my sister, Pam, said we should change his name to Scratchy because he liked to scratch things. Mom and Dad agreed. Scratchy especially liked to claw at the brown Turkish goatskin rug that covered the cracked tiles of our kitchen entryway. We didn't know it then, but Scratchy's habit would soon lead my father toward a critical discovery for the navy.

Over the next year, Scratchy grew to be the size of a small dog. He chased my sister and me around the yard like a bounding puppy and learned how to climb up the slide by lying flat with all four paws outstretched and gliding down from the top. Often he pushed us out of the swing so he could try. When he couldn't manage that, he hit the seat with a paw and sent it swinging while he let out a muted growl, which sounded more like a baby's cry.

Every day, after feeding Scratchy six or more bottles of milk on the back porch, Dad pulled on his uniform and left for work. Mom readied Pam and me for school—kindergarten in my case—and then helped us board the small bus to the base. There we spent our day among other English-speaking military brats learning our Ps and Qs. After school, Scratchy leaped for joy when Pam and I came home, and we put on his collar to take him on a neighborhood walk.

We loved our baby brother, but our Turkish neighbors did not. Most of them muttered and complained. Their animals went crazy when we walked Scratchy through the village. Cats, dogs, horses, goats, burros, chickens, and ducks scrambled for safety. Dad finally built a six-foot concrete-block wall around our backyard to keep Scratchy penned in. He hated it, but we knew it was for his own protection.

Months later, after several incidents when the neighbors complained about our pet bear to the local police, my father was forced to make a decision that crushed his heart. He took me sailing that day. Dad and several of his navy buddies co-owned a twenty-two-foot Marconi-rigged boat that they kept at the Seaside Club near Karamür-sel. They sailed her in the Sea of Marmara, through the Bosporus, and around a sprinkling of small Greek islands. Every now and then, Dad took me along. Sometimes we'd talk, and other times we'd sit in silence and listen to the waves lap against the wooden boat. Today we talked, and he told me that it was time for Scratchy to go. I started crying and begged him to reconsider.

Tears filled my dad's eyes, too. He placed his arm around me and pulled me close. "I found Scratchy a home where he can be happy," he said. "We can see him there whenever you want, okay?"

It wasn't okay, but in my shattered heart I knew we had no choice.

His hands shaking, my dad maneuvered the boat back to the dock. We saw my dad's boss, Captain Frank Mason, standing on the pier. Mason always made my dad a little nervous, but even more so around boats because the captain was an expert sailor. Still a bit green at sailing, Dad crashed the boat into the dock. Mason started laughing and suggested sailing lessons. The next day my father talked his friends into selling the boat. The day after that I watched my dad put Scratchy's collar on and take him away forever.

One of my father's friends was the founder of the American Seaside Club, a private home on the outskirts of Değirmendere rebuilt into a restaurant for the American community. An American oil company, Caltech, operated a facility on the other side of Marmara Bay, and some of their personnel came over in outboards to the Seaside Club. We often brought Scratchy past the club during our afternoon bear walks. Some of the wives of the Caltech employees fell in love with him and came out to pet his soft fur. When they heard we could no longer keep him, they promised my parents that they'd build a mansion of a cage and feed and care for him like a spoiled child. I knew Scratchy would be happy there, but that didn't lessen the ache.

When the Caltech people tried to put Scratchy into their boat and take him away, he struggled and whined. As the boat pulled from the dock, I saw him stand on his hind legs and paw at the air in desperation. He wailed in agony, and I knew what his cries were saying: "Why are you sending me away? Don't you love me anymore?" Mom, Pam, and I cried for hours, and my dad was silent and sullen for days.

Winter winds deepened the chill in our home. My parents forgot how to smile. At first I thought it might be because we'd been forced to give away Scratchy, but I soon wondered if something else might be going on. Although too young to understand the import of what was happening in the world around me, I couldn't help but notice the absence of laughter in my dad's eyes and the worry on his face. He started spending more and more time on the base, and most nights I was fast asleep by the time he came home. Neither my mom nor my sister provided an explanation, and I remained unaware that beyond the edge of our quiet existence, my dad's world had just become a living hell.

IN EARLY DECEMBER 1960, THE HUFF Duff in Karamürsel, Turkey, where my dad worked, transformed into the epicenter of the U.S. Navy's underwater battle against the Soviet Union. The United States built the site in 1957 to monitor Soviet radio transmissions using a sophisticated antenna array. Air Force personnel captured communications emanating from various sites in Russia up to thousands of miles away. These trained experts utilized complex processes, deductions, and heavy doses of transmission analysis to predict when a Soviet missile might be launched, as well as the type and probable destination. Similar stations around the globe detected and analyzed communications associated with specific missiles: short, medium, or long range. If an unusual number of long-range missiles were detected in the preparation stage, there might be time to undertake defensive measures, perhaps even launch a preemptive strike.

My father worked for the TUSLOG Detachment 28 at the Karamürsel Huff Duff, which performed an altogether different mission than the Air Force section. Crammed into a small Quonset hut near the Air Force operations building, they were tasked with using HFDF equipment to locate Ivan's sea monsters. Just before Christmas 1960, those monsters became invisible.

High frequency (HF) generally refers to transmissions in the range of three to thirty megahertz. A megahertz (MHz) is a million hertz, and a hertz (Hz) refers to the number of transmitted cycles per second (cps). Most stereo subwoofers put out between 20 and 150 Hz, or really low bass frequencies, whereas a good set of headphones usually tops out at 20 kilohertz (kHz), or 20,000 Hz, which is the top-end range for most human ears.

Since the ionosphere does a nice job of reflecting HF radio waves (a phenomenon called skywave propagation), operators often use HF for medium- or long-range radio communications. Things that can mess with HF transmissions include the time of day or year at the transmission site, sunspot cycles, solar activity, polar auroras, and electrical wires or equipment. Still, the HF band has long been popular with amateur radio operators, international shortwave broadcasters, and seagoing vessels, including submarines.

Since the invention of wireless transmissions, operators pursued the idea of using two or more radio receivers and bearing triangulations to find the location of a transmitter. In principle, the concept seems simple: stick up a reception antenna, and notice the direction of the strongest signal. Theoretically, that should point toward the source. But with only one bearing, the source could be anywhere along that line. That's why you need at least two or more cross-referenced bearings to get a "fix."

To visualize this, imagine that you are in a parking lot, and you can't find your car. You press the emergency button on your key chain and can can hear your car honking but still can't find it in the dark. You cock your head and listen for the strongest sound. You hear it due east on a "bearing" of 90 degrees on a 360-degree compass. You make a mobile phone call to your spouse, who is southeast of you on the other side of the parking lot. She reports that the honking is coming from the north, on a bearing of 0 degrees from her. If you walk east, and your wife walks north, you will run into each other at your car.

Sure, you could have walked east and found your car without her. After all, you were only a football field away. Now imagine trying to find a transmitter in the middle of an ocean thousands of miles away. To do so requires knowing precisely where your two sets of "ears" are located, then drawing straight lines from each. Where they intersect is the location of the transmitter. Sound easy? Now let's make it tough.

What if we throw in inaccuracies and interference? For example, what if the equipment you are using to determine the direction to the transmitter is inaccurate? In the parking lot, if your ears are a few degrees off, you might walk right past your car without seeing it. In an ocean, those few degrees translate to dozens of miles if the source is thousands of miles away. And what if there's a sun storm toying with the ionosphere? Or how about nearby electrical equipment? This interference can easily distort the perceived bearing to the source. This is why one needs multiangulation, or multiple bearings to a transmission.

Let's assume we have a transmitter in Dallas. Now visualize a direction finder in New York and one in Seattle. Using a ruler, draw a straight line from each of those locations to Dallas. You've found the transmitter! Now move the Seattle line upward by a quarter inch to simulate an inaccuracy. You're in Oklahoma, not Dallas. But if you have a few more lines coming from Chicago and San Diego, you'll be much closer to Dallas. Now take away the ruler and let a two-year-old draw the lines to simulate interference. Your transmitter looks like it's in Cuba. These were the problems facing the early designers of Huff Duff systems where inaccuracies and interference equated to bearing "spreads" of up to three or more degrees. Still, they could be reasonably effective for finding transmitting submarines.

That's why, during World War II, in an attempt to thwart detection by Huff Duffs, U-boat captains started shortening their transmissions. They figured that if they were not on the air very long, HFDF operators wouldn't have enough time to get a bearing. Fortunately for the Allies, the U-boat skippers were mostly wrong. The Soviets regurgitated this thinking years later and decided that, if done properly, it just might have some merit. Unfortunately for the United States, they were right.

Describing what happened in December 1960, when the navy Huff Duff stations suddenly could no longer hear Soviet submarine transmissions, requires an understanding of how these sites operated. Navy personnel at HFDF facilities reported to the NSG, which answered to the NSA. NSG divided sailors and officers of the communications technicians rating, who worked at these stations, into separate branches. Radio operators—or R-Branchers—monitored Soviet signals from various platforms and determined an HFDF bearing to the transmitter when they got a "hit." Operations specialists—O-Branchers—assumed the duties of site operations and logistics, while maintenance personnel—M-Branchers— ensured that the station's equipment stayed running at peak efficiency. Russian-speaking intelligence operatives—I-Branchers—listened with trained ears for tidbits contained in Soviet traffic, while "technical" T-Branchers monitored for new kinds of signals and analyzed character- istics to determine the type of transmitter or platform. Being an R-Brancher, my dad's training focused on monitoring Ivan's HF transmissions, with an emphasis on Soviet submarines. As the senior-ranking chief petty of- ficer at the Karamürsel station, however, he assumed an O-Brancher func- tion as the operations chief.

Det 28 hummed twenty-fours a day, manned by four watch sections of four or five people "in the shack" for an eight-hour shift. Each morn- ing, they received a list of "interest" contacts from Net Control (NC), the central command station that coordinated all the Atlantic Huff Duff stations. In those days, Net Control resided in Northwest, Virginia.

NC compiled a catalog of surface and submarine contacts—not all but mostly Soviet—along with their call signs and probable transmit frequencies. R-Branchers set up their equipment to listen on those fre- quencies, usually in the 2–32 MHz high-frequency range. If they got a "hit," they'd contact NC via CW—continuous waves, or Morse code transmitter—and give them a "tip-off." NC then submitted a "flash" to the other Huff Duff stations to tune into, for example, frequency 12465 and take a bearing if they heard anything. If any station did catch something, they'd send a "spot report" with the bearing for that con- tact back to NC. In the early days, those reports—or abbreviated ver- sions called e-grams—came via CW and later on were sent by way of Teletype machines. Because the navy operated as a separate detachment at Karamürsel, someone needed to run the reports over to the nearby Air Force building every hour so they could be sent to NC.

Operators at Net Control collected all the bearings reported by the stations and, before the invention of automated systems, manually plotted a "fix" to the target. This process took several minutes and con- sisted of nothing more than generating "string bearings" with a com- pass rose. The rose used a figure that displayed the orientation of the cardinal directions—north, south, east, and west—on a four-by-six-foot map or nautical chart mounted on a stand. They called this a gnomonic projection. Operators took a line of string and ran it from each Huff Duff map location, represented by drilled holes in the map to a point on the edge of the map along the bearing line reported by the station. The whole thing resembled a large wall-mounted ocean map covered with strands of Grandma's yarn.

At least two intersecting bearings, where the strings crossed, were needed to pinpoint the location of a contact. Only two, however, offered very low accuracy. With three or more bearings, one could multiangulate a more accurate location, but this could still be fifty or more miles off target. Needless to say, the art of Huff Duffing was an inexact science in the early 1960s but at least close enough to point U.S. submarines, aircraft, and ships to the right ballpark.

One fateful morning in December 1960, the HF airwaves went silent. T-Branchers and R-Branchers at the Karamürsel Huff Duff in Turkey, monitoring frequencies for Soviet submarines, spun dials and searched for hours but found nothing. They checked and calibrated equipment. Still nothing. They contacted other Huff Duffs and discovered that the phenomenon existed at every station around the world. Several days passed without a single sniff. Dad decided it was time to tell his immediate boss, Commander Petersen.

My father straightened his back and adjusted the khaki "cover" on his head. He marched into Petersen's office, located in the Air Force building next to Det 28's Quonset hut, and said, "Sir, they're gone."

Petersen glanced up from a stack of paperwork and peered over the top of his thick glasses. "Who's gone?"

"Ivan's boats," Dad said. "Their transmissions have been decreasing over the last several months, as you know, but now they've stopped transmitting anything on HF. We've heard nothing for days."

"Shit," Petersen said as he removed his glasses and massaged the bridge of his nose. "Does NC have anything to say about this?"

"No sir, Net Control is as clueless as we are."

Petersen shoved his glasses back on. "Shit, shit, shit. They're transmitting, all right. We just can't hear them."

Dad relaxed his stance and looked down at Petersen's desk. The commander kept his workspace in the same shipshape condition as his duty section. At a facility saturated with routine and order, like Det 28, no one could trump the man. But when the proverbial excrement hit the blades, Petersen's smooth edges ruffled.

The commander pushed his chair back and stared at the papers on his desk, as if an answer might jump off the pages. His eyes darted from side to side. "Okay, so now what? Do we keep looking or tell Captain Mason? Did NC give us any suggestions?"

The room felt small and hot and smelled of floor wax. Dad removed his cover and backhanded a bead of sweat. "NC said to keep looking, but there's not even a peep in the three to thirty megahertz range. We've scanned every frequency used in the past thirty years by Russian subs, surface ships, and even life rafts with no luck. We're pretty much out of options at this point, sir."

Commander Petersen scratched at his balding head. Dad cringed because the man exhibited a skin disease exacerbated by stress. Losing the Soviet subs qualified as an ulcer-producing disquietude, and for Petersen, incidents like this sent him scratching. When that happened, pounds of dandruff flaked from his head and coated his shoulders. Bets were often taken for how long someone could last in a "flakey" conversation with Petersen. Dad never won. He left Petersen's office quickly with an agreement to keep searching.

Without answers, Petersen's directives offered as good a course as any to follow: keep looking and pray for a miracle. Everyone on the team agreed that Ivan was transmitting, and probably in the HF range, but the Soviets must have found a way to mask their transmissions. Dad knew that Admiral Sergei Gorshkov, commander of the Soviet fleet, had dark red "control freak" blood running through his varicose veins. He insisted on maintaining constant communications with his fleet, especially with his attack and missile-firing submarines. Soviet subs always checked in with their command stations at least once, sometimes twice or more, per day. With hundreds of submarines operating on a continuous basis, Det 28 often sent dozens of tip-offs per day to NC and received hundreds of flash reports with monitoring assignments based on tip-offs from other stations. Adding up all the hits coming from Soviet ships and subs, some of the larger stations handled up to 3,000 flashes per day.

More days passed without a single submarine tip-off or flash. Dad decided to bounce a few ideas off Captain Mason. Although he considered

Mason a friend, since he'd worked for the captain in Guam years earlier, my father still felt a bit nervous in the man's presence. Mason's graying hair and wise eyes complemented his friendly tone and engaging smile. He stood six foot three, about an inch taller than my father, and commanded a quiet respect. With matching crewcuts, square jaws, and deep baritones, the two had a lot in common. Both men were born with strong demeanors and "take command" attitudes, which is why they often played from the same song sheet when it came to military matters.

For reasons unknown, however, Mason sometimes reminded Dad of his stepfather, Lon Reed. Three months after Billy Joe Bowles came into this world, in Konawa, Oklahoma, on February 4, 1929, my real grandfather, Hoyle Bowles, died in a train accident. My grandmother, Ethel, met and married Lon Reed a few years later, and Lon adopted my dad and his two sisters.

"Drill Sergeant" Lon probably didn't intend to be an evil man, but ignorance blinded him to the kindness of the wise. Trapped in the mold of his forebears and smugly confident of his rightness, Lon played the role of king in a pauper's court. Ignorant men are often haunted by the reflection of their own hatred, and within Lon Reed's small frame walked a man who despised almost everyone. Crude remarks and bigoted bias my dad could endure, for these were passive shortcomings, but when Lon's cruelty turned active, Dad harbored no guilt in wanting his stepfather removed from the planet.

Lon's worst show of spite happened when my father was a boy. Intolerant of animals, especially young ones that barked when hungry, the drill sergeant stuffed Dad's first puppy, along with its little brothers and sisters, into a gunnysack. He then hurled the bag into an irrigation canal and laughed while the puppies drowned. Dad once told me that their pathetic cries and whines, as the waters swept them along, left him with violent nightmares and an obsession to adopt all of the world's helpless animals. I'm certain that losing Scratchy hurt him much deeper than he dared show.

Thoughts of his childhood plagued my dad as he stood outside Frank Mason's office in the Air Force building that day. The reasons for this were unclear, but he suspected that, despite how much he liked the captain, the smell of the man's Old Spice aftershave always reminded him of Lon Reed.

Dad tapped on the open door and stepped inside. "Got a minute, sir?"

Still on the phone, Mason nodded and pointed at a chair. Dad sat.

Mason reeled off a few more commands, then hung up the phone. "What's on your mind, BJ?"

Having been renamed Billy Joe Reed after Lon Reed adopted him, Dad preferred to be called BJ by his friends. "I think I know why we can't find the Soviet subs."

Mason sat forward in his chair. "Go on."

"Do you remember what the Germans did in the war to keep us from DFing their CW transmissions?"

"Yeah," Mason said. "They recorded their Morse code communications, then sped up the recorder before transmitting. That let 'em send out a shortened message on a specified frequency at a set time. Those bursts were so short that we couldn't get a good bearing—" Mason stopped midsentence. He stood up and brought a hand to his chin. "I'll be damned. Ivan's using a burst signal."

"I'd bet my stripes on it," Dad said, "and that means we'll be lucky to find it. And even if we do, how the hell are we going to get a bearing? The duration of the signal will be too short."

"All good questions, Chief, and I wish I had the answers. For now, let's focus on the first step first."

Dad nodded. "We need to find the damn thing."

"And fast," Mason said as he moved to the side of his desk. "Right now we're one of the closest stations to the Soviet backyard. If we can't find the burst, nobody can. And I've got NSA breathing down my neck every day. If Ivan thinks he's invisible, he just might get cocky and start firing missiles."

Dad stood up. "Understood."

As my father turned to leave, Mason called after him. "One more thing, BJ."

Dad turned and cocked an ear. "Sir?"

"You gotta find this thing in less than a week."

"NSA?"

"No, Petersen. I'd say he's got about five days before he scratches himself into the base hospital." Mason flashed a brief smile and turned back toward his chair.

Dad returned the smile and walked out of the captain's office. About a minute later, his smile faded as he thought of the impossible task he'd just been given. He spent the next week monitoring every HF known, along with a wide assortment of frequencies outside the normal range. He heard nothing except static and an occasional pop and scratch. He also employed a sonograph to analyze any suspect signals.

Back then a sonograph was a three-foot-long by eight-inch-wide machine that made sound waves visible. The unit housed a large drum around which operators wound photographic paper for each signal analyzed. On playback of a recorded signal, a stylus imprinted an enlarged image of the signal for inspection by analysts. Unfortunately, given the dilapidated condition of his years-old navy-issued sonograph, Dad saw no signs of a burst signal.

His shoulders slumped, his eyes red and swollen, his smile gone, my father walked through the door of our apartment and plopped onto the couch. My mom tried to console him, but to no avail. My sister and I also did our best to make him feel better, but our attempts at levity went down in shambles. Dad just opened a beer and sat staring at the wall.

The next morning, while getting ready for school, I noticed a worn spot in the rug by the door. Then I remembered. Scratchy earned his name by scratching at that spot, like he was trying to dig a hole to China. I knelt on the rug and scratched at the carpet, just like Scratchy once had, trying to remember his hairless little bear body.

Still sitting on the couch, his chin blackened with stubble, Dad said, "Billy, would you please stop that? You sound just like—" He sat up on the couch and opened his eyes wide. "Oh my God. That's it! The burst signal sounds just like Scratchy's carpet scratching."

Dad jumped from the couch, ran over, picked me up, and gave me a hug. Smiling, he bolted into the bathroom to shower and shave. I couldn't explain why he'd suddenly transformed from depressed to ecstatic, but I figured it must be a grown-up thing. I learned later that our pet bear's scratching probably saved the navy's ass.

Dad sped to the base in our Volkswagen in search of an Air Force colleague named Jimmy Hensley. He bypassed his shabby Quonset hut and charged into the plush concrete-and-steel air-conditioned building next door. There he flagged down Airman Hensley. "I need a big favor."

Noticing the excitement on my father's face, Hensley said, "Does it involve a woman?"

Dad frowned. "No, it involves being a sneaky little thief. Think you can handle that?"

Hensley inched the corner of his mouth into a wry smile.

Whether Air Force or navy, everyone knew that maneuvering the military system to obtain supplies, equipment, or parts required a master's degree in procurement manipulation combined with borderline "cumshaw" thievery. British sailors coined the word *cumshaw* from one they heard from Chinese beggars that meant "grateful thanks." And in the art of cumshaw, Hensley boasted a Ph.D. Often compared to Milo Minderbender, the mess officer glorified in Joseph Heller's *Catch-22*, Hensley could darn near find anything, for a price.

Dad heard that Hensley acquired, on behalf of Master Sergeant Rich Cousins, three brand-new sonographs for the Air Force unit. Dad's section, Det 28, employed an old unit held together with bailing wire, tape, and a wad of bubblegum. Since the Air Force got all the best supplies, Cousins could offer "cumshaw guy" Hensley a few pieces of unused equipment in exchange for the new sonographs, while Det 28 had nothing to trade. With an epiphany running around in his head, Dad knew that he could find the Soviet burst signal, but not without at least two of those sonographs. Getting them, however, would be a major challenge.

"I heard you found three new sonographs for Sergeant Cousins," Dad said as he cornered Hensley near an office doorway.

"Did indeed," Hensley said, standing next to the door.

"I need two of them ASAP," Dad said, feeling like a beggar.

"Well now, that's gonna cost you. I don't think Cousins is going to—"

"How much?"

"More than you can afford, Chief," Hensley said, as he leaned against the doorframe.

Dad's temples throbbed. He needed those units. Lives might be at stake. At the very least, a few careers. Hensley was right, however. Dad could tap everyone at Det 28 for a loan and still not get enough to buy two new sonographs. He had to devise a way to borrow the damn things—indefinitely and for free.

"There must be something we can trade," Dad said out of desperation.

Hensley rubbed his chin. "Don't think so. You boys ain't got nothing anybody needs."

Hensley had a point there. Most of the Air Force personnel called Det 28 the "Orphan Annie" of Karamürsel. Relegated to a Quonset hut on the other side of the tracks, they received none of the perks afforded their Air Force counterparts.

Dad noticed that he was standing on new linoleum tile and that the walls smelled of fresh paint. Probably Hensley's doing. My father searched his brain for inspiration. What could he offer that Hensley might need? Probably nothing. But maybe there was something Cousins needed. "Does Sergeant Cousins know how to use those new sonographs yet?"

Hensley tilted his head to one side like a dog training his ear on a sound. "What do you mean?"

"Those new sonographs aren't anything like the old ones. You did give Cousins operating manuals, didn't you?"

"Well, no, I don't think so."

"So what happens when his men can't get the things to work? Who's he going to blame?"

"I . . . I don't know. He wouldn't blame me, would he?"

Still blocking the doorway, Dad stood up tall. He now hovered a good four inches above Hensley. Furrowing his brow, he stepped up close to the airman and produced a slow, deep bass. "You mean you got him three new pieces of equipment without operating manuals? You might as well have given him boat anchors."

Hensley took a step backward. His lower lip quivered. "This is not good. Cousins will be pissed."

"And he's not a man you want to piss off," Dad said, knowing that Cousins could be meaner than a bulldog on steroids when angered.

"I'll just have to find him some manuals."

"Weeks," Dad said. "That'll take you weeks. Then he'll *really* be angry."

Hensley looked at Dad with wide eyes. "So what should I do?"

My father placed a sympathetic hand on Hensley's shoulder. He softened his tone and said, "Maybe I can help. I was trained on those units in Guam. I could teach Cousins's team how to use them. That'll buy you some time to get those manuals."

Hensley looked relieved. "You'd do that for me?"

Dad smiled. "Sure, for a price."

That afternoon Hensley delivered two new sonographs to Det 28's Quonset hut. Everyone whistled approval as my father opened the boxes and removed the units.

Scratching at his head, Commander Petersen came out of his office and stared at the early Christmas presents. "Where'd you get *those* babies?"

"Borrowed 'em," Dad said as he plugged one in. "Kind of indefinitely."

"Why do we need them? We have one already."

"It's old and worthless. I needed two new ones. I have a hunch."

Petersen scratched his scalp. "A hunch?"

"Yeah," Dad said, trying to ignore Petersen's dandruff. "A *scratchy* hunch."

Over the next several days, my father listened to a series of frequencies on the HF band. Days earlier, before his epiphany, he'd heard nothing but pops and scratches. But one scratch differed from the rest. When I started scratching at the rug in our house, the sound reminded my dad of the "scratch" he heard at around 345 Hz. At first, he passed the burst of static off as an anomaly. Now he listened for the sound with unequaled intensity. Hours passed with no joy. Then, suddenly, he found it again.

Like an excited kid on Christmas Eve, my father used the two new sonographs to make an enlarged picture of the signal. He needed both of the units so he could record numerous signals quickly and compare them side by side. While studying "sound pictures" of the signal, my father noticed something odd. He squinted, then pulled out a magnifying glass. To no one he said, "I'll be damned, this thing has bauds."

He grabbed the sonograph printout and ran over to Mason's office. Bursting through the door without knocking, he threw the printout on the captain's desk. "It has bauds."

Mason sat forward and stared. Dad handed him the magnifying glass.

Mason studied the printout, then smiled. "Best Christmas gift I've ever gotten, BJ."

The term *baud*, named for the French engineer Jean-Maurice-Émile Baudot, became the primary yardstick for measuring data transmission speeds until it was replaced years later by a more accurate term, bits per second (bps). As far as Mason and my dad were concerned, a baud equated to something man-made, and that translated to "Gotcha!"

In similar fashion to the burst signal used by the Germans in World War II to thwart direction finders from locating Morse code transmissions, the Soviets invented their own burst signal for HF communications. The bauds they used were the most compressed ever and represented a huge leap forward in radio technology.

After studying, recording, and confirming the burst signals, Det 28 sent copies of their findings to the NSA in Maryland. The agency assigned their best analysts to the case and instructed all stations to obtain as many recordings of the new burst signal as possible. Soon every Huff Duff started finding them.

My father spent the next several weeks thanking me for helping him solve a major problem. I had no idea what he was talking about, but I rejoiced in the fact that he seemed happy again. I still didn't see him much, as he spent most of his days and nights at the base analyzing the burst signals. He discovered that they came with a "trigger" at 345 bits per second (bps), followed by a series of bauds at 142 bps. He figured that the trigger probably started a recorder at the receiving station. The series of bauds were followed by a short message burst. Dad knew that the NSA might never decipher the contents of those messages, but that was not the mandate of a Huff Duff. These stations were designed to find and analyze, not decode. Unfortunately, that still left them with an impossible task: how to get a good bearing to a burst signal.

These signals were so short that determining an accurate bearing could not be done. Even getting an inaccurate one posed a significant challenge. With the Soviet navy launching hundreds of new submarines, many capable of wiping out dozens of cities in the United States within minutes, the NSA pushed the program up several rungs on the priority ladder. If they couldn't find a way to DF the Red Bear's new burst signal, they couldn't find its submarines. And that made the world a very scary place to live in.

For his team's diligence in finding and analyzing the new Soviet burst signal, Captain Mason received a letter of commendation from the National Security Agency. In turn, he handed my father a letter of appreciation and recommended him for limited duty officer (LDO). My dad was on his way to leaving the ranks of the enlisted and becoming an officer in the U.S. Navy. That winter he flew stateside to undertake one of the most important assignments of his career.

CHAPTER FOUR

The beginning is the most important part of the work.
—PLATO

WHEN MY DAD, WILLIAM J. REED, reported to NSG headquarters at 3801 Nebraska Avenue in Washington, D.C., in early 1961, he couldn't keep his hands from shaking. The shivers were caused in part by the snow falling on the shoulders of his navy jacket, in part by the anticipation of what lay ahead. After discovering the Soviet burst signal, and the fact that this thing contained data of some sort, Reed received temporary orders to return to the States.

Master Chief Reed knew that, in March 1959, the NSA combined all military electronic intelligence (ELINT) programs under one roof, and that Howard Lorenzen's group supported the Advanced Signals Analysis Division of the NSA's Office of Collection and Signals Analysis headed by John Libbert. He also knew that the bulbs glowing in the heads of these engineers were brighter than ship-borne searchlights. And while Reed considered himself a pretty smart guy, he figured that chess matches with these geniuses wouldn't last more than two minutes.

In the world of electronic countermeasures, few icons commanded more respect than Howard Otto Lorenzen. In July 1940, after five years of designing commercial radios, he launched his career in ethereal warfare at the Naval Research Laboratory working for the brightest minds in radio engineering. During World War II, he developed a system to analyze German aircraft radio signals that controlled glide bombs. This allowed NRL's Special Projects Section to create intercept jammers for enemy aircraft bomb controllers that rendered the things useless. German Luftwaffe engineers thought the problem was their fault and dismantled the systems.

Lorenzen worked on similar projects throughout the war, eventually overseeing a dozen small groups tinkering in various fields of radio engineering. After the war, he invented the term *electronic countermeasures,* defining ECM as a "discipline that first detects, then interferes with or analyzes for intelligence purposes any electromagnetic energy emanating from the enemy." The Bureau of Ships concurred with his definition and sponsored ECM projects at NRL for intercept, direction finding, radar jamming, and decoy systems.

Lorenzen and other key members of his team remained government employees after the war, pulling apart captured German electronics like excited kids in science class. He managed to convince the Brits to lend his group a key piece of German technology—used in the Wullenweber (pronounced *VOOL-in-veber*) antenna sites—that eventually helped redefine HFDF forever. Lorenzen's HFDF expertise brought him to NSG's headquarters to meet with Reed and others and help solve the problem of gaining accurate, after-the-fact bearings to Soviet burst signals.

Still nervous, Reed introduced himself to the team. Lorenzen and a dozen engineers grilled him for hours about the nature and characteristics of the burst transmission that he'd discovered and analyzed. They examined the sonograph printouts and grilled him some more. When not grilling, the narrow-tie-wearing team swapped theories and ideas using complex phrases that, as far as Reed was concerned, were akin to Latin. Techno, technara, technatus, technodom. Now and then Reed managed to grab hold of a concept and attempt to bring the cloud-dancing scientists back down to earth, where submarines transmitted and Huff Duff stations listened.

Over the course of several days, the team determined that the burst signal lasted no more than seven-tenths of a second. That posed a huge problem. Equipment at Huff Duffs was designed to locate and determine bearings to ordinary high-frequency transmissions, most of which lasted several seconds or even minutes. Now they needed not only to hear a signal that short but also to determine an accurate bearing to the submarine long after the transmission ended. That's like trying to find your car in a parking lot when the horn honks for only a half-second after you push the key-chain button.

Using recordings captured by Reed and others, the team analyzed the signals and determined that each transmission consisted of a two-tone alert designed to trigger an automatic receiver/recorder. A short encrypted data stream followed that contained message information. The Soviets probably figured that no one could direction-find such a truncated transmission. After a week of analyzing signals and bantering over ideas to solve the bearing problem, Lorenzen figured the Soviets just might be right. He threw up his arms in frustration, stating that the Russkies may have finally found a way to trump American engineering. Another engineer, Robert Misner, then asked if it might be possible to create a device that "triggered" a switch after a burst signal was picked up by an antenna. Lorenzen pondered the question and answered yes, that a trigger might be possible, but to what end?

Misner flashed a smile. He reminded Lorenzen about their work together, several years earlier, on a magnetic tape recorder. When the Soviet threat escalated during the Korean War, Lorenzen's efforts, in collaboration with others, led to a new ECM system installed into antisubmarine aircraft. While gaining operational feedback on this system in 1949, Lorenzen had an epiphany.

That's when he contacted Robert Misner, and together they created the first magnetic tape recorder for intercept work. They called this device the Radio-Countermeasures Sound Recorder-Reproducer, dubbed the IC/VRT-7. After that project, Misner did some research on after-the-fact transmission analysis in 1958. He now thought that by combining what he'd learned from the two projects, perhaps they could create a trigger that started a magnetic recorder and determined a bearing to the transmitter based on the recording.

Other engineers on the team scoffed at the idea, and Reed leaned in favor of the skeptics. Finding an accurate bearing to a live, longer-lasting transmission posed enough of a problem. Misner dismissed the naysayers and sketched his concept on a blackboard. As the white chalk revealed dozens of boxes, lines, and arrows, Reed's eyes slowly opened. Misner's concept started to make sense. If they could engineer a way to record the time at which a transmission was detected, along with the strongest bearing to the signal, then compensate for inaccuracies and other conditions, they just might be able to find Ivan in a haystack.

To achieve this, the NRL engineers needed to overcome a big issue with Lorenzen and Misner's magnetic recorder: the thing didn't have enough capacity to store the hours and hours of recordings needed for after-the-fact analysis. Today we have programs on iPods that can record a song playing on the radio for a few seconds, then upload the recording over the Internet, where it's analyzed to determine the artist and song. We take such a feat for granted, forgetting that this requires gigabytes of storage capacity and superfast microprocessor speeds. In the early 1960s, there weren't microprocessors with billions of bytes of storage and memory capacity. Storing burst transmission recordings magnetically required "out of the box" thinking that pushed engineering envelopes. Matching these recordings to accurate time signals down to the millisecond raised the bar even higher. Months passed before the team could overcome the limitations and build something that actually worked.

At the time, no one on the team, least of all Reed, imagined their groundbreaking new device, christened the AN/FRA-44 recorder/analyzer, might one day earn a place in history as one of NRL's top seventy-five inventions, with Robert Misner accepting the prestigious award. They also did not expect their new system to play an integral part in thwarting a world war less than a year later.

Using innovative technology, the FRA-44, called "fraw forty-four" by operators, allowed the U.S. Navy to record a Soviet microburst, analyze the signal after the fact, and determine a bearing to the source. One major problem in countering the Soviet stealth innovation appeared solved, but an equally daunting one remained: designing a way to receive the burst signals in the first place.

Current antenna and receiver technology used at Huff Duffs already lagged behind the Soviets, who had deployed twenty Krug Wullenweber sites based on captured German designs. The United States still used an antiquated AN/GRD-6 antenna array and HF receiving system, which was no match for microbursts. Although the new recorder/analyzer designed by the team could work at such a site and provided a short-term fix for locating Soviet submarines, the DF accuracy would be worse than a World War II Huff Duff.

With Reed's help, Lorenzen's team, led by Bob Misner, determined that a new type of antenna/receiver was needed, one that increased reception and accuracy by an order of magnitude. Fortunately, such a device had already been constructed by the NRL a few years earlier at the Hybla Valley Coast Guard Station in Alexandria, Virginia. The United States built this station using German Wullenweber technology in 1957 to track the Soviet Sputnik's transmission signal and determine its orbit. While Reed returned to Turkey to test the new recorder/analyzer in the real world using older antennas, Misner commandeered the Virginia station to test the makings of a new electronically steerable array that could find Ivan's silent boats.

The NSA funded both projects on a high-priority status and gave them top-secret code names. They called the project using the recorder/analyzer and related equipment to find the burst signal Boresight and the new Wullenweber antenna project Bulls Eye.

Before Boresight could become operational, however, Reed needed to gather some critical field data. To do so, he'd need to sneak onto Ivan's back porch without getting caught.

MORE THAN FIFTY YEARS AFTER THE invention of Morse code, Guglielmo Marconi launched radio communications by sending the letter *S* across the English Channel in December 1901. The science of locating radio transmissions came to light a few years later when John Stone used radio direction finding (RDF) techniques in 1904 to locate transmitting sources. The team of Ettore Bellini and Alessandro Tosi improved on Stone's designs, and Marconi acquired the patents to this technology in 1912. He then mounted the newly acquired RDF equipment on commercial ships.

During World War I, Captain H. J. Rounds of the Royal Navy installed a series of RDF stations along the east coast of England for Room 40, the British Admiralty's code-breaking intelligence branch. These stations came in handy during the battle of Jutland in the summer of 1916, a World War I clash between battleships in the North Sea near Denmark that to this day is considered the largest naval battle in history.

Captain Rounds ordered his stations to monitor the movements of the German battleship *Bayern*. Using RDF, operators reported that the *Bayern* steamed some distance north during the night. Using this information, Vice Admiral David Beatty, commander of the First Battle Cruiser Squadron, avoided the U-boat threat and caught the Germans off guard. The Brits engaged Franz von Hipper's battleships long before the German admiral expected. While the battle proved costly for both sides, the advantages of RDF were solidified in the minds of military experts.

That same year, under the direction of Commander Laurance F. Safford, head of OP-20-G (20th Division of the Office of Naval Communications) and "father" of the navy's communications intelligence unit, the navy built an Atlantic arc of twenty-six HFDF stations. These Huff Duffs stretched from Britain to Iceland to Greenland, across the eastern states, and down to Brazil and Africa. German submarine tactics mandated frequent radio contact between U-boats and headquarters. When these skippers called home, they were unaware that a giant ring of Huff Duffs was capturing these signals and finding a direction to the source. By cross-referencing bearings from multiple Huff Duff sites, the Allies could multiangulate approximate locations for the transmitting submarines.

Stations reported bearings to Net Control in Virginia, which forwarded the same to head of Naval Communications Intelligence Commander Knight McMahon's staff in Washington, D.C. Fixes were then flashed to the Atlantic Section of the Combat Intelligence Division, which shot them out to U.S. antisubmarine warfare forces. Unfortunately, the system's accuracy left something to be desired, and the definition of a good "fix" equated to fifty miles from the target. Despite this limitation, for more than a decade after the war, the navy did little to upgrade its twenty-six Huff Duffs.

When the NRL team officially launched Project Bulls Eye in 1961 they radically upgraded the ability of HFDF sites to detect weak HF signals and improve DF accuracy. To accomplish this, they needed help from the Germans.

During World War II, German engineers invented the Circularly Disposed Antenna Array (CDAA) as a way to improve their own Huff Duff capabilities. They built the first site at Joring, Denmark. The German CDAA used forty vertical antennas placed in a circle with a diameter of 360 feet—about the same diameter as the average baseball stadium. Forty more antennas, designed to reflect signals from the first circle, were suspended on a circular wooden support structure just inside the outer ring. From the air, the entire affair looked like two giant Ferris wheels, one inside the other, turned on their sides and missing all the seats.

The Germans built only two CDAA arrays under the code name Wullenweber—a name prompted by the exploits of Jurgen Wullenweber, who became mayor of Lübeck in 1531. This iconic figure gained a reputation as a fighter against injustice and the wealthy class, much like Robin Hood. The story of his adventures prompted Dr. Hans Rindfleisch, the group leader of the German navy's communication research command, to use his name for the CDAA program.

After the war, the Brits studied the Wullenweber design in Denmark, then destroyed the array in accordance with Geneva Convention mandates. Some of Rindfleisch's engineers were captured by the Soviets and taken to Russia. The Red Bear's Defense Ministry soon erected its first Wullenweber site at Khabarovsk Krai under the code name Krug, which means "circle" in Russian. The massive antenna array spanned a diameter of more than a half mile. The Soviets built nineteen more sites throughout the 1950s, with many installed in pairs within a few miles of one another for navigation purposes. Four Krugs were installed near Moscow, and some were used to track Sputnik satellites via 10 and 20 MHz beacons.

Although the Allies snatched up their own Wullenweber engineers after the war under Operation Paperclip, they were slow to the game. Antenna researcher Dr. Rolf Wundt, along with his wife and parents, arrived in New York City on the same ship as Wernher von Braun in March 1947, but he did not work on this technology until many years later. The Air Force, and later GT&E Sylvania Electronics Systems, made some progress on Wullenweber antenna technology, but more than a decade passed before the first site became operational.

Professor Edgar Hayden, a bright engineer at the University of Illinois, under contract to the U.S. Navy, led the charge to build America's first Wullenweber. He studied the German design and analyzed potential performance possibilities against current Huff Duffs. That's when he got excited. His calculations concluded that inaccuracies could be reduced from as high as three percent down to one-half of one percent. That small change could be the difference between sending navy aircraft to find a sub off New York City versus Long Island. Hayden also found that Wullenweber arrays could select desired signals and reject interfering signals or noise detection. This helped extend detection ranges out to several thousand miles away—four times that of current antennas. With the Soviets extending the range of their ballistic missiles, and hence their submarine patrol distances away from U.S. shores, longer range capability held a high degree of importance.

Blessed by the navy, after reporting the good news, Hayden assembled a team to build a Wullenweber array at the university's Road Field Station near Bondville, Illinois. The array contained a ring of 120 vertical pole antennas that "listened" in the HF range of 2 to 20 MHz. Tall wooden poles, comprising a hundred-foot-diameter circle, supported a screen of vertical wires located within the ring of monopoles. From a distance, the site looked like a giant circular cage large enough to keep elephants from escaping, which spurred the term *elephant cages* often used by operators.

Based on lessons learned from the Bondville experimental array in 1959, the Air Force awarded a contract to GT&E Sylvania Electronics Systems to build a larger Wullenweber elephant cage—the AN/FLR-9—at RAF Chicksands in the U.K. This "Flare-nine," along with a sister site at San Vito, Brindisi, Italy, was not scheduled to light off until late 1962. The Air Force used these arrays for airborne tracking and not HFDF although the navy planned to borrow these antennas for such by stationing NSG personnel nearby.

In mid-1961, when Robert Misner installed the newly invented Boresight AN/FRA-44 recorder/analyzer, the navy's plan expanded. With help from Stanford Research Institute, the original Wullenweber designs were improved upon, resulting in something more advanced called the Wide Aperture Receiving System (WARS). Since the Air Force owned AN/FLR-9 as its official designator for the new CDAA antenna and systems, the navy named its design AN/FRD-10. Operators called them "Fred Tens."

While these sites were designed to conduct some of the most sophisticated radio interception work ever, much of the equipment used, aside from the special fraw forty-four Boresight recorder/analyzer and related systems, came from "off-the-shelf" sources. Each site contained an abundance of such gear, and even small failures or calibration errors could badly degrade bearing accuracy. With the Bulls Eye and Boresight programs underscored by massive budgets, most everything ordered for these facilities arrived in baker's dozens, from antennas to multicouplers to receivers. Miles of cable snaked through, under, and around the build-ings, ending in hundreds of coaxial connectors for coupling to various devices. Only one special device held the honor of being installed as a dynamic duo: the goniometer.

Used by the Germans in their Wullenweber designs, the goniometer owes its name to the Greeks. *Gonia* translates as "angle," and *metron* means "to measure." A spinning goniometer became the backbone to a functioning Fred Ten by refining the process of searching various fre-quencies. Not unlike a carnival wheel on which various prize amounts are indicated, a goniometer rotates around various frequencies by "touch-ing" the pole antennas in a Wullenweber array. Recall that our array consists of a bunch of tall antennas positioned in a big circle. So, if the strongest signal from a transmitting submarine is coming from due north, as the goniometer spins, it will measure a higher signal strength coming from the antenna pole positioned at zero degrees in that circle. After compensating for inaccuracies, time delays, atmospheric conditions, and so on, via lots of sophisticated equipment and analysis, we can determine a bearing to our contact of, say, 358 degrees—roughly in the direction of Santa's house at the North Pole.

Original Fred Ten designs consisted of two independent goniome-ters that were later replaced by a single ten-foot-long dumbbell-shaped unit with four-foot-diameter router housings on each end. These resem-bled the spinning "g-force" simulators used to train pilots and astro-nauts, only smaller. Since these sites were built prior to the invention of uninterruptible power supplies (UPSs), engineers installed electric mo-tors driven by generators with large flywheels. Diesel engines spun the flywheels during power outages, which took over for the electric motor when the primary power failed.

The U.S. Navy contracted with ITT Federal Systems to deploy a worldwide network of more than a dozen Wullenweber elephant cages for HFDF operations. The Fred Ten near Okinawa, Japan, became the first installation, but it did not come up to full speed until the second half of 1962. An elephant cage near the Scottish village of Edzell also came on line that year. Nestled in a farming area in the foothills of the Grampian Hills, some thirty-five miles south of Aberdeen, that site re-placed less sophisticated listening posts in Germany and Morocco. The navy erected another elephant cage in 1962 at Skaggs Island,

California, not far from San Francisco. Each of these facilities cost just shy of $1 million and employed dozens of navy and civilian personnel. At the time, operators at the Skaggs Island Bulls Eye site were unaware of their destiny to play a significant role in the Cuban Missile Crisis.

IN EARLY 1962, REED RETURNED TO Turkey. Within hours of his return, he jetted to the Karamürsel base to integrate the new Boresight technology into the existing DF systems. Although the Air Force had not yet installed a Wullenweber elephant cage there, which meant that bearing accuracies would be poor, the objective now focused more on getting something working versus working well. Reed was also tasked with writing an installation and operations manual that could aid other DF sites in implementing the new systems.

With the help of his colleagues at Karamürsel, under the watchful eye of Captain Mason and Commander Petersen, Reed installed the new Boresight receiver/recorder and related equipment developed by the NSA team. Now, if he could only get the damn thing to work.

The theory seemed simple: When a receiver encountered a "trigger heading" on a burst signal, a sixty-inch-per-second recorder with two-inch-wide tape automatically switched on. The recorder captured the signal, along with a marker indicating the time to the millisecond that the signal was intercepted. Because the Boresight system enabled operators to also capture directional signal strength and other parameters, synchronized by the time marker, they could now determine, after the fact, the probable bearing to the transmitting sub.

In order for Net Control to get a reasonable fix on the sub's location, additional bearings were needed to create a multiangulation. So until more stations came on line, Boresight remained useless. As such, while Lorenzen's team tackled the enormous problem of building more Wullenweber sites to improve accuracy, Reed received orders to help get other sites—most equipped with older GRD-6 antennas—up and running. The navy hoped that if enough of these sites were operational, they could at least achieve a ballpark fix good enough for ASW forces to have a fighting chance.

For the next several months, Reed flew around the world to install systems and train operators at sites along the Atlantic and Pacific

Ocean perimeters. Operators and station chiefs were excited about the possibility of finally hearing the Soviet subs again, but they were not so thrilled with the amount of work and resources required to become operational.

The space required for the reception and recording equipment covered an area as large as a typical living room and needed to be air-conditioned since the receivers in those days still used vacuum tubes that generated considerable heat. After installation, days of calibration and testing were needed, along with many long hours of troubleshooting to ensure that everything worked properly. R-Branchers needed to be trained on the equipment, what to listen and look for, and how to properly analyze the burst signals. Reed usually spent weeks at each facility before certifying them as Boresight operational.

Back in the States, Howard Lorenzen and his team of geniuses went to work on a jamming system. Using similar technology to that used in intercept jammers developed by NRL's Special Projects Section in World War II to hamper German aircraft bomb controllers, Lorenzen's team built systems that could send out false signals on the same frequencies used by Soviet burst transmitters. This made it a little harder for Moscow to communicate with its subs and vice versa. Reed took several trips to England to help engineers there install the burst signal jammers, but these devices came with a limited range and were effective only when the Soviet subs passed near the British Isles.

In the spring of 1962, William J. Reed found out that he'd been selected for a commission in the U.S. Navy. All those years of correspondence courses, night school, and hard work finally paid off. Commander Mason informed Reed that he'd earn his ensign bars in July, and he and his family would be leaving Turkey that same month. After his arrival in the States, he'd head to LDO School in Newport, Rhode Island, for "knife-and-fork" training in August, then to the NSA facility at Fort George G. Meade, Maryland. Until then, several more months of grueling travel lay ahead.

A key ingredient to ensuring that Boresight could obtain an accurate bearing to a transmitting submarine entailed calibration and signal analysis. Using the example of finding one's car in a parking lot, two things are taken for granted with human hearing that are not prevalent in the world of HFDF. One of these is that we know what a car horn sounds like. The other is that, for most of us, our ears are also familiar to us, and over many years we've learned how to discern from which direction sound is traveling. In other words, we're pretty sure that our horn is the one blaring at us from an easterly direction.

This was not the case for the systems used to detect locations for Soviet burst signals. There were just too many unanswered questions about the characteristics of this new type of signal, and before Boresight could be made fully operational, more information was needed. Someone had to undertake the job of finding a Soviet sub or two and get them to transmit while analyzing and calibrating signal location, strength, type, frequency, and time on the air. Using these parameters, operators could test and properly calibrate Boresight systems to be sure they were not providing false hits.

When Reed was ordered to ride on a Turkish sub to see if he could capture a burst signal from a nearby Soviet boat, his heart raced. He'd never been on a submarine before, let alone an old smoke boat that appeared to be missing half a lung and one eye. The Turks called her the *Birinci Inonu,* which loosely translated as "First Prize" or "Number One"—hardly an apt description befitting this blue-haired geezer in an Istanbul harbor that oozed the foul scent of diesel fumes. Holding his nose, Reed crossed the wooden gangway and boarded the sub.

The *Birinci* once served the U.S. Navy in World War II as the USS *Brill* (SS-330). She launched from Groton, Connecticut, on June 26, 1944, and the Turks bought her after the war on May 23, 1948. A slick film of oil surrounded her 312-foot black hull, where ten torpedo tubes, six forward and four aft, had fired MK-14s at the German navy seventeen years earlier. Reed once read that the *Birinci* could hit around twenty knots on the surface and ten submerged, driven by a couple of large diesel engines and electric motors.

The topside watch saluted as Reed approached. He handed over his orders and in Turkish asked to see the skipper. Long minutes passed before a stocky barrel of a man emerged through the hatch. He displayed short-clipped hair and a tight mustache and carried a stern "I'm in charge" look. He introduced himself as Captain Celik and motioned for Reed to follow. Grabbing his seabag, Reed descended the ladder into the belly of the dragon.

Below decks, the *Birinci* smelled even worse than she did topside. So did her crew of eighty-five. They paid Reed little attention as they prepared to get under way. Captain Celik escorted Reed to his stateroom, which was also a misnomer. The small space housed two bunks and a curtain. No door. Another officer who shared the space—introduced as the navigator—smiled and shook Reed's hand. They talked briefly, then walked to the wardroom for the mission briefing.

Captain Celik greeted Reed near the wardroom and handed him a cup of black coffee and a pastry. He smiled and said, "A cup of coffee commits one to forty years of friendship."

Recognizing the Turkish proverb, Reed returned the smile and said, "A hungry stomach has no ears."

Captain Celik cocked his head, offered a friendly Turkish hand gesture, and entered the wardroom. Reed said little during the briefing, as a majority of the crew did not have a "need to know" the details of this mission. Such was the tacit agreement between the two navies: we cooperate like allies; we defend our secrets like enemies. Three U.S. Navy technicians trained on Boresight and ESM equipment were also present. The ESM equipment had been installed days earlier by those technicians.

The *Birinci* sputtered and belched as she edged away from the pier. The diesels vibrated and hummed, and the saliva in Reed's mouth disappeared. Standing in the control room, one level below the conning tower, Reed watched a Turkish seaman attempt to repair a leak in a hydraulic line—with a hammer. Any doubts that Turkish submariners were the most dangerous species of mammal on the planet evaporated.

Captain Celik steamed the *Birinci* into the Black Sea and submerged. The world turned quiet as the batteries spun the boat's propellers, and Reed spent most of his time in the conning tower working with the technicians to test and calibrate the ESM systems. The Turkish sailors gazed at the U.S. techs with curious eyes, but having been briefed by their CO regarding secrecy, they refrained from asking any questions.

While patrolling near Sevastopol, days passed without a contact. Then the sonar operator heard the muted chugging of a snorkeling submarine. Captain Celik steered toward the contact. Chatter in the boat ceased. Faces turned serious as they closed to within a few nautical miles. Although the *Birinci* was an old girl, she'd been upgraded with reasonably decent sonar gear. The same could not be said for the sonar operators. They were clueless as to the possible contact type. Reed asked for permission to take the headphones. The captain nodded agreement, and a Turkish sailor handed Reed the phones. He sat near the sonar stack and listened. His face wrinkled with concentration. Then he heard it: the distinct diesel engine chug of a *Foxtrot* submarine.

Although Captain Celik remained in charge of the boat, Reed assumed command of the mission. Once contact was made with the enemy, the mission commenced, and Celik now technically reported to Reed. Technically.

Reed ordered Celik to close the distance so the ESM gear could get a signal. Reluctantly, the captain issued the proper orders to his crew. The *Birinci* turned, slowed, and inched toward the Soviet boat. Sweat dripped from faces and soaked coveralls. Reed stood behind the U.S. Navy techs and gave directions regarding the signal types and frequencies to listen for. If they could capture a burst transmission and match that against the actual position of the transmitting target, that would go a long way toward accurately calibrating the newly installed Boresight systems. They could only hope that the Russian submarine transmitted before she left the area and went deep, and that might be a long shot at best.

"Now we wait?" Celik said in Turkish.

"*Evet,*" Reed said. "Now we wait."

"Do you gamble, Mr. Reed?" Celik said.

Reed flashed a puzzled look. "Gamble?"

"*Evet.* Gamble. Poker, blackjack, you know, gamble. Do you not understand this word?"

"Yes, I understand. Why do you ask?"

"We have an old saying, perhaps you've heard this. 'The wind that the sailor likes does not blow at all times.'"

Still perplexed, Reed said, "What's your point?"

"I like to gamble, but not in a house of bad odds. If I had to gamble now, I'd bet that your mission fails."

Reed removed twenty American dollars from his pocket and waved them in front of Celik. "I'll take that bet."

Celik smiled and removed some bills from his pocket.

Hours passed with no joy. The Soviet sub continued to snorkel without transmitting. Reed started to wonder if he'd just lost a day's pay to Celik. His mouth dry, his armpits moist, his temples throbbing, Reed knew that the success of their mission depended on getting that Russian boat to send out a burst. But how?

Reed's mind scrambled for an answer. At first he refused to listen to his own thoughts, as to do so meant risking more than he cared to, more than he knew Celik would accept. Reed walked over to the sonar console and asked the operator to let him take the stack. He pulled on the headset and listened. Celik watched from the other side of the conning tower, his forehead forming curious lines above his thick black eyebrows. Reed ordered Celik to slow and pull to within 4,000 yards—less than two nautical miles away.

"I will not!" Celik said.

"You will," Reed said, "or I'll see to it that you lose your command."

Celik glared. "Two captains sink a ship." He gave the order to the helmsmen, and the *Birinci* moved closer.

Several minutes later, the diving officer reported that they were now 4,000 yards away. Celik ordered all stop and raised the ESM mast and the attack periscope. He swung the scope left, then right. He marked two bearings and lowered the mast. "*Foxtrot* at two-three-five and a *Skory* at one-eight-nine."

"A *Skory*?" Reed said. "We never heard her."

"She's not moving. She's just sitting there about 2,000 yards behind the *Foxtrot*."

Reed searched his head for stats and recalled a few. The Soviet *Skory*-class destroyer carried a slew of ASW equipment and weapons, including four depth charge racks on her afterdeck. No doubt the *Skory*'s captain longed for the chance to use them against a macho Turkish sub skipper. That ship wouldn't sit still forever, and if she came their way, that just might end the mission. They couldn't chance having an extended ESM mast popped up while trained Soviet eyes scanned the seas for intruders.

Reed sat at the sonar console and stared at the active sonar key. That key, when pushed, sent a focused beam of sound into the water that bounced off nearby objects, like ships and subs. The active ping returned distance and bearing information to those objects, along with visual outlines displayed on a screen in a similar fashion as radar. That was good. On the other hand, the loud ping could be heard by anyone in the area, thus alerting them to the sub's location. That was bad.

From the other side of the conn, Celik watched Reed like a shop owner monitoring a potential thief. The muscle in Reed's chest thudded like a Turkish ramazan drum as he moved his hand closer to the active sonar key. Celik's eyes shot open when he detected the move. He ran toward the console but did not get there in time.

Reed hit the key. One loud active ping blasted the water.

Celik arrived at the console, pulled out his sidearm, and placed the cold steel against Reed's temple. "The cock that crows at the wrong time is killed."

Reed said, "One hand does not clap, two hands do. Maybe now he'll transmit."

Celik pulled the gun away. His lips formed a half smile. "If he does and we live, I will not make good on our bet."

"Why not?"

"Because you cheated."

"Fair enough."

One of the navy technicians raised an excited hand. "The *Foxtrot* just sent out a burst!"

Reed ran over to the ESM equipment. "What do you have?"

The lanky tech pointed at a spinning recorder. "We got her nice and clear."

Another tech looked up from a console. "Can we go home now?"

Reed smiled. "Absolutely."

His smile disappeared as the sonar operator reported the sound of a killer on the move. Had Reed not pinged the water, the Soviet *Skory*-class destroyer might never have heard the *Birinci*'s quiet, battery-powered propellers. She might never have seen the ESM masts or periscope smoothly gliding through the Black Sea. But now, as she eased toward her prey, this formidable hunter/killer knew that something lurked under the waves nearby. The *Skory*'s active sonar lit up the ocean as she neared. The ringing vibrations penetrated the hull, and a dozen men in the conning tower recoiled with each ping.

Captain Celik glared at Reed. "You've killed us."

Reed said nothing.

A loud explosion rocked the boat. Sailors in the control room, one deck below the conning tower, yelled obscenities as they struggled to maintain depth and course. More depth charges shattered the silence.

Celik ordered a dive to test depth—about 400 feet—and all ahead full. Reed figured he was probably trying to find a thermal layer to hide under. It didn't work. The *Skory* kept rolling cans off her deck, and the explosions got louder. And closer. The hydraulic pipe the Turkish seaman earlier fixed in the control room with a hammer sprung a leak. Hydraulic fluid shot out from the pipe like water from a pinched garden hose.

Reed thought about his home, his wife, his children. He recalled that years earlier, on board his first ship, the PCS-1380, he'd held Bible studies and Sunday church services. He even bet some of the atheists on board that if he bested them in the boxing ring, they had to attend the following Sunday. He never lost. Since then his faith had diminished to an ember, but as another depth charge rattled his teeth, he whispered a silent prayer.

Celik took the boat deeper and slowed to a crawl. The *Birinci* moaned and shrieked. Despite the slower speed, the batteries would be depleted in less than a dozen hours. Reed's lungs heaved as the carbon dioxide buildup made it hard to breathe. The heavy air smelled of sailor stench.

The boat leveled off at 475 feet. Pipes sprang leaks, and the Turkish crew scrambled to make repairs. The depth charges crept closer, along with the *Skory*'s incessant pinging. If neither stopped soon, Reed swore to himself that he'd grab Celik's sidearm and end the ordeal on his own. Thankfully, he didn't have to. The *Skory* passed overhead and moved away. She did not return.

Celik ordered a turn in the opposite direction, looked at Reed, and said, "Dogs bark, but the caravan goes on."

Reed smiled and said, "If a dog's prayers were answered, bones would rain from the sky."

After another four hours, with the *Skory* now far enough away, Captain Celik brought the boat shallow and snorkeled. Having survived her brush with death, the *Birinci* ran for home.

A few days later, Reed walked through the door of his apartment near Karamürsel and held his children in his arms longer than he ever had before.

CHAPTER FIVE

We don't receive wisdom; we must discover it for ourselves after a journey that no one can take for us, or spare us.
—Marcel Proust

During the month of August 1962, while a small window of summer warmed the city of Moscow, Vice Admiral Leonid Rybalko sped down the Kutuzovskv Prospekt in a black Volga sedan. Summoned to a last-minute meeting with Sergei Gorshkov, the fleet admiral of the Soviet Union, Rybalko ruminated over the reasons for the urgency. Through the windshield of the vehicle, driven by an enlisted man with peach fuzz on his face, the walls of the Kremlin reflected the morning sun and splashed the Arbat with a blood red hue. Vendors along that ailing street unpacked their goods and looked up briefly as the Volga passed by.

The driver turned the Volga onto Yanesheva ulitsa and pulled to a stop in an annex parking lot. Rybalko stepped from the car and bounded through the arched tunnel toward the Ministry of Defense building. Military police, adorned in leather boots and white guard belts, popped to attention as Rybalko approached. The admiral returned their salutes and entered the building through the main door. Ordinarily, he entered from the side, along with the senior operations and intelligence staff, but today was no ordinary day.

Rybalko vaulted up the steps and paused for a moment at the top to catch his breath. At fifty-three years of age, he could no longer ignore his limitations. Socialist paintings lined the walls of the hallway. Most depicted Soviet supremacy over Nazi fascists during the war, as if winning those battles validated the Communist way of life. Reaching his destination, Rybalko entered. Defense Minister Rodion Malinovsky waited two steps inside the large wooden door. Rybalko had met the barrel-chested Malinovsky during the war when the field marshal commanded the Soviet Sixth Corps on the southern front. Malinovsky received two decorations for bravery and became a close friend of Joseph Stalin during the war. That friendship eventually led to his selection as defense minister in 1957, trumping more senior officers, including Admiral Gorshkov.

"To your health, Comrade," Rybalko said.

"And to yours," Malinovsky said.

The defense minister guided Rybalko to a seat, whereupon he proceeded to reminisce about their escapades during the war. Rybalko survived that time partly by fate and partly by luck. He recalled the siege of Leningrad in 1943, when his submarine sent torpedoes into the sides of two Nazi troop ships before they unloaded reinforcements. While other boats suffered from mechanical failures and personnel issues, Rybalko's luck steered him clear of those sandbars.

The two shared a few laughs, then Malinovsky's smile faded. His large eyes narrowed. "I'm not going to sugarcoat this, Leonid. What we're going to discuss today could change the balance of world power. Based on our actions over the next few months, the outcome could go either way."

"I see," Rybalko said, though he really didn't. An orderly brought a tray with two cups of bitter tea and handed one to Rybalko.

As Rybalko sipped his tea, Fleet Admiral of the Soviet Union Sergei Gorshkov burst through the door and strutted toward him. Gorshkov's round red cheeks and down-turned mouth made him look permanently angry. With Admiral Vitali Fokin in tow, Gorshkov's short legs carried his stocky frame across the room at a fast clip. He pulled up a chair and sat. Admiral Fokin did the same.

Famous for his direct style, Gorshkov hurled a question at Rybalko. "Have you heard of Operations Kama and Anadyr?"

Rybalko recalled hearing rumors but nothing more. "Yes, sir, I've heard the names but not the details."

Gorshkov leaned back in his seat. A slight smile played on his lips, as though he were about to impart gossip to his grade-school buddies on a playground. "As you know, on May 12, Premier Nikita Khrushchev finalized his decision to deploy strategic weapons to Cuba under the cover of a humanitarian aid program."

Rybalko said nothing.

Gorshkov continued. "After the first Soviet delegation visited Havana later that month to consult with the Cubans, and Fidel Castro agreed to the plan, the Soviet General Staff Directive devised Operations Kama and Anadyr."

When Gorshkov took a breath, Admiral Fokin said, "The name *Anadyr* came from Stalin's plan to attack Alaska in the fifties with a million-man army. Obviously, he never executed the plan, so we took the name."

Gorshkov sneered at Fokin for the interruption, then said, "On July 10, General Issa Pliyev, our Cuban forces commander, along with his staff, flew from Moscow to Havana on a transport plane. They were disguised as engineers and agricultural experts offering humanitarian aid. In July the *Maria Ulyanova* became the first of eighty-five cargo ships bound for Cuban ports. Do you know what these ships carried in their holds?"

Rybalko did know, but he again feigned ignorance. "I have heard speculations, sir, but no confirmations."

Gorshkov's eyes lit up. "Long-range nuclear missiles. On that day Operation Anadyr began. Now the world will never be the same."

Inside, Rybalko shuddered. Outside, he remained stoic. His patriotism and love for his *Rodina* ran deep, but his respect for some of his country's leaders often waned. This was especially true when it came to the premier. Party First Secretary Khrushchev had insisted on nosing his way into the navy's postwar naval construction programs. He ordered Gorshkov to dismantle all large ships, claiming that these behemoths were "good only for carrying heads of state on official business." Now, with a potential conflict brewing near Cuba, the navy could not even muster two cruisers. Plagued by reactor problems, the long-promised fleet of nuclear submarines remained nothing more than a pipe dream. The party's Central Committee could not find enough raw materials to build much more than a rowboat, so the Soviet Union found itself staring at the backside of American ingenuity and production. If Khrushchev's Anadyr were indeed destined to change the balance of power, it would have to include a way to create resources from thin air.

Minister Malinovsky leaned forward in his chair. "Here's where you play a key role, Leonid."

Rybalko held his chin steady. "What's my assignment?"

"You will lead Project Kama," Gorshkov said.

Rybalko recognized the title of the river that ran from Siberia to the Volga, but he'd heard almost nothing about the operation bearing the river's name.

"Kama is the naval segment of Operation Anadyr," Fokin inserted. "This plan calls for the permanent relocation of the seven missile submarines of the Eighteenth Division from Polyarny to Mariel, Cuba. Accompanying those submarines will be two Project 68 *Chapayev*-class gun cruisers, two squadrons of mine warfare craft, and two missile destroyers."

"There's more," Gorshkov said, again displaying agitation at Fokin's interruption. "Four Project 641 diesel boats from the Sixty-ninth Brigade will also transit undercover to Cuba, but these boats will carry special weapons."

"Special weapons?" Rybalko asked.

"Very special," Fokin said grimly.

"Each submarine," Gorshkov said, "will be issued one nuclear-tipped torpedo."

Rybalko's eyes opened wide. "Nuclear? But . . . our 641 boats aren't trained for such weapons."

Gorshkov waved a hand dismissively. "Captain Shumkov of B-130 earned the Order of Lenin award for firing two live nuclear torpedoes near Novaya Zemlya last year. That should be sufficient."

"That's true," Rybalko said, "but these torpedoes have a sixteen-kilometer kill zone. Getting close enough to hit an American ship could put our submarines at great risk."

Gorshkov remained silent for a moment, then drew his lips tight and said, "Hopefully, your boat commanders will never need to fire one. In the event they are forced into a corner, they will be guided by clear rules of engagement. Is that understood?"

Reluctantly, Rybalko nodded. "Understood."

Fokin piped up again. "Your submarines will transmit position reports daily at midnight Moscow time using their SBD high-frequency transmitters. We will broadcast updates in parallel using low-frequency and high-frequency single sideband. To receive these broadcasts, one boat must remain near the surface to monitor the HF band."

"That will make them vulnerable to American ASW forces," Rybalko said. He also knew that the new "burst" transmission radios, dubbed SBD, for ultra rapid activity, were not very reliable. Due to natural and manmade interference, including a new jamming signal used by the British near their coastline, Soviet boats often needed to stay near the surface and transmit dozens of times to ensure receipt by Moscow.

"We appreciate the dangers," Fokin said, "but the mission's importance takes precedence."

"We have limited acoustic and sea condition knowledge for the Sargasso Sea," Rybalko said. "We're also not certain how effective the American hydroacoustic array is now, and avoiding enemy ASW aircraft may be difficult. Also, the warmer tropical waters could cause living conditions in these boats to become unbearable."

"No one said this mission would be easy," Gorshkov said. "That's why we selected you to lead the charge."

Rybalko wanted to voice further concerns, including the possibility that firing a nuclear torpedo at an American ship could cause World War III, but he realized that his admonitions would be lost on deaf ears.

Admiral Fokin offered further instructions, including details about store loads, crew preparation, and the planned departure time. The four then rose, shook hands, and departed. As he left the Ministry of Defense building, Rybalko thought about his wife, Galena, and his mother, Natasha, who lived with his sisters and their families north of Moscow in a small village called Klin. For a fleeting moment, he pictured their pained and twisted faces as they turned to ashes in the fiery center of a mushroom cloud.

ON AUGUST 17, 1962, ON BOARD the spy ship USS *Oxford* (AG-159), an R-Brancher heard something strange, not too unlike the faint sound of tires screeching in a parking lot. Instantly, he recognized the electronic chirp of a Soviet radar code-named Whiff. The R-Brancher informed the officer in charge, and the OIC radioed Net Control, which sent a

CRITIC (critical) message to the National Security Agency. Russian-speaking I-Branchers assigned to the A Group Soviet signals intelligence desk at NSA headquarters in Maryland ran down hallways and out doors. Within minutes they reported to the office of the operations chief, Major General John Davis. Most were ordered to assist the B Group Spanish linguists listening to intercepted traffic coming from Cuba. Previously, all transmissions from the island came from Cubans speaking Spanish. Within the past week, however, much of that banter changed to Russian—or Spanish spoken with a heavy Russian accent. In response to this unprecedented change, the NSA set up its first around-the-clock SIGINT command center, establishing the foundation for the National Security Operations Center (NSOC).

While Russian-speaking I-Branchers at NSA strapped on headphones and listened, officials in Washington, D.C., hurried to meetings. CIA Director John McCone insisted that the detection of Whiff radar signals and other collected data supported only one conclusion: the Soviets were installing offensive ballistic missiles in Cuba, possibly even nuclear. Secretary of Defense Robert McNamara and Secretary of State Dean Rusk dismissed McCone's "Chicken Little sky is falling" concerns, believing the military buildup to be only defensive. Still, under direct orders from President John Kennedy after he received the news, the NSA established FUNNEL as the new top-secret code word restricting access to information related to Cuban SIGINT—especially anything containing evidence of Soviet offensive weapons.

Thousands of miles away, in the silent cold of the Arctic Ocean north of Russia, the USS *Nautilus* crept along at three knots. Her periscope peaked above the icy sea near the remote island of Novaya Zemlya. Thirteen miles from ground zero, T-Brancher "spook" John Arnold, a communications technician chief, waited for a nuclear explosion. Arnold knew that the Soviets had detonated a fifty-eight-megaton bomb—the largest thermonuclear beast extant—at this very location the previous year, and a *Foxtrot*-class submarine also shot two live nuclear torpedoes into the harbor around that same time. In fact, the Soviets had conducted so many nuclear tests near Novaya Zemlya that they took to calling the island Black Harbor.

A seasoned submariner, Arnold had previously served aboard the USS *Scorpion* (SS-278), a diesel boat that almost collided with a Soviet

November-class nuclear submarine. When he received orders to report to the *Nautilus* for a special mission, he envisioned a technically advanced underwater marvel. He soon found that low tech still ruled the day when he learned that an ordinary cardboard toilet paper roll played a critical part in conducting periscope photographic intelligence. The crew placed the lens of a Canon camera on one end of the roll, with the other end fitted to the periscope's eyepiece. The jury-rigged setup remained in place with a double helping of black electrician's tape.

While on station at Black Harbor, Arnold witnessed more than a dozen spectacular explosions through the periscope in which the Soviets filled the sky with crimson mushroom clouds. During each test detonation, while the *Nautilus* rocked back and forth, bright flashes could be seen through the toilet paper roll, despite the heavy coating of tape. Sonic booms clapped in ears, and fluorescent lights shattered.

Several weeks later, Arnold transferred to a spy ship operating just off the coast of Cuba, where he strained to hear the signals emanating from nearby Soviet radar and missile guidance systems. He and other spooks monitored signals from SA-2 surface-to-air missile (SAM) and other conventional weapons platforms brought to the island by the Soviets on merchant ships. At 2:00 A.M. on the morning of September 15, 1962, Arnold detected something that made the hairs on his neck stand at attention. He checked and double-checked his readings. Without a doubt he was listening to the tone of a Soviet Spoon Rest radar system, indicating that the Soviets had completed the construction of the SA-2 missile platforms. These conventional SAM sites were now fully operational, and from now on, any U.S. aircraft flying over Cuban airspace could be shot down within seconds.

Other R-Branchers located at a Huff Duff high-frequency direction-finding station in Homestead, Florida, and on board the spy ship USS *Oxford* operating in Cuban waters also heard the signals and multiangulated the source. They estimated the location of the SA-2 battery as three miles west of Mariel. The navy ordered the *Oxford* to move in closer and gave her a new set of orders: start listening for signals that indicated the presence of nuclear missiles in Cuba.

In August 1962, CIA Director McCone advised President Kennedy about the Soviet SA-2 conventional SAM batteries in Cuba and the possibility that nuclear missiles might be present, though they were yet to be verified. Kennedy sanctioned a U-2 spy plane flight over the island, which confirmed eight conventional SAM sites. He voiced a strong protest to Khrushchev about the sites and further warned that the United States would not tolerate nuclear missiles in Cuba. The Soviet premier denied any such intentions, claiming that only a few conventional weapons and "agricultural equipment" would be shipped to the island.

By early September 1962, dozens of Soviet ships had delivered spare parts and munitions to Cuba. Secretly, these ships also unloaded several *Komar*-class missile-firing patrol boats designed to thwart amphibious landings, which Gorshkov warned could happen within weeks if the Americans discovered nuclear missiles in their backyard. Already the United States had increased surveillance flights and eyed Soviet merchant ships suspiciously. Two of those ships, the *Indigirki* and *Aleksandrovsk,* departed Severomorsk and carried a cargo of nuclear missiles into Cuba's Mariel harbor eighteen days later. The *Aleksandrovsk* transported fourteen warheads, which would later be married to R-14 missiles after they arrived on another ship. Each missile could hit targets as far away as San Francisco, California, and packed more than sixty times the destructive force that leveled Hiroshima.

The first indication that Khrushchev might be lying about sending nuclear arms to Cuba came on September 18 when, off the coast of Tunis in the Mediterranean, a U.S. Navy frigate confronted a Russian merchant ship and inquired about its cargo. The ship reported that she was carrying only agricultural machinery. Binoculars aboard the frigate indicated otherwise, as the deck was covered with large crates of irregular sizes, which appeared to be the kind that carried disassembled military aircraft.

By the third week of September, U.S. warships and aircraft were intently watching thirty-five Russian merchant ships en route to Cuba. All told, the United States counted 129 ships leaving Russian ports and 94 arriving at Mariel. Due to frequent overflights by U.S. surveillance planes, Soviet personnel on the ground in Cuba worked only at night, unloading ships and assembling missile silos. The Cubans nicknamed their new allies "night crawlers." Though the U.S. government suspected foul play, it had no proof. At least not yet. Fearing the worst, the U.S. Navy planned for a potential future blockade of Cuban waters. Naval aircraft and "tin can" destroyers increased patrols, and personnel were put on high alert. Suspecting that Gorshkov would not send so many merchant ships through the Sargasso Sea unprotected, the navy issued instructions to search for possible Soviet submarines. Those orders were also given to every Huff Duff station within range of the Atlantic.

ON SEPTEMBER 30, ABOARD THE SOVIET diesel submarine B-36 harbored in Sayda Bay, Captain Second Rank Aleksei Dubivko examined a suspicious bundle of ocean charts. They lay against one corner of the chart room, a small, highly classified enclosed area in the port front corner of the control center that only a few on board were allowed to enter. The fleet headquarters duty officer had brought the charts on board a few days earlier. The large stack of nautical maps covered the Caribbean and North Atlantic seas. One chart provided channel approaches to enter several Cuban ports, including Mariel, a small harbor west of Havana. No more than a few seconds elapsed before Dubivko added up the clues—including a recent overload of stores—and guessed where they might be headed. Why they were being sent on the longest deployment ever made by Soviet submarines remained a mystery. Those details would be revealed only after they submerged in the Barents Sea and opened their sealed orders.

Dubivko had no doubt that, regardless of where they were going, he and his crew would execute those orders efficiently. An aggressive commander in his early thirties, he demanded top performance from his seventy-eight officers and men. His motivation to achieve perfection often led to top grades for operational and engineering tests. Captain Nikolai Shumkov, commander of the submarine B-130, was the only peer who had ever bested him in a competition, and only in the weapons department. And this was because B-130 was the only boat in their group to have fired live nuclear torpedoes into Black Harbor one year earlier. Although Dubivko had always longed for that opportunity, he hoped that the need to fire a nuke on this mission would never occur.

After graduating from the Vladivostok Higher Naval School, Ukrainian-born Dubivko originally served on board a "skimmer" surface ship. He later transferred to the submarine fleet, and as a senior lieutenant, accepted command of his first boat out of Gorky on the Volga in 1953. Dubivko learned a great deal under the leadership of Fleet Commander Admiral Chebanenko and Commander of Submarine Forces

Vice Admiral Orel by participating in scores of exercises held by the Northern Fleet. These maneuvers, as a rule, were conducted in the Norwegian and Greenland seas and the northern part of the Atlantic so boat skippers could polish submarine tactics and antisubmarine defenses. Dubivko's drive earned him considerable recognition and opened the door in 1960 to his selection as commander of B-36, a new Project 641 boat fresh out of Leningrad's construction halls.

Dubivko and his crew turned the key on B-36 in 1961 and ran her around the track on sea trials. A year and a dozen runs after that, he received his orders to transfer the boat to Sayda Bay. There, on a cold September afternoon, he stood on the bridge of his award-winning race car and watched two officers from the weapons facility walk across the gangway. One officer, adorned with the Northern Fleet headquarters staff emblem, clutched a briefcase in his right hand. B-36's topside watch clicked his heels, saluted, and pointed at Dubivko high up in the sail. The visiting officers nodded and walked toward the side hatch. They undogged the hatch and entered the confined space that led to the bridge. Dubivko looked down and watched the staff officer, with the second officer following, swing onto the lower rung of the ladder leading up to the bridge platform.

The heavyset staff officer scrunched his broad shoulders, stepped from the ladder, and squeezed onto the bridge. He told Dubivko that he was the Northern Fleet special weapons directorate. He turned and waved a hand at the baby-faced man behind him and introduced Alexander Pomilyev, who he said was a lieutenant assigned to B-36 as a special weapons expert. He cracked opened his briefcase, removed a copy of the lieutenant's orders, and handed them to Dubivko, who read the papers and asked for a definition of "special weapon."

The weapons directorate said Dubivko would find out soon enough.

After the directorate left the bridge, Dubivko stared at Pomilyev and asked, "What do you know about this special weapon?"

"Everything," Pomilyev said.

"Is it nuclear?"

"I'm not at liberty to say."

Dubivko rubbed his chin. "Am I delivering this to our mission destination?"

"I'm not at liberty to say."

"What are you at liberty to say?"

Pomilyev's face softened. "Nothing, I'm sorry. I understand that you'll be receiving more information about this at the briefing, and detailed instructions are included in your sealed orders. All I can tell you is that the weapon must be stored as a service-ready torpedo in the forward compartment and loaded into the number two tube once we've crossed the Iceland gap area."

Dubivko frowned. "You mean after we're in waters patrolled by enemy forces."

Pomilyev nodded.

Dubivko narrowed his eyes. "Are we going to war with the Americans?"

"I don't know," Pomilyev said.

"Are you qualified in submarines?" Dubivko asked.

"No, sir, but I'm a fast learner and will study under way."

"You're damn straight you will," Dubivko said as he turned to watch the weapons directorate cross the gangway and waddle toward the pier. Still staring at the directorate, Dubivko said, "That will be all, Lieutenant. Report to First Officer Kopeikin for your berthing and watch assignments."

As Pomilyev left the bridge, Dubivko desperately wanted to see his wife and children and tell them good-bye before he left on the most important mission of his career.

STANDING ON THE DECK OF HIS submarine, staring at a strange-looking torpedo, Captain First Rank Ryurik Ketov flipped up the collar on the back of his navy blue overcoat to shield his neck from the cold. A fading September sun coated the waters of Sayda Bay and reflected remnants of orange and yellow from the sides of a floating crane. The crane hovered over Ketov's boat and lowered a purple-tipped torpedo through the loading hatch. Within minutes the long cylinder disappeared into the forward torpedo room. Blowing into his gloved hands to keep his nose warm, Ketov glanced at the submarine's conning tower. Three large white numbers were painted on the side, but Ketov knew this label held no meaning, except to serve as a numerical decoy for enemy eyes. The boat's real designation was B4—B as in *Bolshoi,* which means "large."

The handsome, blue-eyed Ketov inherited his B-4 Project 641 submarine—known as a *Foxtrot* class by NATO forces—from his former commander, who was a drunk. Tradition dictated that submarine captains who were too inebriated to drive their boats into port should lie below until they sobered up. First officers took charge and positioned a broomstick on the bridge in their captain's stead. Atop the handle they placed the CO's cap so that admirals on shore peering through binoculars would raise no eyebrows. Ketov stood watch with a broom more times than he could recall. He didn't dislike vodka, nor did he disapprove of his CO's desire to partake, but Ketov felt that a man must know his limits and learn to steer clear of such rocks when under way. He demanded no less of his crew. Unfortunately, as his appointment to commander required the approval of the dozen sub skippers in his group, and all of them drank like dolphins, Ketov's stance on alcohol held him back for a year when he came up for promotion.

The Soviet navy formed the sixty-ninth Brigade of Project 641 submarines in the summer of 1962. Ketov and his comrade captains were ordered to prepare for an extended deployment, which they suspected might be to Africa or Cuba. Some wives, filled with excitement, anticipated a permanent transfer to a warm locale.

The four subs arrived in Gadzhiyevo at Sayda Bay a month earlier and were incorporated into the Twentieth Submarine Squadron along with the seven missile boats. Vice Admiral Rybalko assumed command of the squadron, and over the next thirty days, each boat was loaded with huge quantities of fuel and stores.

Now, aboard B-4, Captain Ketov coughed into the wind and turned to stare at the weapons security officer. Perched near the crane, the man shouted orders and waved long arms at the fitful dockworkers. The officer's blue coveralls and *pilotka* "piss cutter" cap signified that he belonged to the community of submariners, but Ketov knew better. The shape of a sidearm bulged from under the man's tunic, and his awkwardness around the boat made it obvious that he was not a qualified submariner.

Ketov also knew that the security officer came from Moscow with orders to help load, and then guard, the special weapon. Although he'd not yet been briefed about the weapon, Ketov figured this torpedo with the purple-painted nose, which stood in sharp contrast against the other gray torpedoes on board, would probably send a radiation Geiger counter into a ticking frenzy.

Ketov looked down at the oily water that slapped against the side of his boat. Attached by long steel cables, three sister boats of the Soviet Red Banner Northern Fleet floated nearby. If one approached these late-model attack subs from the front, their jet-black hulls, upward-sloping decks, and wide conning towers with two rows of Plexiglas windows might look menacing. The silver shimmer of their sonar panels, running across the bow like wide strips of duct tape, might appear odd. The reflective panels of the passive acoustic antenna, jutting from the deck near the bow, might look borrowed from the set of a science-fiction movie. But the seasoned sailors on the decks of these work-horses were unmistakably Russian, and undeniably submariners.

Ketov strutted across the wooden brow that connected B-4 to the pier. Two guards, with AK-47 assault rifles slung on their shoulders, snapped to and saluted. Ice crunched under his boots as he walked toward a small shed less than a hundred meters away. Captain Second Rank Aleksei Dubivko, commander of B-36, matched his stride and let out a baritone grunt.

"Did they give you one of those purple-nosed torpedoes?"

"Yes," Ketov answered, "they did."

Although the round-faced commander was about Ketov's height of five foot seven, Dubivko's stocky frame stretched at the stitches of his overcoat. He let out another grunt and said, "Why are they giving us nuclear-tipped weapons? Are we starting a war?"

"Maybe," Ketov said. "Or maybe we're preventing one."

Dubivko's boots clicked on the ice as he hurried to keep up with Ketov. "We haven't even tested these weapons. We haven't trained our crews. They have fifteen-megaton warheads."

"So?"

"So if we use them, we'll wipe out everything within a sixteen-kilometer radius. Including ourselves."

Ketov neared the door of the shed and stopped to face Dubivko. "Then let's hope we never have to use them."

Dubivko let out a low growl and followed Ketov into the shack.

Inside, Captain First Rank Nikolai Shumkov, commander of submarine B-130, stood by the door. Only a few stress lines underscored his brown eyes and marked his boyish features. Next to Shumkov, Captain Second Rank Vitali Savitsky, commander of B-59, appeared tired and bored. None of them had slept much since their trip from Polyarny to Sayda Bay.

The tiny shed, once used for storage, offered no windows. A single dim bulb hung from the ceiling and cast eerie shadows inside. Someone had nailed the Order of Ushakov Submarine Squadron flag on one wall. The unevenly placed red banner, fringed in gold and smeared with water stains, appeared as if hung by a child in a hurry. In one corner sat a small stove that flickered with yellow sparks but offered little warmth. The air smelled of burnt coal.

One metal table graced the center of the room, where the squadron commander, Leonid Rybalko, sat with his arms crossed. Ketov noticed that the vice admiral shivered, despite being bundled in a dark navy greatcoat and wool senior officers' *mushanka* cap. The tall, broad-shouldered Rybalko had a reputation for analytical brilliance and a smooth, engaging wit. A dedicated performer, Rybalko exuded the confidence and mastery of a seasoned leader.

To the side and behind Rybalko, the deputy supreme commander of the Navy Fleet, Admiral Vitali Fokin, fidgeted with his watch. Thin and lofty, Fokin kept his back straight. Ketov deduced that Fokin, given his close relationship with Fleet Admiral Sergei Gorshkov, held the reins of whatever mission they were about to undertake. A slew of other officers filled the room, including Anatoly Rossokho, the two-star vice admiral chief of staff. Ketov suspected that Rossokho was here to define their rules of engagement about using the special nuclear torpedoes.

Vice Admiral Rybalko motioned for everyone to find a seat. He coughed and brought a handkerchief to his lips to spit out a clump of mucus. His face looked pale and sickly. He locked his eyes on each submarine commander one at a time. When he looked at Ketov, those few moments seemed like days.

"Good morning, Commanders," Rybalko said. "Today is an important day. I'm not going to discuss mission details, as we've included those in your sealed briefings, which you will open under way. So instead we will focus on other aspects of your mission."

Metal clanked as an attendant creaked open the front panel on the hot stove and dumped in another can of coal pellets.

Rybalko continued. "I'm sure you all know Admiral Fokin. He asked me to emphasize that each of you has been entrusted with the highest responsibility imaginable. Your actions and decisions on this mission could start or prevent a world war. The four of you have been given the means with which to impose substantial harm upon the enemy. Discretion must be used. Fortunately, our intelligence sources report that American antisubmarine warfare activity should be light during your transit."

Ketov hoped that the ASW intelligence report was correct but feared that optimism probably overruled reality. He glanced at the other sub commanders. Dubivko and Shumkov wore excited smiles. Savitsky, who'd earned the nickname "Sweat Stains" because he was always perspiring about something, wrinkled his brow. Ketov, who received the title of "Comrade Cautious," shared Savitsky's angst. As adventurous as this might seem to Dubivko and Shumkov, Ketov knew Project 641 submarines were not designed for extended runs into hot tropical waters and had no business carrying nuclear torpedoes.

Rybalko imparted more information, concluded his speech, and asked if anyone had questions.

Ketov raised a hand. "I do, Comrade Admiral. I understand that our sealed orders provide mission details, but we share concerns about our rules of engagement and the special weapon. When should we use it?"

Vice Admiral Rossokho broke in. "Comrade Commanders, you will enter the following instructions into your logs when you return to your submarines: Use of the special weapons is authorized only for these three situations—One, you are depth charged, and your pressure hull is ruptured. Two, you surface, and enemy fire ruptures your pressure hull. Three, upon receipt of explicit orders from Moscow."

There were no further questions.

After the meeting, Ketov followed the group out into the cold. A witch's moon clung to the black sky and hid behind a dense fog that touched the ground with icy fingers. Ketov reached into his coat pocket and took out a cigarette. Dubivko, standing nearby, held up a lighter. Ketov bent down to accept the flame. Captains Shumkov and Savitsky also lit smokes as they shivered in the dark.

Between puffs, Ketov posed the first question to Captain Savitsky. "How are your diesels holding up?"

Savitsky cringed. "No problems yet, but I'm still worried about what might happen after they've been run hard for weeks. If they fail on this mission . . ." Savitsky's voice trailed off as he shook his head.

Ketov knew that shipyard workers had discovered flaws in B-130's diesel engines during the boat's construction. The shipyard dismissed the hairline cracks as negligible, and Savitsky did not press the issue, as to do so would have resulted in his sub's removal from the mission. Still, he fretted endlessly about the consequences.

Sensing his friend's distress, Ketov changed the subject. "Have you seen those ridiculous khaki trousers they delivered?"

"I'm not wearing those," Savitsky said.

"I wouldn't either," Shumkov said, "if I had your skinny duck legs."

Savitsky snorted and threw his head back. "I'd like to see how you look in those shorts, Comrade Flabby Ass."

"Right now," Dubivko said as he pulled his coat tighter, "I'd rather look like a duck in shorts than a penguin in an overcoat."

Ketov smiled and shook his head. "I'm going back to my boat, try on those silly shorts, and have a long laugh and a can of caviar."

"And maybe some vodka?" Shumkov said.

"I wish," Ketov said. "We cast lines at midnight."

Shumkov nodded and said nothing.

Savitsky raised his chin toward Ketov. "Do you think we're coming back or staying there permanently?"

Ketov shrugged. "All I know is that we can't wear those stupid shorts in this weather."

Back on board B-4, Captain Ketov sat on the bunk in his cabin and stroked the soft fur of the boat's cat. "It's time to go, Pasha."

Over the past year, the calico had become a close member of B-4's family. Like many Russian submarines, B-4 enlisted the services of felines to hunt down rats that managed to find their way on board, usually by way of one of the shorelines. Boats often carried at least one or two cats on board, and the furry creatures spent their entire lives roaming the decks in search of snacks and curling up next to sailors on bunks. Unfortunately, for reasons unknown, headquarters decreed that cats were forbidden on this journey. Given no choice, Ketov found a good home for Pasha with a friend who could care for her and keep her safe.

As Pasha purred by his side, Ketov reached for a can of tuna. "The least I can do is give you a nice snack before we leave."

Ketov thought about his mother, still living in the rural Siberian village of Kurgan. She'd lost her husband to one war; would she now sacrifice her firstborn son? When Ketov was thirteen, his father, who was an accountant with bad eyesight, was forced to fight in the battle at Leningrad. He was killed in his first engagement. Ketov became the man of the house and helped support his younger siblings and his mother, who earned a meager teacher's salary. He could still not explain why, but the day he turned eighteen, one year after the war ended, he took the train to Moscow and enrolled in the naval college. He also had no explanation for why he'd jumped at the chance to serve aboard submarines. He only knew that, despite the sacrifices and often miserable conditions on the boats, no other life could fulfill him like the one under the sea.

A few minutes past midnight on October 1, 1962, Captain Ketov stood on the bridge of B-4 and watched Captain Savitsky cast off lines and guide B-59 away from the pier using her quiet electric motors. Captain Vasily Arkhipov, the brigade's chief of staff, stood next to Savitsky in the small cockpit up in the conning tower. A flurry of snow mingled with the fog and dusted the boat's black hull with streaks of white. Thirty minutes later, B-36, commanded by Dubivko, followed in the wake of her sister sub and disappeared into the darkness of the bay. After another thirty minutes, Shumkov, in B-130, followed by Ketov in B-4, maneuvered away from the pier. Ketov stared into the blackness as the three subs ahead of him, all with running lights off, vanished into the night. Then he heard the low rumble of B-59's diesel engines, signaling that Savitsky had cleared the channel and commenced one of the most important missions undertaken by the Russian navy since World War II.

A SHIVER OF EXCITEMENT RAN DOWN his spine as Captain Dubivko stared at the radar repeater in the bridge of B-36. Though he could not see the other boats in the dark, he knew that he was second in line with B-59 ahead, B-130 behind, and B-4 at the rear. The narrow space, high up in the boat's sail, enclosed all but the top part of the bridge with a blanket of steel. The brisk, cold air swirling through the space bit at Dubivko's hands and nose. One lookout, a few feet above and behind him, stood on the bridge platform with binoculars planted against his frozen face. Sea foam dotted the deck as the submarine's bow carved through the Barents Sea.

Dubivko looked down and to his right. He studied the green glow of the radar repeater screen that painted a picture of the channel. The dense fog lowered visibility through the scratched Plexiglas bridge windows to no more than a few meters. The radar, despite its eighty-kilometer range limitation, offered the best means to avoid running aground or into the other three boats. Metal clanked and groaned as Arkadyi Kopeikin, the boat's *starpom* first officer, scrambled up the ladder to the bridge.

Kopeikin settled in next to Dubivko, remained silent for a long moment, then said, "We're ready to dive, sir."

Not speaking, Dubivko offered a short nod.

"The charts are spread on the nav table, and the officers are anxious," Kopeikin said.

Dubivko let the hint of a smile play on his lips. "Let's take her down, Comrade. Then we can open our orders."

"Yes, Captain." Kopeikin bent down and yelled into the voice tube, "All hands clear topside, prepare to dive the boat." He then dropped through the hatch leading to the conning tower.

Dubivko imagined a scurry of feet and voices below in the command center as the crew prepared for entry into the silent world of moaning whales. Blue-green ocean shot through the limber hole vents. He took one last scan of the horizon, pulled the smell of salt and sea into his lungs, then yelled down to the watch officer. "*Pogruganye!*"

The submarine angled down at the bow as the ballast tanks flooded, forcing air to bubble and hiss through the limber holes. The lookout secured his binoculars, scrambled down the platform ladder, and disappeared through the bridge hatch. Dubivko captured a final glimpse of the blue sky, shimmied down the ladder, and with a loud clank, pulled the spring-loaded hatch shut above him. He slid down past the conning tower periscopes and entered the CC in Compartment Three. He removed his *kanadka* and handed the fur-lined coat to a *michman* (warrant officer) as he entered. In the dim blue light, surrounding the two cylindrical periscope housings in the center of the room, more than a dozen men manned various stations, including helm and diving control, radar, navigation, ballast tank operations, and torpedo fire control. The crowded compartment offered the dank smell of hydraulic fluid, diesel fumes, and something that could only be described as mechanical.

The low thrum of the diesel engines ceased as the boat angled downward and switched to battery power. Dubivko pictured B-36 in his head, from her silhouette to her soul, her long black hull, systems, schematics, and statistics, her crew, capabilities, and compartments. He saw her as more than just a tube made from steel and sweat. More than a tube full of engines, motors, batteries, bunks, and torpedoes. Much more than a scattering of pipes, valves, switches, and gauges. He saw her as a woman imbued with a loyal heart and stunning beauty. And like any woman, she deserved a great deal of patience and care.

Dubivko shot an order to the duty officer, who relayed the same to the helmsman. "Right full rudder, come to course two-five-five, speed nine knots."

The helmsman, seated on a small bench in the front right corner of the CC, acknowledged the order and pulled a black knob to the right. The navigator, Captain Lieutenant Sergei Naumov, from the navigation table on the port side near the center of the CC, sounded off the next course change as the boat leveled off at one hundred meters.

Dubivko turned over CC command to the electronics officer, Yuri Zhukov, and took a half-dozen steps toward the nav table. There he met First Officer Kopeikin, as well as Navigator Naumov and the political officer, Captain Third Rank V. G. Saparov. The three men were positioned around the small rectangular table. Five times the typical number of chart cases filled the small corner near the curved white bulkhead. The extra cases were brought on board to maintain complete secrecy about their final destination—despite the many rumors and clues that alluded to Cuba.

Dubivko's eyes found Kopeikin's. They had shared many missions together on this boat, and while most of them began just like this one, few were similar. "Comrade Kopeikin, you may open the safe."

Kopeikin reached toward the bulkhead safe and dialed. The square box clicked open, and the first officer removed a large manila envelope. He handed it to Dubivko. Wide red "top secret" stripes ran diagonally across the package. Dubivko opened the outer seal, then the inner one. He removed a booklet, examined the cover, flipped to the first page, and read. The single word *Kama* headlined the top of the page, followed by small type outlining their mission orders.

Dubivko read in a low voice, so as not to be overheard by others in the CC. "Operation Kama tasks the submariners with performing reconnaissance of all seaward approaches to Mariel, Cuba. Acoustic area conditions are to be logged for port entry in preparation for seven ballistic missile submarines."

Dubivko took a breath and glanced at his senior officers. He could tell by the look in their eyes that they shared his anticipation. He read further about the additional gun cruisers and destroyers joining them in Mariel and concluded with "This internationalist intervention mission by the Soviet Socialist Republic is designed to equip the Socialist Republic of Cuba with sufficient resources and support to undermine further Western aggression. Our brigade is tasked with a special mission for the Soviet Union, which includes transiting the Atlantic in secret to a new home port in an allied country. This transit must remain undetected by enemy forces, and the submariners must arrive in Mariel, Cuba, by October 20."

Dubivko pondered the expectations implied in his orders. Reaching Cuba by October 20 without being detected was practically impossible. They'd have to run near their top speed submerged, snorkel at night, and by sheer luck try to avoid American sea-mounted sonar arrays and ASW ships and planes. Dubivko held his doubts and anger in check and removed another envelope. As he read, the officers standing near the nav table leaned in closer. "These are the rules of engagement for the use of weapons. One, while in transit, all weapons will remain in combat-ready condition. Two, conventional weapons will be used as directed by the Main Navy Staff. Three, the use of nuclear torpedoes is allowed only as directed by the Ministry of Defense of the Main Navy Staff."

Silence descended on the tiny cubicle.

Dubivko cleared his throat and said, "We have our orders. Let's make the appropriate preparations. I will tell the crew what they need to know."

Dubivko left the cubicle and walked to his stateroom. His head spun as he lay in his bunk and digested the information he'd just read. Their mandated arrival date translated into a fast and dangerous transit time. To maintain high speed, they'd have to run on the surface for as long as possible and hope that Mother Nature sent storms to shroud them under dense clouds. After they neared Cuba, where they'd be required to stay submerged, they needed to remain near the surface every day at midnight Moscow time to receive and acknowledge transmissions. Midnight in Moscow equated to late afternoon in the Caribbean, and that meant risking exposure during daylight hours.

Each boat carried a special OSNAZ group of nine young men trained in signals intelligence—similar to "spooks" in the American navy. OSNAZ was short for *osobennogo naznachneniya,* which in Russian means "specialized designation." Five of these were English-speaking communications experts tasked with monitoring HF and UHF bands to determine where the Americans might be concentrating their ASW efforts. Each captain needed this intelligence not only to remain undetected, but also to determine if, why, and when they should fire their nuclear torpedoes at the enemy. Dubivko thought about his family back in Russia. By now they had been informed about the mission and were preparing to join him after B-36 arrived in Cuba. He thought of the special weapon resting in a torpedo tube only fifty feet away from his stateroom and wondered if his boat would be vaporized long before he ever reached the distant tropical island.

CHAPTER SIX

The ability to get to the verge without getting into the war is the necessary art. . . . If you try to run away from it, if you are scared to go to the brink, you are lost.
—JOHN FOSTER DULLES

IS ARM CHAINED TO A BRIEFCASE, William J. Reed counted the Florida palm trees through the car window as they whisked by at high speed. His eyelids sagged from jetlag, and his fingers tingled from lack of circulation. He'd been a man with a portable toothbrush through most of September, and now, in early October 1962, his final destination lay just ahead. This after more than a dozen stops in foreign countries around the Atlantic and Pacific rims. This after attending limited duty officer's school in Newport, Rhode Island, where he studied navigation, chart reading, ship handling, and how to be a "politically navy correct" officer. This after taking his family to see the 1962 America's Cup Yacht Race in the summer and then reporting to Section 22 of the Soviet SIGINT A Group at the NSA facility in Fort George G. Meade, Maryland—commanded by Operations Chief Major General John Davis.

Reed was the only person on the team with Soviet burst signal field experience, so his boss, Commander Jack Kaye, and his peers at A22, descended upon him like squawking geese. They made him the "point man" for completing the technical and operational manuals for Boresight. They also locked a briefcase on his wrist and filled his schedule with site visits around the globe.

Reed's driver, a twenty-something petty officer named Smith, glanced in the rearview mirror and with a touch of a Southern accent said, "Kinda humid today, ain't it?"

"Kinda," Reed said. His tired brain cells allowed only one word at a time.

"That's what happens after it rains real hard like it did last week. Ever been to Homestead before?"

"Nope."

"Great duty station, at least *I* think so. The enlisted and officers' barracks are actually in town, and the ops building is about five miles away down Card Sound Road."

"That so," Reed said, having mustered the mettle to form two words.

"Most everybody takes the bus to the base, but since I'm one of the drivers, I get to take this here car."

"That's nice." Still on two words.

Smith wrestled with the wheel as the car stuttered across a road with potholes that looked like they'd been made with C-4 explosive. "They built the station in 1957, and now NSG has, like, forty or fifty people here at Site Alpha, so that means we have good duty rotations and plenty of time off, so we can go have fun up in Miami and stuff. At least, we did until last month, when some kinda shit happened, and they told us to double up our watches. Do you know what's going on?"

"A little."

"You could tell me, but then you'd have to kill me, right?"

"That's right."

"You can't tell me even if I ask real nice and say pretty please with a cherry on top?"

"Not even then."

"That's a bummer."

"I know." Reed smiled and turned again to look out the window. A flat plain baked by a tropical sun and mottled with green vegetation stretched to the horizon. The barest hint of pineapple and coconut drifted by.

Smith turned onto an even more dilapidated road. A small building, surrounded by mud craters, sat alone on the rain-pummeled ground. The car pulled to a stop, and Smith said, "This is it."

Still clutching his briefcase, dressed in khakis, Reed stepped from the car. Brittle dirt cracked beneath his brown naval shoes. The gray cinder block structure displayed no windows and appeared to contain no life. At the entrance, a uniformed marine saluted Reed, checked IDs, and opened the door to an all-too-familiar sight. Racks of equipment beeped, whirred, and blinked; seated operators monitored, logged, and chatted. The piña colada odor vanished, replaced now by something indefinable that smelled electronically bitter. A tall man in his early thirties approached and reminded Reed of Commander Petersen from Turkey.

The man offered a smile and a palm. "I'm Lieutenant Clower, the facility OIC. Welcome to Homestead."

Reed shook Clower's hand and forced his head to allow more than three words. "Ensign Reed, NSG Maryland. Have my techs arrived yet?"

Clower pointed to a far corner. Reed squinted and recognized his two cohorts, their hands buried in an equipment rack.

"I'd like to chat with you before you join your team," Clower said.

Reed nodded but did not reply.

Clower motioned for Reed to join him a few feet away from listening ears and said, "I feel like I'm in the dark here. I was hoping you could fill me in."

"On what?" Reed asked.

Clower crossed his arms. "Homestead's Strategic Air Command Bomb Wings are on heightened alert, we've been ordered to double up on watches and keep an eye on every merchant ship headed to Cuba, and now you guys show up with some new DF toys. What the hell's going on?"

Reed shrugged. "I'm not exactly sure, sir; I've only gotten tidbits myself, but it probably has something to do with former vice president Nixon's suggestion for a quarantine to keep the Soviets from shipping more arms to Cuba."

"I heard that the Senate Foreign Relations Committee passed a near-unanimous resolution to spank the Cubans pretty hard if they decided to get nasty with their shiny new Soviet weapons," Clower said.

Reed wiped a sleepy wink from the corner of his eye and wondered how many of those new weapons delivered by the Soviets might be nuclear versus conventional. "There's more; that's why I'm here."

Clower stood up straight. "Go on."

"As you know, our guys are upgrading your DF systems so you can hear Ivan's new burst signal."

"Yeah."

"A couple of our Atlantic stations got a hit on some *Foxtrots* leaving Sayda Bay."

"Yeah, I know. We got the flash."

"What you don't know is that it looks like those boats are headed toward Cuba. That means something big is about to happen, and that's probably why your watches were doubled."

Clower's face lost half its color. "God help us."

Reed removed his cover and backhanded his forehead. His shoulders ached, and he fought off a splitting headache. "We can't expect God to do all the work. We need your team to pitch in."

"Absolutely. What do you want us to do?"

"As soon as my guys get you up and running, we'll need to calibrate the systems for a couple days while I train your team. Then you'll have to play bird dog like you've never played before. Since you're still running a GRD-6, even with the new Boresight equipment, detecting a burst will be like finding a needle in a haystack, and any fixes you get will be way off."

"But better than nothing."

"Let's hope so," Reed said. "If not, things could get pretty ugly in the Caribbean."

After his conversation with Clower, Reed and his team spent the next few days installing and testing the Boresight receiver/recorder and related equipment for capturing, recording, and analyzing the Soviet burst signals. Reed's team consisted of Master Chief Carl Odell and Second Class Petty Officer Tommy Denofrio, both M-Branchers. The Midwest-born Odell looked like Jackie Gleason and apparently copied the comedian's diet. He piled on bacon and biscuits for breakfast like a last-meal death-row inmate. Reed often wondered if Odell's heart would give out halfway through a job.

Denofrio was wired differently. An Italian from New York and a consummate lady's man, he counted calories and pumped iron with the dedication of Jack LaLanne. The twenty-something petty officer displayed a talent like no other when it came to smooth-talking women but flashed the temper of a runaway train when someone disagreed with his conservative views. Both sailors displayed real genius on the job, which is why Reed selected them to help get the Boresight stations up and running.

Unfortunately, at Homestead, interference problems made that task quite a challenge. Odell and Denofrio worked day and night to correct the issues, while Reed trained the operators in small groups so as not to pull anyone away from current operations.

Meanwhile, outside the small building twenty miles south of Miami, the world's superpowers edged closer to a showdown.

ABOARD B-36, SECOND IN THE LINE of four *Foxtrot* submarines heading to Cuba, Captain Dubivko rested his round cheeks on a pillow in his bunk. He curled his fingers into rock-hard fists and reeled off a stream of obscenities in his head. He cursed Premier Khrushchev, threw harsh words at Admiral Sergei Gorshkov, and blamed Admiral Rybalko for sending him toward failure. How could he possibly arrive in Cuba by the date ordered while maintaining complete stealth? The boat heaved to port, and Dubivko's stomach knotted. A day earlier they'd run head-long into a massive storm that still lingered. After he'd spent two days awake battling pounding waves as the boat snorkeled, Dr. Buinevich ordered him to rest.

Unable to sleep, he instead launched into another imaginary attack on his superiors. While mentally plotting Khrushchev's demise, Dubivko's stateroom phone rang. He answered, listened, hung up, and questioned whether there was a God or just a universe full of demons with the single-minded purpose of punishing him for the rest of his life.

Dubivko hurried aft down the passageway to the wardroom, where he saw Dr. Buinevich hovering over Sublieutenant Pankov, head of the hydroacoustic (sonar) group. The room smelled of fresh alcohol, obviously used to sterilize the wardroom table. The high-intensity light, placed in the overhead for surgeries, bounced off Buinevich's balding head like a searchlight. The captain of medical service filled a syringe, while Pankov moaned in pain. Four sailors held the man in place on the small table as the boat rocked to and fro in the storm. The moaning sublieutenant's feet dangled over the edge of the table, and the doctor fought to keep his hand steady as he shoved a needle into Pankov's arm. The moaning stopped.

"What's going on here?" Dubivko asked.

"He has appendicitis," Buinevich answered. "I need to operate."

"Now?"

"If I don't, he'll die."

Dubivko bit his lip. "Do it."

They were now transiting through the Norwegian Sea, past an imaginary line known as the GIUK gap, which stood for Greenland, Iceland, United Kingdom. The line represented a gauntlet laid down by the Americans. Any Soviet submarines venturing past that line entered into a narrow fishbowl less than one thousand miles across where sub-hunting NATO aircraft and ships patrolled in droves. Fortunately, a large gale developed as they sped past the Faroe Islands off the coast of Britain, and it now masked their diesel engines and hid their surfaced vessel under dark clouds.

The storm above worsened, however, forcing the boat to dive. Now the crew could feel its anger at a depth of more than twenty meters. Dubivko held no doubts that when the boat surfaced again during the night to charge batteries, the weather would still be turbulent. A Project 641–class submarine snorkeling on the surface was no match for a force of this magnitude, and the two running diesel engines would certainly whine and choke every time a wave crashed over the snorkel mast, causing the flapper to slam shut to avoid sucking in salt water. Dubivko looked at his watch. Two hours to sunset. The doctor could barely keep his hands steady now. Once the boat surfaced, and ran at a higher speed in the storm, performing an appendectomy would be nearly impossible.

Dubivko walked over and clasped Pankov's hand. Bleary-eyed, the sailor tried to smile. Dubivko smiled back. "Hang in there, Pankov. You're in good hands."

Pankov managed a quarter nod just before the injection sent him under.

Dubivko let go of Pankov's hand and turned toward Buinevich. "How long will this take?"

"Several hours," the doctor replied.

"We don't have several hours. If you can't complete the operation by sunset, we will risk missing our snorkeling window, and we'll fall too far behind schedule."

"What are you saying?" the doctor asked as his scalpel penetrated Pankov's skin.

"I'm saying," Dubivko said, a lump forming in his throat, "you must finish before nightfall, or we'll be short one acoustic sublieutenant."

Buinevich looked up from his incision, his scalpel dripping blood. He stared at Dubivko for a long moment, shook his head, then returned his attention to Pankov. Dubivko clenched his teeth, and left the wardroom. Back in his cabin, he cursed his superiors for forcing him to kill a crewman.

Two hours later, a sailor shook Dubivko awake. He walked across the passageway and splashed his face with a sprinkling of tepid water. Like all submariners, Dubivko learned early on to conserve water use, especially on extended patrols. B-36 left port with thirty-six tons of fresh water, and submarines of this type did not have a condenser to produce more. Cooling the water was simply not possible, and when they entered tropical zones, tea became the preferred beverage.

Dubivko dried his face and, with the waves high above still rocking the boat, returned to the wardroom. The doctor stood near the table, his feet dancing to maintain balance while B-36 rolled ten degrees to starboard.

"Doctor?" Dubivko said.

Buinevich looked up from the table and grumbled. "I never wanted to go to that damned hospital for special training, you know."

"I know."

"But I went anyway, and I guess that decision paid off, Captain. The boy will live."

Dubivko let out a slow breath. He did not need to assassinate Premier Khrushchev after all. He offered a job well done to Buinevich and walked toward the CC. There he assumed command and issued an order. "*Vspletye!*"

The boat angled toward the surface. Several minutes later, Dubivko, along with Political Officer Saparov, climbed up the ladder into the tiny conning tower. There Dubivko depressed a switch and said, "*Podnyat periscope.*"

The mast shot toward the sky as the boat slid to the surface, where the storm tossed her about like a tiny balsa wood model. Seawater splashed through the open bridge hatch above and soaked Dubivko's shirt. Steadying himself, he slapped down the handles on the navigation periscope. He swung the scope left and right, then focused on a small moving speck against the sunset-filled sky. At first he thought it might be a bird, but he then switched the scope to high power and recognized the shape as a British Shackleton ASW aircraft circling in the distance. He contemplated quick diving the boat but decided to risk staying on the surface. Without fully charged batteries, they'd have no hope of remaining on schedule. He also figured that the Shackleton's sonobuoys and radar would not be able to discern a snorkeling submarine from a crashing wave in this storm.

As Dubivko continued to watch the aircraft fly figure eights, something bothered him. He took a bearing to the plane and turned the scope over to Political Officer Saparov. He slid down the wet ladder into the CC and took three steps over to the navigation table. With Navigator Captain-Lieutenant Naumov staring over his shoulder, he bent over the nav table and silently ran a finger backward along B-36's track since leaving Sayda Bay. In a low whisper, he said, "The Americans are following our course. How the hell do they know where we are?"

Dubivko had little time to ponder this further as a giant wave slammed into the boat's side. The snorkel's flapper valve closed tight. The diesel engines continued in vain to suck in air that caused a high-pitched vacuum whistle in the boat. Dubivko's ears popped, and he swallowed hard. The control room filled with the rank odor of diesel fumes as the exhaust backed up from the engine room. A watchstander coughed and vomited onto the deck, and another grabbed his chin and groaned as the vacuum pressure made his eyes bulge and threatened to suck the fillings from his teeth. The boat pitched again, and men in Compartment Four ducked as canned meats flew from overhead cubbyholes like small metal missiles. Dubivko's ears finally cleared in time to hear the watch officer scream from up in the bridge.

"Right full rudder!" Dubivko yelled.

The boat turned toward the waves, and the flapper valve popped open. Someone relieved the officer on the bridge, who, clutching his chest, stumbled down the ladder into the CC, then hobbled toward the wardroom. Dubivko later learned that Brigade Engineer Captain Second

Class Lyubimov broke three ribs when he crashed against the side of the bridge gyrocompass.

Dubivko climbed up the ladder into the conning tower and relieved Saparov on the periscope. A splash of fading sunlight snuck past a storm cloud and glinted orange-red off the Shackleton's wings as the plane turned toward B-36. Batteries charged or not, he knew they were out of time as he ordered the boat beneath the roiling sea.

Thirty meters deep, running silent on electric power, with the storm now above them, Captain Dubivko returned to his cabin. He was haunted by what might happen if they couldn't return to the surface soon to snorkel. Or if they lost more than one engine while thousands of miles away near Cuba. Or if the Americans ruptured their pressure hull and forced them to use the nuclear weapon.

And he wondered how the Americans appeared to know their general course. He closed his eyes, but sleep did not find him.

ON OCTOBER 10, BASED ON R-BRANCHER input from the USS *Oxford,* and the listening station at Homestead, Florida, the NSA advised the White House that the Cuban air defense system appeared complete and armed. The Cubans were now relaying their radar tracking information between their headquarters and jetfighter bases using Soviet standard procedures.

Boresight stations continued to track the four *Foxtrot* submarines as they sped toward Cuba, while U.S. naval forces maintained a watchful eye on the Soviet oiler *Terek,* which they knew was there to resupply the *Foxtrots.* The navy also kept track of the electronic eavesdropping ship *Shkval,* which had a reputation of collecting intelligence information from U.S. warships and feeding it to nearby submarines. The presence of these two vessels indicated that the Soviets were planning something for the four *Foxtrots,* and American forces patrolling the Sargasso Sea east of Cuba remained alert and on edge.

ON OCTOBER 14, A U-2 SPY plane piloted by Major Richard S. Heyser flew over Cuba on a course that placed him sixty miles west of Havana. Heyser snapped 928 pictures in less than six minutes, covering an area seventy-five miles wide. The National Photographic Interpretation Center in Washington, D.C., examined the pictures the following day. Wide-eyed analysts confirmed that a series of nuclear launch sites now existed in Cuba that were capable of targeting most major cities in the United States. These included Soviet SS-4 *Sandal* medium-range ballistic missiles with forty-two projectiles that carried two- and three-megaton nuclear warheads. Also on board were twenty-four SS-5 *Skean* intermediate-range ballistic missiles that could go twice as far and kill eighty million Americans in less than five minutes. Throughout the United States, fallout shelters could hold forty million people at best.

All six Polaris submarines based in Holy Loch, Scotland, pulled out of port on October 16 and aimed their nuclear missiles at the Soviet Union. Two days later, McGeorge Bundy delivered the bad news about the nuclear missiles in Cuba to President Kennedy, who called a meeting with his high-level executive committee advisers just before lunch.

At Section A22 in Maryland, William J. Reed's boss, Commander Jack Kaye, assembled his team in the conference room. For the next several hours they discussed the Cuban situation and the need to locate any and all Soviet submarines operating in the area as soon as possible. To that end, they reviewed which stations had received the Boresight system upgrades and the operational status of each. Only a half-dozen contained the equipment, and none were fully operational. Most still used old GRD-6 antenna arrays, which meant their ability to locate and accurately pinpoint Soviet submarine locations were shaky at best. Only Edzell, Scotland, in the Atlantic, along with Hanza, Japan, and Skaggs Island, California, in the Pacific were equipped with Boresight and the new Wullenweber elephant cages.

Even with the more advanced antenna capabilities at those stations, the maximum range for detection was around 3,200 nautical miles, with one ionospheric "hop" of 2,700 miles as more typical due to interference and weather conditions. An ionospheric hop is what happens when ionized atmospheric gases reflect high-frequency radio energy and "bounce" the transmitted signals back down to earth. These signals are often reflected back into the ionosphere for a second bounce, or hop. The upshot was that the optimal listening range for the arrays left Hanza and Edzell out in the cold and Skaggs right on the ragged edge of hearing anything near Cuba. The team concluded that, for now, they'd have to find a way to tweak the five Atlantic GRD-6 sites enough to give U.S. ASW forces a snowball's chance of keeping American ships from becoming sport-diver relics on the ocean floor.

CAPTAIN DUBIVKO'S B-36, SECOND IN A lineup of four Soviet *Foxtrots,* made good time for several days under the cover of bad weather. They arrived at the edge of the Azores—in the middle of the Atlantic northwest of Africa—on October 15. After several more sleep-deprived days, Dubivko's eyelids twitched as he stood near the number two tube on the port side of the forward torpedo room. He hated the involuntary reaction to stress. Staring at Alexander Pomilyev, he said, "Step away from that."

Pomilyev, the young special weapons officer (Weps) brought on board in Sayda Bay by the Northern Fleet special weapons directorate and who'd yet to qualify in submarines, removed his arm from the yellow rectangle adjacent to the number four tube. He turned to see what "that" might be.

Dubivko pointed to the silver lever on the front of the rectangle. "That's the emergency bow plane operating mechanism. If you were qualified, you'd know not to rest your arm near that lever."

Pomilyev nodded. "I'll make a note of that."

"See that you do," Dubivko said. "You wanted to see me?"

"Yes, sir. I wish to inform you that none of your crew are sleeping in the torpedo room."

"I'm already aware of that, Comrade Pomilyev."

"Yes, of course, but I'm concerned that this will diminish response time in the event of a weapons emergency."

"There are plenty of men in Compartment Two that can respond in time, but your point is duly noted."

"Perhaps the men could be persuaded to return if we convinced them that the special weapon will not cause radiation poisoning."

"Will it?"

Pomilyev's perplexed look made his face look like a used dish towel. "Of course not! That weapon is perfectly safe."

"Perfectly?"

"Well . . . not perfectly, but close enough."

Dubivko glanced at the three torpedo tubes on the starboard side of the boat. Each housed conventional weapons, as did two of the tubes on the port side. Only the number two tube carried megaton destruction. "Comrade Pomilyev, most of the men assigned to sleep in the forward torpedo room are young sailors. They have wives or girlfriends and aspirations of raising children. Unless you can guarantee that your purple-nosed weapon will not render their family jewels useless, they will probably continue to bunk in the aft torpedo room."

Pomilyev shook his head from side to side. "There's absolutely no proof that—"

"Proof?" Dubivko said. "Until you are qualified, your credibility on this boat is less than that of bilge slime. Earn your qualification, Comrade. In the meantime," Dubivko waived his arm around the torpedo room, "enjoy your solitude."

Dubivko turned and headed toward Compartment Two. He undogged the hatch, grabbed the bar above the opening, and shot his legs through. On the other side, a sailor with a large bayonet guarded the hatch—a mandatory requirement for any boat carrying nuclear weapons. All persons entering the torpedo room were stopped by the guard and required to surrender sharp objects, tools, matches, lighters, or anything that could be used to sabotage the weapon. Although everyone on board had survived intense background checks prior to selection for this mission, none were trusted near the nuclear torpedo. No wonder they didn't want to sleep up front.

Dubivko popped into his cabin and glanced at the picture of Khrushchev hanging above his bunk. He grumbled once, turned toward his wooden cabinet, and rummaged through a drawer. After finding his toothbrush, paste, and hand towel, he strode across the passageway to the officer's washroom. There he splashed some water on his face, brushed his teeth, and tapped at the flickering light above his head.

As he returned to his cabin, he noticed the assistant navigation officer coming out of the four-man cabin just aft and to starboard. Seventy-eight men shared this home under the waves, and most of the noncommissioned sailors berthed in the "sleeping wagon" area in Compartment Seven back by the aft torpedo tubes. One shower and toilet in Compartment Six, used exclusively when submerged, serviced most of the crew. When B-36 surfaced, only the toilet and saltwater shower in the bridge were used, and the crew could enjoy an ocean shower without restrictions. Each compartment was issued a metal token, and only one person who held that compartment's token could use the facilities at a time—similar to the key-on-a-stick system at an American gas station. On board Soviet diesel submarines, this system also ensured that everyone was accounted for when the boat submerged.

Once underwater, the boat's freshwater supply had to be allocated for washing, cooking, and drinking. Showers were limited to only two per week, so the doctor dispensed wash towels daily to maintain hygiene. Unfortunately, the towels did little to control body odor.

Dubivko stored his sundries and walked a few feet over to the acoustic room. A thin operator wearing headphones sat in front of a rack of metal rectangles covered with small airholes. Each box contained an indented section that housed several black control knobs, dials, and indicators. B-36 still used the older Herkules medium-frequency active/passive and Feniks passive search/attack acoustic arrays that could hear contacts up to twenty-nine kilometers away. Only Captain Ketov's B-4 had an upgraded RG-10 passive system that offered greater range and accuracy.

The acoustic sublieutenant looked up, smiled, and pulled the headphones off one ear. "Captain?"

Dubivko smiled. "How are you feeling, Pankov?"

"I'm still a little sore but doing fine, sir."

"Glad to hear," Dubivko said. "It's a good thing our doctor is highly trained in appendectomies."

"Yes, sir."

Dubivko pointed to the headphones. "What are you listening to?"

"Mostly whales, sir. There's nothing much else out here right now."

"No sonobuoys from that Shackleton we saw a few days ago?"

"No, sir. Not even a peep."

Dubivko stepped back into the passageway and said, "Well, let's hope it stays that way so we can stay on schedule."

Pankov nodded and replanted the headphones.

Dubivko walked past the wardroom, where Pankov had received

his operation, and undogged the Compartment Three hatch. The CC hummed with activity as he entered, and he reveled in the vibrancy. Standing just below the ladder that led up to the conning tower, he glanced at his watch. At this time of day, high above them, a blazing sun warmed the sea and transformed wave crests into reflective shards of glass. Although they needed to snorkel to recharge batteries, they dared not do so in broad daylight, but capturing the required daily radio broadcast from Moscow could not be avoided.

Dubivko held back an expletive as he thought about the mental midgets at Fleet HQ who ignored the fact that midnight in Moscow meant daytime this far from Russia. Their orders complicated things in two ways. First, the submarines needed to slow down to capture a broadcast. Second, receiving a signal required approaching near the surface, which left them vulnerable to detection. Apparently the mission planners were not qualified submariners.

"Watch Officer, make your depth twenty meters," Dubivko ordered.

"Yes, Captain," the watch officer acknowledged. He turned to the planesman, seated starboard near the front of the compartment. "Bow planes up fifteen degrees."

"Prepare to raise the HF antenna," Dubivko said.

Another acknowledgment.

In the CC, a dozen faces instinctively glanced upward as the needle in the main depth gauge moved counterclockwise.

The planesman, sitting forward and to the right of Dubivko, barked off a reading. "Passing forty meters."

The watch officer, so designated by his blue and white elastic armband, echoed the planesman's report as the boat's hull creaked like arthritic bones in response to the change in ocean pressure.

"Raise antenna," Dubivko said as the boat neared periscope depth.

Lieutenant Zhukov stepped through the hatch from Compartment Two. As the boat's electronics officer, he was responsible for all of B-36's electronically operated equipment, including acoustic, radar, weapons control, and radio. Without saying a word, Zhukov pointed toward the radio room and continued aft. The familiar routine happened daily, and Dubivko hoped that this time the outcome would be different.

For all the years Dubivko had operated aboard Northern Fleet submarines in the Barents and Norwegian seas, he had experienced the curse of the Arctic. This high-latitude region created sporadic interference like Dubivko's mother-in-law spewed insults. Virtually all transmissions were sent and received via HF or UHF, and during the winter and parts of the summer, magnetic storms and interference were constant concerns. But of all the transits Dubivko could recall, this one, so far, held the record for the most transmission problems, far surpassing anything the Arctic could muster.

The planesman sounded off another report. "Passing thirty meters."

Having crossed the Faroe–Iceland line, B-36 descended into a proverbial radio vacuum. All Northern Fleet stations were masked by static, and the only audible voices came from fishermen on trawlers near Murmansk. For two days Zhukov and the radio operators tried to find clear frequencies but never succeeded. They analyzed signals based on different times of day, weather conditions, and other factors and tried to make radiogram sending and receiving adjustments but without any luck.

"Steady at twenty meters."

"Very well," Dubivko said as he prepared to enter the conning tower.

The main problem they faced with radio communications centered around power. Communicating with Moscow could only be kept secret by using the low-power fifteen-kilowatt transmitter. The antenna for that system required air drying for almost twenty minutes before sending or receiving transmissions. In submarine time, that meant forever, especially while trying to stay undetected and on schedule. The only way to mitigate the problem required using a slightly wet antenna and re-transmitting the same burst message as many as thirty times to ensure receipt by Moscow.

Now up in the conning tower, as he swiveled the periscope back and forth, Dubivko contemplated the endless obstacles before him. Through the scope he spotted an American P2V ASW aircraft on the horizon. His cheeks turned hot as he wondered how they could possibly be following B-36's course so accurately. He contemplated whether it might have something to do with the Americans' new underwater sonar arrays but then quickly dismissed that thought. Unless the United States somehow learned how to defy the laws of physics, given B-36's distance from any known array, and given that they only snorkeled at night, they couldn't possibly achieve this close a fix with that system. No, either the Americans were damn lucky, or they had deployed some new unknown technology. But what?

LAST IN THE LINE OF FOUR Soviet submarines headed to Cuba, B-4 cruised past the Azores on October 15 and entered the Sargasso Sea. A few days later, his nerves on edge, his armpits soaked, Captain Ryurik Ketov thought about Pasha, the boat's cat they'd left behind in Sayda Bay. He hoped she was happy in her new home with plenty of caviar, or at least some nice tuna. He peered over Vladimir Pronin's shoulder in the radio room as the young electronics officer stared at the short burst data radio with frustrated yet hopeful eyes.

The SBD radio consisted of a yellow box covered with dials, switches, and indicators seated next to a silver device that resembled a typewriter. Ketov didn't like the new burst radios. The concept, stolen from the Germans after World War II, was forced upon the submarine fleet in late 1960 and added to B-4 prior to her launch in 1961. The SBD's encoder—the typewriterlike device—suffered from a limitation of having only seven groups of symbols that severely truncated what could be sent to or from HQ. Ketov always harbored concerns that this could lead to misinterpretations of important orders or rules of engagement. What if they received updated instructions on when to fire their nuclear torpedo, but the ultra rapid activity radio turned these into something vague and unintelligible? That could make for a very unpleasant day.

Pronin licked his upper lip and rechecked the settings on the SBD. Still nothing. The burst itself took only seven-tenths of a second, but the wait to receive a transmission could take seven hours, or so it seemed to Ketov.

Finally, the SBD received an order update just as the political officer approached the radio room. Ketov read the strip of paper and shook his head from side to side. They were being redirected from their transit to Mariel. The new orders told them to assume combat readiness in the Caribbean Sea, south of Jamaica, and wait. Wait for what? Ketov wondered if imminent war with the Americans caused the change in plans. He ordered Pronin to have the OSNAZ group begin monitoring civilian radio transmissions to find out what might be going on out there.

After changing course, the next several hours bordered on boring until they caught up with a hurricane. That monstrosity dredged up giant waves and hurled them onto the decks of helpless vessels. One of these was a merchant ship that let out a desperate cry for help. Pronin heard the plea on the radio and reported it to Ketov, who stood watch on the bridge. In the dead of night, B-4 barely held her own against the hurricane while snorkeling on the surface. In rough seas she became a black spec of metal toyed by a force older than time and stronger than Neptune. Ketov bit his lip when he heard Pronin's report. Though he desperately wanted to, he knew he could be of no help to the floundering merchant ship.

Pronin pleaded over the communications circuit to do something, but Ketov knew they could not. The distressed ship lay twelve kilometers to the east, and if B-4 turned away from its southern heading, that might cause them to take on too much water to snorkel. They needed to charge batteries to stay on schedule. They were also under strict orders not to reveal their position to anyone, not even a friendly vessel. Last but certainly not least, what assistance could a tiny submarine possibly offer to a large merchant ship?

Ketov stood on the bridge and watched the waves form crests and troughs of foam and spray. With a dense scud strafing the wave tops, some as high as seventeen meters, he could see no farther than ten meters to either side. For all he knew they might never spot that ship and might run right past her in the dark, or worse, smack hard into her side and end their mission. Pronin called up again from radio and tried a different tact. Perhaps they could launch a flare and try to pick up survivors? Ketov knew it must be agonizing for Pronin to listen to the repeated SOS, but he had to decline his electronics officer's request yet again. They had no room on board and could ill afford to take merchant seamen on such a secret mission.

A blast of salt water doused Ketov's face. The wet cold reminded him of a decade earlier when Ketov's former commander on the SS-26, Captain Second Rank Abram Tyomin, taught him a valuable lesson. A giant wave struck the SS-26 from the side and rolled her almost ninety degrees. Water flooded through the open bridge hatch and cascaded into the CC. Electronics shorted. Helm control died. As watch officer, Ketov stood motionless, lost in a daze of panic. Tyomin ran into the CC, saw the look on Ketov's face, and backhanded him hard across the cheek. He told Ketov to calm down, remember his training, and deliver clear and deliberate orders to the men.

Ketov did just that and directed the crew to manually control the helm from the aft torpedo room. They survived, and from that day forward, he recalled the incident during times of trouble. Now he recalled Tyomin's hard-learned lessons to make the right decision about the distressed merchant ship. That, unfortunately, meant standing by while dozens of fellow countrymen drowned.

Or did it?

Ketov called down to radio. He told Pronin to relay the distress call on the HF band, but to send only a few transmissions and then stop. Hopefully, someone might hear and respond. With relief in his voice, Pronin thanked Ketov for taking this risk. And risk it was, for Ketov knew that transmitting on an open channel could end his career. But if he violated the seamen's code by letting those men die, he'd have to avoid mirrors for the rest of his life.

ON OCTOBER 19, DEEP IN THE middle of the Sargasso Sea, B-36 swayed to the beat of a weather-tossed ocean. Captain Dubivko's fingers tingled like they always did when he neared the surface. For when a submarine sheds her deep ocean security blanket and ventures near the domain of surface ships and aircraft, she becomes vulnerable and takes one step closer to death.

"Twenty meters," Zhukov reported. As the on-duty watch officer, he stood to Dubivko's right near the planesman and helmsman.

"Very well," Dubivko said, leaning against the conning tower ladder. "Raise the HF antenna, and open the main induction. Engine Control, start the diesel engines and battery charging."

Although they had received no communications from Northern

Fleet HQ since their departure weeks earlier, they were nonetheless required to slow, surface, and check for transmissions every evening. Brigade Commander Vitali Agafonov, riding on B-4, sent a daily radio check "click" over the airwaves during that time but never sent anything more tangible. B-36's radio operator simply responded with one "click" to verify reception. All four boats orchestrated this dance in a successive relay to ensure that only two submarines were near the surface at a time.

Assistant captain, Lieutenant A. P. Andreev entered the CC from Compartment Four and relieved Zhukov of the watch. After a two-minute routine, Zhukov returned Andreev's salute and handed him the watch armband. He then looked at Dubivko, and the two headed aft toward the radio room. Dubivko caught a whiff of something that smelled like soap and wondered if Andreev had just showered. He also wondered when he'd last enjoyed that luxury himself.

Dubivko wanted desperately to receive a transmission from Moscow. Anything would be better than nothing. He took a deep breath and forced himself to relax. As the captain of B-36, he could not afford to display signs of anxiety or nervousness. That could undermine his ability to command respect from the crew. If they started to doubt his confidence, they might also doubt his orders. Still, he couldn't help but wonder why Moscow remained silent, and what might be happening in the world around them. Were they headed toward a war with the Americans? Absent communications from HQ, he felt like a blind man in a room full of sword-wielding Cossacks.

Dubivko stuck his head through the radio room door. First Officer Captain Third Rank Kopeikin and Political Officer Saparov peered over the shoulder of the radioman, who was seated near a rack of equipment. Their eyes remained transfixed on the rectangular teletypewriter that sat on a white metal shelf. The typewriter clicked away as the carriage moved left, then right, then left again. Nothing appeared on the paper ribbon except repeated groups of seven letters that spelled nothing.

"It's just the carrier tone," Kopeikin said.

Dubivko said nothing as his heart sank into his gut. He knew Soviet procedures dictated that fleet broadcasts sent from the large antenna farm southeast of Moscow should remain on air for no more than ten minutes. Either you received the transmission on time or you didn't. There were no repeats. Dubivko glanced at his watch. They were on time, but so far they'd been greeted by nothing but a carrier signal. The four men continued to stare at the Teletype, as if their combined sheer will could produce a message. Long minutes passed in silence.

Dubivko checked his watch again and turned to leave. Though he fumed with anger, he dared not show it. Not more than two steps away, he heard Zhukov's excited voice.

"Something's coming in!" Zhukov said. He tapped the radioman on the shoulder. "Test the synchronization line."

The young radioman reached for a few knobs on the SBD.

Dubivko stared at the Teletype as an unreadable message appeared. Talking aloud, Zhukov counted backwards from ten down to zero and then hit the encryption key. The seven-letter groups of gibberish transformed into Cyrillic sentences. The clatter of the teletypewriter ceased, and Zhukov tore off the message ribbon. He handed the piece of yellow paper to Dubivko.

Holding his hand steady, Dubivko read the message, frowned, and said nothing to the others. He left blank stares behind and hurried back through the hatch into the CC, with Kopeikin in tow. They reached the navigation table a minute later.

Dubivko tapped Navigator Naumov on the shoulder. "Show me charts for the Bahamas and Sargasso Sea."

"Which ones?" Naumov said.

"Southern entrance."

"Yes, sir."

Naumov combed through a stack of charts, removed one, and placed it on the nav table. The map displayed a colorful lined drawing of the Bahamas Islands chain near Florida. To the south lay the Turks Islands and Cuba. Dubivko traced a finger across the chart. The paper felt bumpy and coarse as he neared Miami. He thought of the millions of people lying about on sandy beaches there, drinking margaritas, and having fun while completely unaware that a Soviet submarine might be watching them from afar.

Dubivko glanced again at the typed message in his hand. He pictured Ryurik Ketov, on B-4, and wondered if his friend now shared similar concerns and confusion.

"What does it say?" Kopeikin asked.

Dubivko handed over the paper ribbon. "Read it aloud."

Kopeikin grabbed the message and read with a soft voice. "Secret modification of operational orders to follow. The Sixty-ninth Brigade of Submarines are ordered to change course, assume combat readiness, and form a line west of the Caicos and Turks Island passages in the Caribbean Sea."

Dubivko pictured the location in his mind. Just southeast of the Bahamas and north of Puerto Rico, around 500 miles from the southern tip of Cuba and 600 miles south of Miami, dozens of tiny islands dotted the area, and water temperatures soared into the eighties. B-36 was now south of the other submarines, closer to the Turks Island passage. For submariners, transiting this narrow passage was a highly dangerous prospect when detection *wasn't* an issue.

Kopeikin lifted his eyes from the paper. His face registered bewilderment as he stared at Dubivko. "What does this mean? Are we not going to Mariel as originally ordered? What's going on? Are we at war?"

"I don't know," Dubivko said, "but I'm going to find out."

"How?"

"Follow me." Dubivko moved away from the table and walked toward Compartment Two. Kopeikin followed. Dubivko entered his cabin and sat on his bunk. The first officer stepped inside and shut the door.

Dubivko grabbed the phone on the wall and brought the black device to his ear. "Zhukov, join me and Kopeikin in my cabin." He hung up the phone.

"Sir?" Kopeikin queried.

"I'm tired of being deaf, dumb, and blind."

"What do you have in mind?"

"We have five English-speaking OSNAZ communications experts on this boat. It's time we put them to good use," Dubivko said.

"I don't understand. What—"

A knock on the cabin door interrupted Kopeikin's question.

"Enter," Dubivko said.

The door opened, and Zhukov squeezed inside. He bent down slightly as the curving bulkhead ran just above his head. "Sir?"

"Has our OZNAZ group heard anything on open frequencies?"

"No, sir. Our schedule has not permitted more than a few minutes of HF antenna time."

Dubivko stared at the picture of his wife and children tacked to the bulkhead and said, "Comrades, we're going to remain near the surface for another two hours."

Kopeikin's eyes extended. "But that's a direct violation of orders!"

"I will take full responsibility. I want our communications experts to scan American broadcast frequencies and record what's being said."

"Military frequencies?" Zhukov said.

"And commercial," Dubivko said. "Voice of America, for example. They even have a broadcast in Russian. We need to know what's going on out there, and Moscow is not in a position to tell us."

"What about the *Zampolit*?" Kopeikin said. "He'll be obligated to report this breach of protocol."

"I've thought of that," Dubivko said. "We'll tell Political Officer Saparov that our new orders require us to assume combat readiness. As such, we must ascertain the intentions of our enemy. Listening to commercial radio broadcasts is essential in accomplishing this task."

Kopeikin smiled. "Brilliant. He will be required to listen in. And when he does, he will be complicit. That places him squarely on our side, whether he wants to be or not."

"Precisely," Dubivko said. "We have a nuclear torpedo on board. We've been ordered to assume a combat stance in enemy waters. We need to know what the hell is going on out there before we ready tube number two. Understood?"

"Yes, sir," the two men said in unison.

"Zhukov," Dubivko said, "prepare three sets of headphones. I want to listen in as well."

"Yes, sir."

Thirty minutes later, Dubivko pulled on a set of phones and listened to the Russian version of Voice of America. A woman's voice came across clearly and with no accent. She reported that three days earlier, on October 16, 1962, the New York Yankees beat the San Francisco Giants to win the fifty-ninth World Series four games to three. Dubivko breathed a sigh of relief. The Americans were still talking about baseball instead of hurrying toward fallout shelters. That meant they were not at war. At least not yet.

CHAPTER SEVEN

It's not the size of the dog in the fight, it's the size of the fight in the dog.

—MARK TWAIN

O N SUNDAY, OCTOBER 21, ABOARD THE USS *Oxford* off the coast of Cuba, T-Brancher Aubrey Brown came to attention in his chair. Other operators inside the darkened room did likewise. Most of them squinted at an assortment of green screens and indicators. Racks of receivers hummed, while reels of tape spun on six-foot-tall 3M reel-to-reel tape recorders. Other T-Brancher technicians monitored the flickers running across the scope of an X-band receiver. The screech of an unknown radar signal, captured by the receiver, blurted from a speaker mounted on the unit. The technicians grabbed stopwatches, which dangled on shoestrings wrapped around their necks. They timed the intervals between the whooping sounds to measure radar scan rate and cataloged the highs and lows of the signal spikes showing up on the screen.

Once certain of the signal's characteristics, they checked what they'd found against the NSA's super-classified TEXTA (Technical Extracts of Traffic Analysis) manual to confirm the radar's identity. Judging by the parameters captured, the T-Branchers verified that the Soviets were testing radar systems on fully operational offensive nuclear weapons systems in Cuba. The station then reported this to their navy command, who immediately dispatched a helicopter from Florida to retrieve the tapes for verification.

President Kennedy received the news about the operational nuclear missile systems in Cuba during a meeting with Secretaries Dean Rusk and Robert McNamara at 10:00 A.M. After a brief discussion, Kennedy approved the final plan for a quarantine of the island. Later that morning, the president met with General Walter Sweeney, the commander of the Tactical Air Command (TAC), to review the plan for an air attack. Sweeney admitted that at best they might destroy ninety percent of the missile sites. Kennedy expressed concern that the remaining ten percent could still kill hundreds of thousands of Americans and ordered Sweeney to prepare for a potential strike on Cuba within twenty-four hours.

That afternoon the president convened a formal meeting with the National Security Council in which the chief of naval operations, Admiral George W. Anderson, briefed the group on quarantine rules of engagement. Anderson announced that every Russian ship approaching the line would be signaled to stop for potential boarding and inspection. Should a ship fail to halt, they'd fire a warning shot across its bow. If that didn't work, a U.S. destroyer would cripple the merchant ship by demolishing the rudder with cannon fire. Kennedy reacted to this announcement with unease that such a provocation could unintentionally sink the ship and trigger a war. Although Anderson provided assurances that gun-crew accuracy should ensure that no Soviet vessels sank, Kennedy remained dubious. Nonetheless, he agreed to the plan when Anderson stated, "The biggest danger lies in taking no action."

On Monday morning, October 22, at NSA in Fort George G. Meade, Maryland, William J. Reed sat in a chilled conference room and listened intently as his boss, Commander Kaye, read the daily intelligence report. The scent of steaming coffee and fresh aftershave drifted through the large room that held around a dozen members of the Boresight A22 section. Kaye, with an abundance of Texas accent and attitude, launched the meeting by describing a series of events that had transpired over the past several days.

Kaye said that Kennedy had met with Soviet Ambassador Anatoly Dobrynin and Foreign Minister Andrei Gromyko on October 18, and the president received false assurances that the Cubans had received no offensive missiles from Russia. Reed heard a rumor that Kennedy privately called Gromyko a "lying bastard."

Halfway through the meeting, Kaye glanced at Reed and said, "What's the latest on those Soviet submarines?"

Reed frowned. "Well, as you know, we got a few good hits on the four *Foxtrots* when they were near the North Cape, but then four *Zulu*-class submarines showed up out of Gadzhiyevo and created some interference. We haven't seen the *Zulus* since. The only Boresight hits we're getting now are from the *Foxtrots,* which we believe have approached to within five or six hundred miles of Cuba."

"How's our bearing quality holding up?" Kaye said.

"Not very well," Reed said. "We get rough bearing hits on three or four of the *Foxtrots* in the afternoon, which is when we believe Moscow requires them to transmit a burst signal update. Good news is they're retransmitting up to thirty or more times, probably because their antennas are wet. That really helps us get a bearing, but then they go dark for a day. We've been sending P2Vs to the area, but right now our bearing quality is worse than a World War II Huff Duff. Maybe fifty or sixty nautical miles at best. Still, we're starting to get enough data to formulate some interesting conclusions."

Kaye cocked his head to one side. "Such as?"

Reed said, "Well, for one, they're still in the Sargasso Sea northeast of Cuba and averaging almost ten knots, which is damn aggressive for a diesel boat. Two, they don't appear to be shadowing any of the merchant ships, so they must have other orders."

"Like what?" Kaye asked.

Reed shrugged. "We can only assume they're posturing for a fight."

"What about SOSUS?"

"Not much help yet," Reed said as he glanced at the wide-eyed faces seated around the table. "Given that our underwater sonar arrays can't hear beyond 150 nautical miles, they're only occasionally catching a whiff of a snorkeling boat, and the bearing accuracy is worse than Boresight. Also, those subs only snorkel at night. During the day, when they're running on battery, SOSUS can't hear them at all."

"Shit," Kaye muttered. "This is starting to feel like a train wreck in the making."

A female yeoman walked into the room and handed Kaye a printed message. Kaye's face went white as he read it. He looked around the room and said, "I'll deny I told any of you this, but this CIA memo just frosted my balls."

"CIA?" someone said.

"Yeah," Kaye continued. "I probably wasn't supposed to be briefed on this yet, but I have a few friends in low places. The memo says that they verified sixteen of the MRBMs and four of the longer-range IRBMs are already saddled up and out of the barn."

"Damn," someone else said. "That means they can actually launch now."

"And that means they can vaporize millions of us within minutes," Reed said. He knew that when the U-2 spy planes originally photographed the medium- and intermediate-range ballistic missile sites in Cuba, the launchers were still under construction. None were fully operational. Now that they were, the end of the world was only a button push away.

"The Cubans also have twenty-two Soviet IL-28 bombers and thirty-nine MIG-21 jet fighters now," Kaye said. "What's worse, they have a bunch of short-range Frogs with megaton warheads. That means if we try to send in an invasion force, they'll nuke thousands of our boys into dust clouds."

Reed's stomach turned sour as he pictured that scenario. A mixture of fear and excitement shot a bolt of current through his body—fear that perhaps he might be witnessing the end of humanity, and excitement at being center stage for one of the most important events in human history. "What do you think Kennedy's going to do?"

Kaye pursed his lips. "According to this memo, the CIA concluded that he's basically fucked like a junkie whore on Main Street no matter what he does."

One of the officers stifled a laugh. Another asked a question. "They don't think the Soviets are doing this as a bargaining chip to get Kennedy to pull our missiles out of Turkey, do they?"

Kaye ran his tongue over his upper lip and shook his head no. "It's a lot more serious than that. The Soviets want to show the world that if we can waltz into their neighborhood with big-ass weapons, they can do the same in our neck of the woods. Problem is, if we just let them get away with it, we open the door for the Russkies to arm every commie pinko bastard country that might want to point a six-shooter at us. But if we confront Khrushchev about this, that won't stop him from shipping more nukes, and he'll just stall us with bullshit negotiations and U.N. red tape. A confrontation would also show our hand and take a surprise invasion of Cuba off the table."

Reed shifted in his seat. "What about a blockade? I heard that the *Enterprise* got marching orders along with the Eleventh and Thirty-second Air Wings out of Puerto Rico. And with the *Essex* leaving Gitmo Bay with Task Force Bravo, I gotta believe they're planning something."

"Could be, but a blockade won't help us get rid of the dozens of missiles that are already there. Also, we probably can't muster more than sixty ships, and the ocean around Cuba is a lot bigger than the Ponderosa. The Soviets could probably still sneak in some nukes on submarines."

Reed thought about the four *Foxtrots* they'd detected heading for Cuba and let out a soft whistle.

Kaye stood up straight. "What?"

"I just had a holy shit thought."

"What kind of holy shit thought?"

A dozen eyes stared at Reed.

"What if," Reed said, "those *Foxtrots* are already carrying nukes?

"You mean like IRBM parts?" an officer asked.

"The CIA cautioned about that in a recent memorandum," Kaye said.

"No," Reed said. "I mean like nuclear-tipped torpedoes."

A dozen officers sucked in a collective breath.

"Jesus," Kaye said. "That'd mean fifteen-megaton bad boys with a blast radius of—"

"Ten nautical miles," Reed finished. "That could put an end to any blockade Kennedy might be planning."

Kaye's brow furrowed as he shot Reed a stern look. "You need to find a way to tweak those Boresight stations to get us better bearing fixes. And you've probably got no more than a week to do it. Otherwise, we might wind up like the dead guys at the O.K. Corral."

Reed's throat tightened. He knew the odds were against him, but he and his team would just have to find a way to get the job done. There were two GRD-6 Boresight stations that, given the substandard range and accuracy of those antennas, might be close enough to get good fixes if their systems were optimal. Those were the sites in Northwest, Virginia, and Homestead, Florida. There was also one Wullenweber elephant cage close enough that might be able to lend a hand. Unfortunately, that facility, located at Skaggs Island, California, sat right on the edge of the system's maximum range. Somehow they'd have to get all three sites up to optimal capability, then hope that God sent them some good weather and minimal interference. Reed whispered a silent prayer that God didn't prefer the color red.

BY THE AFTERNOON OF OCTOBER 22, T-Branchers aboard the USS *Oxford* spy ship reported that at least five Soviet missile regiments were operational or nearly so, each with eight missile launchers and sixteen missiles. Cuba now possessed the ability to fire a salvo of forty missiles that could devastate dozens of targets in the United States. In response to this threat, the Strategic Air Command initiated a massive alert for the entire B-52-bomber strike force, guaranteeing that thirteen percent of all aircraft be airborne at any given time. The plan ensured that every time a bomber landed, another one took to the air. SAC also dispersed almost 200 B-47 nuclear bombers to thirty-three civilian and military airfields, with another 161 aircraft delivered by the Air Defense Command to sixteen bases within nine hours. For the first time in history, all these planes were armed with nuclear bombs.

THAT SAME AFTERNOON, CIA DIRECTOR JOHN McCONE informed President Kennedy that the four Soviet submarines were positioned to reach Cuba within a matter of days. He received that information from Chief of Naval Operations Admiral George Anderson. Neither McCone nor Kennedy was informed that the original source for that estimation came from Boresight Net Control in Maryland.

In light of the Boresight location estimates for the four *Foxtrot* submarines, Admiral Anderson issued a warning to the blockade fleet commanders: "I cannot emphasize too strongly how smart we must be to keep our navy ships, particularly carriers, from being hit by surprise attack from Soviet submarines. Use all available intelligence, deceptive tactics, and evasion during forthcoming days. Good luck."

The navy positioned hunter/killer Group Bravo, headed by the aircraft carrier USS *Essex*, 200 miles northeast of Caicos Passage, just outside and in the center of the Walnut Line—a boundary arcing through the ocean in a semicircle 500 miles off Cuba. This "do not cross" line that started at the southern tip of Florida and ended east of Haiti represented the outer boundary of Kennedy's quarantine area. Three sub-hunting destroyers escorted the *Essex*, including the USS *Blandy*. The destroyer USS *Cony* (DD-508) took up station near the carrier USS *Randolph* (CVS-15) farther northeast of the *Blandy* (DD-943), and the destroyer USS *Charles P. Cecil* (DDR-835) accompanied the USS *Enterprise* (CVN-65) to a position southeast of the other two groups. Sonar and radar operators on all three destroyers listened and watched for signs of Soviet submarines approaching Cuba. None were aware that they'd soon come within a breath of nuclear annihilation.

ON BOARD B-36, NEAR THE TURKS Island Passage at the southern tip of the infamous Bermuda Triangle, Captain Dubivko's eyelids started twitching again. Only forty kilometers of ocean in that passage separated East Caicos from Grand Turk. Although depths in the center of the channel, which lay just north of Puerto Rico and east of Cuba, plunged to greater than 2,200 meters, shallow waters and sandbars on both sides made the area extremely dangerous for maneuvering. With American ASW planes and ships smothering the entrance and exit routes day and night, remaining undetected required a lot of caution, skill, and luck.

Dubivko envied Captains Savitsky and Shumkov on B-130 and B-59, respectively. Both were transiting through the wider, less treacherous Caicos Passage to the north. Captain Ketov, on B-4, had circled around Puerto Rico to take up station near Jamaica. Dubivko contemplated his unlucky orders and decided to try an old trick. If successful, they just might make it through the passage undetected and unscathed. If not, given the difficulty of his planned maneuver, they could take up permanent residence at the bottom of a foreign sea.

That evening, on October 22, Zhukov relayed the report that the OSNAZ team had heard President Kennedy address the American nation. He stated that the U.S. Navy had initiated a quarantine around Cuba to block Soviet merchant ships carrying nuclear weapons. They still heard nothing but truncated order changes from Moscow and so could rely only on commercial traffic intercepted over the airwaves to give them a hint of what might be happening in the world. After hearing the news about the quarantine, Dubivko wondered if he'd soon receive a message from Moscow ordering him to use the purple-tipped weapon to rip a large hole through Kennedy's blockade. He took a deep breath and patted a handkerchief against his sweat-soaked forehead.

The mediocre air-conditioning unit on B-36 had served them well while operating in cold northern waters. But once they reached the Sargasso Sea, where temperatures exceeded fifty degrees centigrade even at depth, the unit could not keep up, and the crew started to swelter. Dubivko dreaded the thought of being pursued in these heated waters by a swarm of ASW aircraft for hours on end while running slow and deep. He moved B-36 into the passage and descended below the thermal layer to thwart possible sonar detection. He then ordered all stop and neutral buoyancy to keep the boat still and level while they waited. Nerve-grinding hours passed as Pankov, in the acoustics cubicle, listened with trained ears to a distant contact.

Dubivko's mind plagued him with disastrous scenarios as his submarine sat motionless hundreds of feet below the surface. What if an ASW plane picked up their scent and started dropping sonobuoys? Where could they hide? There seemed to be hundreds of planes flying around. He figured American destroyers could not be far away.

Standing in the CC, Dubivko queried a question over the comm. "Acoustic, is that merchant ship still there?"

"Control," Pankov said, "the merchant ship is still making turns for ten knots."

"Acoustic, range and bearing?"

"Control, 3,000 meters bearing three-five-eight and closing."

Dubivko glanced to his left. "Navigator, plot an intercept course for nine knots speed."

Naumov acknowledged the order. Two minutes later he said, "Turn right on my mark to course zero-six-zero." Five more seconds passed. "Mark."

The boat's deck tilted a few degrees as the helmsman pulled the steering lever to the right.

"Intercept time?" Dubivko asked as he wrapped his hand around a pole on the conning tower ladder to steady himself during the turn.

"Intercept in ten minutes," Naumov said.

"Watch Officer, make your depth sixty meters."

Captain-Lieutenant Andreev, the current watch officer, echoed the order. Those standing in the CC shifted their stance as the bow of the boat tipped downward.

Five minutes passed. Dubivko knew that if he miscalculated the merchant ship's depth or his angle of approach, he could crash B-36 into a pair of massive propeller blades. He recalled that another boat, while operating in the Barents Sea, suffered an almost life-ending collision while attempting a similar scheme. The captain of that submarine miscalculated and smacked into the stern of the cargo ship they were planning to hide under. The ship's propeller blades sliced into the boat's bridge and conning tower and caused catastrophic flooding in the CC. After surfacing to make temporary repairs, the ill-fated submarine limped back to Polyarny. Had the weather been rougher, they might have sunk. Dubivko's mouth went dry as he remembered the story. B-36 was now 5,000 miles from home.

"Watch Officer," Dubivko said, "make your depth forty meters, speed seven knots."

Andreev repeated the order, and the bow tilted upward.

"Attention in the CC," Dubivko said. "I intend to close within one hundred meters range and ten meters below the merchant ship's hull. After they pass overhead, we will turn about and match their course and speed. The tanker's propeller noise should allow us to hide in their wake. We will need steady hands on helm and depth control."

Andreev shot Dubivko a concerned look.

Dubivko returned the look and said, "Take us in, Watch Officer." Through the comm he said, "Acoustic, range to tanker?"

Pankov answered from acoustic. "Control, 500 meters and closing."

Dubivko glanced at the shallow depth gauge on the starboard bulkhead. He then moved a few feet to his left and looked over Naumov's shoulder. The navigator drew lines on a small sheet atop the Plexiglas on the nav table. The lines depicted two merging contacts heading southwest. Dubivko studied the contact markings, did a few calculations in his head, and returned to his previous location near the watch officer.

A few minutes later, Dubivko ordered a single-range ping using active sonar on low power. He knew that U.S. destroyers and aircraft could potentially detect even one ping, but running into the blades of a merchant ship posed a greater risk. Once they closed the distance to the tanker, passive acoustic would be almost useless. The ship's massive screws generated too much noise to discern any kind of range estimation, and bearing information would be meaningless. At that point, Dubivko would need to rely on periscope observation from the conning tower.

"Acoustic, range?" Dubivko said.

"Control, 300 meters," Pankov replied from the acoustic room.

"Watch Officer, bow planes up five degrees, come to thirty meters depth."

Andreev reiterated the order, and the boat inched upward. Dubivko took a step toward the conning tower ladder. "Stand by to open hatch."

A sailor darted up the ladder and waited, both hands gripping the cold steel of the hatch wheel.

"Steady at thirty meters."

The sailor muscled the wheel, and a sheet of ocean water splattered onto the deck. Several drops splashed Dubivko's face. He licked his lips and tasted salt.

As the waterfall diminished to a few drops, the sailor stuck his head into the conning tower, did a quick visual sweep, then slid back down the ladder into the CC. After a verbal "all clear" from the sailor, Dubivko gripped the sides of the wet ladder and climbed upward. Behind him followed Saparov, as regulations stipulated that only the political officer could join the captain or authorized relief officer in the conning tower.

Once inside the small enclosure, just above the CC, Dubivko asked for an update. "Acoustic, range to tanker?"

"Control, passing beneath the contact now," Pankov said.

Dubivko had approached the tanker head on and now intended to make a 360-degree turn and follow the ship while hiding behind her.

Kopeikin relayed the order to the helmsman. Dubivko held his breath as he heard propeller blades thrashing above him. He knew that maintaining an exact distance and depth was now critical. The vacuum produced by the surface effect of ocean water rushing beneath the tanker's hull posed an immense danger. Just a few meters too close, and they could be sucked into the ship's deadly propeller blades.

"Acoustic, range?" Dubivko asked again.

"Control, we've passed underneath. Range is opening, now twenty meters."

Dubivko called down to the watch officer. "Watch Officer, come right to course two-five-five, increase speed to ten knots." He then depressed the switch for the approach periscope. "*Podnyat periscope.*"

Saparov said nothing as he stood behind Dubivko and observed. Dubivko knew that the officer's presence in the conning tower had nothing to do with good seamanship and everything to do with politics. No matter the years, sacrifice, or evidence of loyalty, Moscow conservatives never granted full trust to their submarine captains.

Dubivko rested his twitching eyelid against the rubber eyepiece on the scope. An eerie glow from a carved moon shimmered on the wave tops as he spun the scope left and searched for the tanker's stern light post. He knew that if he steered too far to port or starboard, the metal shield around the light would prevent the glow from being seen. Viewing only darkness, Dubivko swung left, then right. Still nothing.

"Control, contact has changed course, now heading two-seven-zero."

Dubivko's ears burned as a wave of panic swept past. He now knew why he couldn't see the ship: they had changed course. He fought to stay calm, to not show signs of concern in front of Saparov as he called down to the CC. "Watch Officer, right two degrees rudder." Again he moved the scope back and forth. Still no light. "Left two degrees rudder!"

B-36 edged to port. Still no sign of the tanker.

Then, suddenly, a twinkle. *Was that a star?* Dubivko squinted and stared. Out of the black, another blink. Then a glimmer. Finally, the stern light came into view, resting on a stanchion about six meters above the waterline.

"Watch Officer, steady on this course," Dubivko ordered.

As his eyes adjusted to the dark, he could now see the vessel's main deck sprinkled with equipment and lifeboats. Judging by how much of the ship protruded above the waterline, he figured she must be running with a light load. The reduced tonnage caused the ship's large propeller to break the surface as it churned the ocean into white foam. Dubivko smiled. A breaching prop translated to more turbulence and noise, making it harder for ASW aircraft sonobuoys to discern B-36's hull from that of the tanker's. The merchant ship's abundance of steel would also help mask the submarine's electromagnetic signature, and the foamy wake made periscope detection almost impossible.

With any luck, the tanker would maintain a steady course and speed, at least long enough for B-36 to make it through the passage. As Dubivko watched for signs of life aboard their newfound friend, he wondered what flag this ship operated under. For all he knew, they could be American. If so, he imagined how the tanker's captain might react should he learn that a Soviet submarine lurked behind him like a silent leech.

The glow of a cigarette came to life on the deck of the ship, illuminating a silhouette near the stern. As he slung his arms over the periscope's handles and played the part of a Peeping Tom, Dubivko wondered if the man might be from Miami. Captain Dubivko was now in his element, in command of his vessel, heading into harm's way while cleverly hiding from the enemy. Excitement, pride, and fear owned equal portions of him, none more so than fear, as this honed his senses, heightened his instincts, and increased his odds of avoiding a fatal mistake. Then Dubivko heard an unwanted report from acoustic.

"Control, contact is slowing."

The sound of the tanker's thunderous screws grew faint. The stern light blinked once and then vanished into the darkness.

"Control, contact has slowed to eight knots, bearing zero-one-zero."

"Shit!" Dubivko said to himself, hoping that Saparov did not overhear. For reasons unknown, the merchant ship was decreasing speed and turning to port. "Watch Officer, slow to six knots."

Dubivko called down to the CC. "Navigator, are there any shallows or sandbars nearby?"

"No, sir," came the reply. "But if she maintains this course, she will run aground."

Dubivko frowned. "Acoustic, report fathometer depth."

"Control, one hundred meters," Pankov replied.

Dubivko quickly weighed his options. Without the tanker's turbulent noise to hide under, B-36 could be exposed within minutes.

"Control, three knots and slowing."

"Watch Officer, slow to three knots," Dubivko said.

"Control, she's dropping anchor," acoustic reported.

"Shit." Dubivko said, no longer attempting to hide his concern from Saparov.

"Acoustic, fathometer depth?"

"Control, eighty meters."

"All stop," Dubivko said.

The boat slowed and crawled to a stop.

"Zhukov," Dubivko said.

"Sir?" Zhukov replied from below in the CC.

"Raise the zenith navigation scope. Check for aircraft."

Hydraulics whispered. Dubivko also heard a few clicks and squeaks as the scope swiveled in its housing.

"Nothing, sir," Zhukov reported.

"Navigator," Dubivko said, "find us a clear patch."

Naumov relayed a course and heading, and Dubivko issued maneuvering orders.

"Watch Officer," Dubivko said once they'd reached the new location, "make your depth eighty meters. Set us on the bottom."

"Yes, sir."

Dubivko still did not know why the tanker came to a stop but surmised she needed to make repairs of some kind. Regardless, for now, with dawn approaching, he had no choice but to sit and wait. Thankfully, they still retained almost a full battery charge. Given their close proximity to the tanker, snorkeling was out of the question.

With Saparov close behind him, Dubivko descended the ladder into the CC and closed the hatch. Hours passed without a sound from the tanker. At 3:00 A.M., while Dubivko studied the nav plot with Naumov, he heard an excited report from Pankov in acoustic.

"Control, new contact bearing two-five-five! Two tandem screws."

"Watch Officer, make turns for three knots, ascend to twenty meters," Dubivko said. From their position on the bottom, they could hear but not see, and Dubivko wanted to know what was going on up there. Who was approaching the tanker? A repair ship? An American destroyer?

The boat angled toward the surface and the answer.

"Stand by to open hatch."

Again a sailor stood poised by the conning tower ladder.

"Acoustic?" Dubivko said over the comm line. "Contact type?"

A few seconds passed, then Pankov said, "Control, American destroyer, speed fifteen knots."

Now what? They could descend back to the bottom and wait, but for how long? On the other hand, if for some reason the destroyer saw their periscope and went active with sonar, they'd be sitting ducks.

At twenty meters, the hatch flung open, and Dubivko climbed back into the conning tower. He peered through the scope and zeroed in on the destroyer's lights. Though he could not make out the enemy ship's bristling antenna, spinning radar, and menacing armament in the dark, he imagined all of these in his mind. He ordered the ESM radar detection mast raised. An occasional beep could be heard from the Nakat ESM panel in the CC as the system detected enemy radar. NATO called this system Stop Light. As the intervals between the beeps decreased, Dubivko knew that the destroyer's radar was starting to lock on to their masts, and he could not afford to keep them up much longer. The ESM antenna array, which sat between the two periscopes on the sail, carried four bands of direction-finding antennas that captured enemy radar signals and fed data to a cathode-ray tube (CRT) screen. Enemy ships and planes used radar to try to spot submarine masts protruding above the surface, and when they ventured too close, the Nakat system beeped with a warning.

The American destroyer closed to within 1,000 meters and stopped. Then a signal lamp pierced the dark as it flashed out a message in Morse code. Dubivko translated the dit-dah flashes in his head. "Alpha, Alpha," he said to himself, feeling Saparov's presence behind him. "What ship?"

Dubivko lowered the scope to avoid detection. He raised the mast again a few seconds later, swung the scope toward the tanker, and watched for a reply. Flashes appeared like shooting stars against the backdrop of endless black. Given the lamp's location on the upper level of the tanker, the large superstructure and bridge blocked most of the light, and Dubivko could see only a few of the flashes.

"Shit!"

Dubivko turned the scope back toward the destroyer and waited. Long seconds ticked by without a response. The ESM beeps increased in frequency. Then, from out of the ink, the flashes came. Dubivko whispered the translation aloud. "Radio frequency three four three point eight." He lowered the masts and called down to the CC, "Zhukov, raise the antenna, and have our English translators listen in. Three four three point eight."

Zhukov complied. Minutes later he called back up to Dubivko. "The tanker is Norwegian. A boiler failed, and they are bringing a spare online. The Americans offered assistance, but the Norwegians declined."

After hearing this report, Dubivko raised the scope and watched as the destroyer's running lights turned to starboard and dimmed. "Acoustic?"

Pankov said, "Control, American destroyer is heading away, making turns for ten knots and accelerating." Chains rattled from outside the hull. Pankov issued another report. "Control, the merchant ship is reeling in her anchor."

Dubivko emptied his lungs in relief.

AT 10:12 P.M. ON OCTOBER 22, a few hours after President Kennedy's speech to the nation regarding nuclear missiles in Cuba, R-Branchers at an NSA listening post picked up a high-priority message sent from the Soviet spy ship *Shkval* on station near the Bermudas. The Russian merchant ship *Alantika* received the message and rebroadcast the same to Murmansk, near the *Foxtrot* submarine's home port of Sayda Bay. Excited R-Branchers informed Net Control, reporting that "this type of precedence is rarely observed. Significance unknown." When the NSA received the flash message, officials there feared the worst. Were the Soviets planning to run the blockade? Were their *Foxtrot* submarines preparing to attack the U.S. fleet? Were they minutes away from launching a nuclear attack?

A few hours after midnight, a flurry of radio signals hit the air as Soviet merchant ships called home to Russia, asking for instructions. One ship sent an urgent plea for help.

Moscow remained silent, and their cargo ships, along with the rest of the world, came to a halt and waited for an answer.

Just before sunrise on October 23, the residents of Palm Beach, Florida, were shocked awake by the rumbling sound of a squadron of P2V ASW aircraft. The P2V's twin-props sliced the tropical air as high-pitched engines gulped gallons of fuel to push through the humidity. Sleepy-eyed spectators watched with curiosity as the planes lowered their landing gear and descended onto the runway at Palm Beach International Airport. Following just behind the P2Vs, a squadron of B-47 bombers lumbered onto the ground, tires screeching and jet engines roaring to slow the aircraft.

Farther south, thirteen attack submarines slid from the docks at Key West Naval Base as dungaree-clad sailors scurried topside to stow lines and gear. With torpedo tubes fully loaded and pointed toward

Cuba, the black silhouettes disappeared beneath the choppy waters of the Florida Straits. A division of *Gearing*-class destroyers followed the submarines, each armed with one ASROC (antisubmarine rocket) launcher, triple torpedo launchers, and two "DASH" antisubmarine helicopters. One lone submarine and destroyer stayed behind to defend the base.

On board the USS *Robert E. Lee* (SSBN-601), on deployment in the Atlantic, Commander Charles Griffiths received a change of orders from CINCLANT (Commander-in-Chief, Atlantic). His original mission orders were to conduct a Follow-on Test (FOT) firing of the Polaris A1 ballistic missile in the wake of the first launch by the USS *George Washington*. Now he was instructed to arm all missiles and ensure readiness to fire on the Soviet Union within a moment's notice. Griffiths knew that four other SSBNs (ship submersible ballistic, nuclear) received the same orders, and eighty nuclear weapons were poised to turn Russia into a radioactive wasteland.

"We were mindful that our loved ones were in imminent danger and that we could be facing an unbelievable future," said Griffiths. "Yet we would have fired as ordered, and no one on board would have tried to prevent it. . . . It was up to the president and God to avoid Armageddon."

IN NAVY FLAG PLOT ROOM 6D624 at the Pentagon, Admiral Anderson paced nervously. In this nerve center of the navy's planned blockade of Cuba, charts and maps of the Caribbean and Atlantic oceans lined entire walls, where personnel meticulously plotted the movements of every warship in the area. Something strange was happening in the Sargasso Sea near Cuba. The HFDF station at Homestead sent a flash message moments earlier reporting that the Soviet merchant ship *Bol'shevik Sukhanov* "has altered course and is probably en route back to port." Yet another report said that "HFDF fix on the Soviet cargo ship *Kislovodsk,* en route to Cuba, indicates that the ship has altered course to the north."

Despite indications that the Soviets might be backing down, or at least taking a breather, Anderson harbored concerns that his sixty ships along the Walnut line were still having trouble finding those four *Foxtrot* submarines. The ASW boys on board ships and flying in planes would get a sniff, run down the track, but then lose the contact. A recent CINCLANT situation report showed only nine ASW hits since October 22. The SITREP (situation report) also revealed that Aircraft TG 136 got a Hot status on two *Foxtrots*, prosecuted, but never found anything. Anderson knew that some of those hits came from SOSUS arrays, but not many.

What did appear consistent were the Boresight tips. Though he'd not yet been fully briefed on the technology, he was familiar enough with standard HF direction finding to understand the concept. When the Soviets sent out a burst, the stations recorded the transmission, then tried to find a bearing after the fact. Sounded simple enough, though he knew a bunch of beacon heads spent more than a year figuring it out.

SOSUS occasionally helped find the *Foxtrots* when they ran on diesel engines at night, but given the long distances from most of the arrays, getting good bearings was tough. When those boats went silent and deep on batteries, SOSUS became useless. Fortunately for the good guys, the *Foxtrots* transmitted multiple times every afternoon, and when they did, Boresight stations could get bearings, even if not very accurate ones.

Still, until ASW forces nailed those submarines cold and forced them to surface, the blockade was at great risk of failure. Anderson knew that Attorney General Robert Kennedy, following a recent intelligence briefing, said that "the president ordered the navy to give the highest priority to tracking the submarines and to put into effect the greatest possible safety measures to protect our own aircraft carriers and other vessels." Anderson also knew that the president sent an ultimatum to Khrushchev stating that any Soviet submarines detected near the quarantine line must surface and be identified. The fleet operated under directives to use international code signals and nondestructive explosive charges to warn the subs, but if that failed, Anderson's orders were clear: use every means possible to sink those *Foxtrots*. That is, if he could find them.

Meanwhile, the defense readiness condition (DEFCON) catapulted to its highest level ever. The military established DEFCON as a measure of the activation and readiness level of the U.S. Armed Forces, and standard peacetime protocol dictated DEFCON 5. DEFCON 1, never formally declared in U.S. history, was synonymous with war. DEFCON 2 ran a close second and also had never been mandated.

A few years earlier, then Secretary of Defense Thomas Gates ordered the DEFCON raised to level three to test the response system. Although he intended at that time to keep the alert secret, the Soviets, while monitoring U.S. force movements, discovered the change. Bristling with anger, Khrushchev called the DEFCON alert a "provocation." When U.S. forces were ordered to DEFCON 3 on October 22, no attempt was made to mask the intention. The message and the response were deliberately transmitted on open frequencies. Three radar bases also activated Operation Falling Leaves to monitor the Soviet response, including any missile launches from Cuba, but Anderson knew that the radar systems were experimental and unreliable.

Earlier that day, on October 23, SAC received orders to invoke DEFCON 2 for the first time in history. Although the rest of the military remained at DEFCON 3, preparations were under way to ensure maximum readiness in the event that conditions changed. Anderson hoped that never happened.

That afternoon, in the third meeting of the Executive Committee (ExComm) of the National Security Council with President Kennedy, Secretary of Defense Robert McNamara reported on the plans for naval interception, noting "the presence of a submarine near the more interesting ships," and warned that radio silence should be imposed. The committee agreed that should a Soviet submarine interfere with the blockade or adopt a hostile posture, the enemy sub must be sunk.

After visiting a Boresight station to try to optimize their systems, William J. Reed returned to Washington, D.C. There he met with a group of NSA engineers tasked with Boresight research and development. They burned midnight oil and chain-smoked while frantically trying to solve the perplexing problems that Reed reported finding in the field. During those infrequent times when the technology worked, it worked reasonably well. Unfortunately, more often than not, the systems were plagued by intermittent errors, annoying interference, and inconsistent results. Reed's boss, Commander Kaye, reminded everyone that unless these issues could be resolved in time, President Kennedy's blockade might become as "useless as tits on a bull."

AFTER SNEAKING THROUGH THE TURKS PASSAGE unscathed, Captain Dubivko hid B-36 in the wake of the Norwegian tanker until they neared 150 nautical miles north of Haiti. He then headed west toward Cuba and slowed. He circled on station until nightfall, then came up to recharge batteries. This night marked the final time they could surface to snorkel and vent exhaust fumes externally on either side of Compartment Five. From now on they'd need to switch to snorting. This operation ensured stealth in enemy waters and could be done at depths up to fourteen meters. Snorting required raising the snorkel mast to suck in air directly to the diesels. Fumes were piped back up through the conning tower and cooled prior to venting into the ocean. Doing so lowered the probability of detection by reducing the boat's heat signature.

"What was her name?" Dubivko asked while peering through his binoculars on the bridge. Two of B-36's engines hummed from below decks, scattering the smell of burnt diesel fumes into the tropical night air.

"Who?" Zhukov asked while studying the gyrocompass to his right.

"The Norwegian tanker. What was her name?"

"*Cretan Star*, I think. At least, that's what the American destroyer called her."

"Too bad," Dubivko said. "I was hoping for a female name."

"Sir?" Zhukov said.

Dubivko lowered his binoculars and smiled. "We hid under her skirt for hours. Shouldn't she be a woman?"

Zhukov shook his head and laughed. Long days filled with tension made that luxury rare, and Dubivko knew that even less levity lay ahead. B-36 received new orders after crossing the Turks Passage behind the Norwegian oil tanker, directing them to assume station to the northeast and monitor U.S. ship movements. On station 200 miles south of Bermuda and 600 miles east of Cuba, they snorkeled at night and ran silent on battery power during the day, still faithfully coming shallow in the afternoon to transmit an update.

ASW aircraft continued to plague them. P2Vs and newer P3 Orions splashed sonobuoys along precise patterns that nearly matched B-36's track, making Dubivko wonder if the Americans had recruited a team of psychics. On the other hand, Zhukov reported that American pilots were lax in their communications protocol, often transmitting unencoded messages over open HF frequencies. This was especially true for planes attached to the carriers *Essex* and *Randolph* operating to the north in the vicinity of B-130 and B-59. Determining which pilots were communicating from which planes became a great pastime for B-36's radio operators, who often made friendly wagers with each other as to which call sign matched which pilot.

Avoiding enemy detection exhausted the crew but took second fiddle to the heat. Deep inside warm tropical waters, given the inadequacy of their air-conditioning system, B-36's insides turned into a blistering sewer pipe. Temperatures surpassed unbearable, especially in the engine room, where they often eclipsed thirty-seven degrees centigrade.

Dubivko ordered the rationing of drinking water, allowing each man to consume no more than one glass per day. He also granted one glass of red wine at dinner. The crew's health deteriorated rapidly, given poor personal hygiene and constant exposure to humidity, high temperatures, and diesel fumes. Most suffered from painful rashes or oozing skin ulcers. To combat this problem, the doctor handed out disposable towels every day. When those supplies ran out, he used alcohol-doused cotton balls. Many of the sailors just stuck the balls in their mouths and sucked out the alcohol instead of using them to treat their rashes.

Zhukov's English-speaking experts, after sucking their alcohol swabs dry, continued to monitor shortwave and high-frequency broadcasts, such as Radio Liberty, BBC, and Voice of America. These broadcasts revealed that Soviet statesman Anastas Mikoyan, one of Khrushchev's closest advisers, was trying to negotiate a compromise with the Americans after Kennedy reacted so harshly to the discovery of nuclear missiles in Cuba. Dubivko could not help but wonder if their mission might soon come to an abrupt end, either by way of war or by recall to Russia. He knew that in the event of a recall, no one in power would admit to failure.

Dubivko informed his crew about the blockade to ensure they stayed alert. He also told them that hundreds of aircraft and dozens of ships from the U.S. Atlantic Fleet were hell bent on finding and possibly sinking their submarine. Zhukov mentioned that an operator heard that the Americans had established prisoner of war camps in Florida, and Dubivko let his mind play on the possibility of meeting someone from Miami.

When the sun rose, bad luck returned. The chief engineer, Captain

Lieutenant Potapov, reported that the upper lid on the VIPS—the imitation cartridge projection device used to fire decoys to ward off enemy torpedoes—had been damaged in the last storm. Potapov insisted that it would be suicide to submerge the boat deeper than seventy meters. The depth limitation posed significant problems. First, most of the thermal layers under which they might hide were below that depth. Second, when they operated at shallower depths, ASW planes could find them more easily with magnetic detectors. Repairing the lid required surfacing, which became impossible once they neared Cuba and ASW activity intensified.

The OSNAZ specialists reported that there were now at least three carrier groups operating in the Sargasso Sea, along with hundreds of aircraft flying about—all intent on finding them. Cloudless blue skies aided the enemy's objective. Dubivko had continued to come shallow at night to snort and descended back to seventy meters during the day to hide.

Five or six times, while snorting under a blanket of dazzling stars, they had spotted a plane through the periscope or detected a radar signal nearby on the Nakat ESM mast. Dubivko then yelled, "*Srochnoiya pogruganye!*" and the boat made a quick dive. Zhukov avoided using any standard HF transmissions, as they knew the Americans were trying to locate them with their Huff Duffs. Still, every afternoon they had been compelled to come shallow again to receive a burst transmission on the SBD from headquarters and send a verification of receipt. Dubivko did not even consider the possibility that a little more than 500 nautical miles away, someone might be listening.

CHAPTER EIGHT

One death is a tragedy; one million is a statistic.
—JOSEPH STALIN

ON THE MORNING OF OCTOBER 23, when Communications Technician John Gurley entered the 1,400-square-foot Boresight building in Homestead, Florida, he did not have an inkling that this day would become one of the most memorable in his life. The day began like most others, with an abundance of routine, coffee, doughnuts, bad jokes, a couple Boresight flashes, and some whining from Lieutenant Clower about staying diligent. But that all changed after lunch.

That tall ensign from NSG had visited them again a few days earlier. What was his name, Reed? The ensign brought a couple of his techies back, and they spent the better part of a day tweaking and testing the Boresight equipment again. Damned if those guys didn't manage to improve the bearing accuracy by half a degree or so. In a big ocean, that could mean a lot. Ensign Reed spent an hour with Clower talking hush-hush about something. After the NSG guys left, Clower called a meeting. He told everyone some harry shit was going on near Cuba, and they needed to crank up the alert factor to full throttle. Clower didn't offer many details but said they should be looking for *Foxtrot* call signs. *Foxtrot* submarines near Cuba? Gurley could only imagine why that might be, as he thought about his trek from unemployed to sub hunter.

When he graduated from high school in 1956, Gurley's parents didn't have enough money to send him to college. Finding a job in Dallas during those days was pretty tough, so he enlisted in the navy to avoid becoming a ground pounder in a rice paddy. During boot camp in San Diego, someone noticed his Texas drawl and said he should learn how to communicate better. He took their advice and struck for communications technician. Weeks later they sent him to radioman school in Imperial Beach just south of San Diego.

After graduating from radioman school, Gurley received orders to Morocco. He spent an exotic three years in the desert, then wound up in the frigid north near Kodiak, Alaska. In between hunting and fishing in heaven's wilderness, Gurley and twenty-six other sailors ran the Huff Duff station there, reporting to a chief warrant officer who let them play as much as they worked to keep morale and efficiency at a peak. After a year in Kodiak, the navy pulled him out of one paradise and plopped him into another: Homestead, Florida. They needed more R-Brancher radio "collection" experts there.

At Homestead, Gurley rode the bus every day from the barracks to the operations building down Card Sound Road. In the ops center, he and a dozen other R-Branchers sat in front of various stations and monitored Soviet traffic, while M-Branchers did their maintenance thing, I-Branchers listened for intelligence tidbits, T-Branchers analyzed signals, and O-Branchers ran the shop. Four R-Branchers usually searched for active targets, while one or two others encoded and decoded messages sent from the crypto center.

The job really wasn't that tough until Ensign Reed and his team waltzed in and set up all that Boresight equipment. Now they'd get flashes from NC that included time windows, so they had to find the related reel-to-reel tape, load it up, and listen on the specified frequencies during specific time slots across maybe thirty or forty bearings. Problem was, they were looking for a burst that lasted only seven-tenths of a second, so they were forced to play the thing back and forth ad nauseam, sometimes for hours.

Four of these systems recorded signals on various frequencies starting at 840 kilohertz. When the systems heard a burst signal, they were supposed to alarm, but the loud bell went off only if they got a strong enough hit. If another station gained a stronger signal to trigger the alarm, they sent NC a tip-off, and NC sent out a flash. That's when Gurley and his colleagues spent hours poring through recording tapes to see if they also picked up the burst, because one bearing to a target could not accurately pinpoint a transmitting sub's location.

"In those days we still used the old compass rose board," said Gurley. "When we got a bearing, we manually ran a piece of string from one end of the board to the other along the bearing. We prepared reports to Net Control on a machine that used a punch tape machine. We'd wrap the yellow paper tape around our finger to form a tight ball and then shove it inside of a Coke can. The ribbon came with two carbon copies that made it thick and hard to handle. We shoved the can into a tube that dropped down to the communications center. The boys downstairs pulled the report out of the Coke can and sent it off to NC. Very high tech."

Operators often played jokes on new guys by ordering them to unravel the three-ply tape, which, if done incorrectly, resulted in carbon-stained fingers and a mess that resembled a ticker-tape parade. "Our job was demanding but mostly routine," said Gurley. "A little levity now and then kept us from going stir crazy."

That day, on October 23, an alarm bell interrupted the routine. Gurley wheeled his chair over to the recorder and started the procedure he'd been taught by Ensign Reed and his team. Check this, verify that, do something else. He and other operators spent the next hour analyzing the recording and checking bearings. The transmission call sign pointed to a probable *Foxtrot*-class submarine.

Gurley went to the compass rose board, grabbed a piece of string, and pulled it down the line of the probable bearing. The string ran just north of Cuba and due east of Florida. Based on the signal strength, he figured the *Foxtrot* was probably less than 1,000 miles away. Gurley wheeled over to the punch tape and generated a report for the Coke can. He dropped the can into the tube and waited.

A half hour later, the station received a copy of the flash sent to the other stations from NC. Gurley waited another hour and then contacted a buddy at Net Control. He asked if any of the other stations had reported bearings to that contact. The answer came back as affirmative. Gurley pulled two more strings across the compass rose board, representing those bearings. When he saw where the lines converged, his heart started pounding so hard he thought he'd pop a shirt button. He didn't know it then, but he and other Huff Duffs had just nailed

Captain Savitsky's B-59, operating near the Bahamas a few hundred miles southeast of Florida.

IN THE FLAG PLOT ROOM, THE navy's command center at the Pentagon, Admiral Anderson's eyes ached from lack of sleep. Too much coffee soured his stomach, but he downed another cup anyway. The black "navy joe" offered that bitter-burnt taste that Anderson liked so well. Just one more reason to go navy. He squinted and stared at the dozens of red tags dotting the large wall chart. Each tag, numbered C1 through C29, denoted a probable submerged contact, most likely a *Foxtrot* or maybe a *Zulu*. Positively identified submarines, like the *Zulu* spotted on the surface days earlier, drew B designations.

Around thirty men and women in Flag Plot kept track of estimated course and speed information for B and C contacts and, most importantly, their probable distance to any U.S. surface ships in the quarantine zone. Another set of flags on the plot followed Soviet merchant traffic as those ships approached the Walnut line, the outer perimeter of the quarantine zone. Anderson suspected that many of these ships carried nuclear missiles and launcher parts destined for Cuba. After Kennedy's quarantine went into effect, around sixty U.S. Navy ships now patrolled along the Walnut line that arced from the tip of Florida to an area just south of Cuba. If Soviet merchant ships tried to cross the line, the navy had orders to stop them. But with four *Foxtrot* submarines lurking nearby, accomplishing that task could prove difficult, if not deadly.

Anderson turned to see Defense Secretary Robert McNamara stride into the room with Roswell Gilpatric, a New York lawyer turned deputy defense secretary, who followed in a "brown nose" position. Behind the two, an entourage of clean-cut press-corps bulldogs came outfitted in suits, ties, and dresses. Anderson scowled. The last thing he needed right now was a bunch of McNamara's White House public-relations questioners, especially when every answer required a heavy dose of sidestepping.

McNamara shook Anderson's hand and asked for an update. The admiral pointed to the set of merchant-ship flags on the plot. McNamara's eyes followed Anderson's finger. While the press dogs scribbled in tiny spiral-bound notebooks, Anderson explained that they expected the first sortie of Russian ships to hit the line at around 10:00 A.M. the following day. He said that things started to change after Kennedy's quarantine announcement a few days earlier. Where once these ships acted as lone wolves heading toward Cuban ports, they now appeared to be forming one large phalanx, with the tanker *Bucharest* leading the charge. They could not explain why.

McNamara asked about the red dots. Were there really that many Soviet submarines out there? Anderson said no, that many flags represented more subs than the Soviet navy possessed. He explained that they plotted each reported sighting but considered most false positives. They believed that less than a half-dozen Soviet subs were in the area, and the navy's ASW forces were most concerned with finding the four *Foxtrots*. An underling from the press office raised her hand. "What's a *Foxtrot*?" she asked. Anderson patiently reeled off a few specifications about that class of Soviet submarine.

McNamara asked how the navy's ships intended to force Soviet subs to the surface once they found them. Anderson took another gulp of coffee and recalled his days as the commander of the Sixth Fleet working for then Chief of Naval Operations Admiral Arleigh Burke. The admiral more than once grumbled that McNamara possessed the steel-trap mind of an encyclopedia but the detail-oriented focus of a micromanager who asked more questions than a five-year-old.

Knowing that McNamara loved the devil in the details, Anderson doused him with a fire hose. He delivered a long diatribe about international signal codes transmitted by the United States on open frequencies. They knew Soviet submarines monitored these frequencies and so should be informed as to the expected rules of engagement during the conflict. That morning the U.S. Naval Oceanographic Office broadcast Mariner's 45–62, Special Warning 32, which spelled out submarine surfacing and identification procedures. The warning stated that U.S. ships and planes would signal Soviet submarines by dropping four to five hand grenade–sized explosives, followed by the international sonar-transmitted signal IDKCA, which means "rise to the surface." Submarines hearing this signal had to surface on an easterly course at once and move slowly away from Cuba.

"And if they don't?" McNamara asked. Pencils poised, the press geeks stared at Anderson with curious eyes.

"We will sink them," Anderson said.

McNamara didn't blink. Anderson hoped that, given the presence of the press staff, the SecDef would not probe for more. Besides, in their meeting days earlier, he'd already briefed McNamara, along with Kennedy and the ExComm group, about these rules of engagement. He'd explained that shots would be fired across bows and then into rudders of noncompliant merchant ships. Neither McNamara nor Kennedy liked the plan, but they conceded that there were no alternatives.

McNamara pointed to a lone flag on the plot, positioned some distance away from the others around the quarantine area. The flag represented the position of a U.S. Navy destroyer off the coast of Florida, almost a hundred miles east of the Walnut line. "Why is that ship out of line?" McNamara asked.

"She's prosecuting a probable submerged contact operating near the Bahamas," Anderson said.

"I don't understand," McNamara said. "I thought you said that most of the sightings were false positives. If you move every ship away from the Walnut line, we won't have a blockade."

"This submarine hit came from a more reliable source," Anderson said.

"What source is that?" McNamara pressed.

Anderson remained silent. Neither the press staff nor many of his own watch officers were cleared for that information.

"Admiral?" McNamara said.

"Mr. Secretary," Anderson said, motioning toward a side room. "Please come with me."

McNamara glanced at Gilpatric. "I'll be right back."

McNamara followed Anderson into the small "inner sanctuary," which was reserved for private conversations regarding sensitive "need to know" information. A few chairs stood guard near a table that held a couple of used coffee cups and a navy regulations manual. Anderson leaned on the back of a chair and proceeded to tell McNamara about Operation Boresight. He explained that SOSUS did a fair job of helping to track the four *Foxtrots* when they snorkeled, but this occurred only at night and only up to 150 nautical miles away. Fortunately, these submarines transmitted multiple burst signal updates to Moscow every day in the late afternoon. When they did, HFDF stations equipped with Boresight technology obtained ballpark fixes.

"Ballpark?" McNamara asked.

"Around fifty nautical miles," Anderson said. "That's why that ship is out of line. She's prosecuting a probable *Foxtrot* hit initially received from our HFDF station in Homestead."

McNamara stood up straight. "I want to know more about this Boresight thing. Send someone to the White House tomorrow to brief me and the president."

"This is a highly classified program, I don't know—"

"Send someone, Admiral."

Anderson nodded, said nothing.

"Now," McNamara said, "about those rules of engagement. I don't give a damn what the rules say. Your boys are not to fire a single shot into a rudder without direct permission from the president or me. Is that clear?"

"Sir, we have sixty ships on that line. They need to be able to act quickly and independently. We need to trust that our captains will do the right thing."

McNamara fumed. "The purpose of this quarantine is to send a delicate diplomatic message to Khrushchev, not start a shooting war!"

"No one's going to—"

"You're damn right they're not!" McNamara yelled, loud enough that Anderson was sure that Gilpatric and the press kids overheard the comment.

Frazzled from days without sleep, his gut churning with coffee acid, his nerves stretched thin, Anderson picked up the thick bound navy regulations manual from the table and held it high. "May I remind you, Mr. Secretary, that the navy has been running blockades by the book since the days of John Paul Jones. I think we know what we're doing by now!"

"I don't give a damn about Jones! I want to know what *you* intend to do. Be a renegade or follow orders?"

Anderson marched to the door and turned the handle. As he stepped back into the Flag Plot room, he said, "I always follow orders, Mr. Secretary. Now, I suggest you go back to your office and let us handle things here."

His face sour, McNamara strutted past Anderson and motioned for Gilpatric and the press nerds to follow. He turned one last time before exiting the room and said, "I want that briefing, Admiral."

Anderson nodded and turned back toward the wall chart. He stared at the flag representing the USS *Cony,* unaware that the navy destroyer "out of line" was running headlong toward a battle with Captain Savitsky's B-59.

AT THE RUSSIAN EMBASSY IN WASHINGTON, D.C., the Soviet naval attaché, Vice Admiral Leonid Bekrenyev, formed his lips into a tight line as he read a recently received diplomatic message regarding the rules of engagement. The message stated that Soviet submarines, if found, would be forced to the surface by American ASW forces. He handed the paper to a radio operator along with instructions to inform the main navy headquarters in Moscow immediately and ensure they transmitted the message to the submarines near Cuba.

President Kennedy met again with members of ExComm to review the latest quarantine intelligence, world reaction to the building crisis, status of negotiations at the United Nations, and potential incidents on the high seas. McNamara provided a detailed briefing on recent reconnaissance photos from Cuba, and the group debated the need to disperse planes at Florida bases in the event of attacks by Soviet MIGs.

McNamara then revealed what he'd learned earlier that morning from Admiral Anderson about the Soviet submarine sighting from a reliable source. He did not reveal the nature of that source. He expressed concern over the "very dangerous situation since [Russian merchant] ships approaching the quarantine line are being shadowed by a Soviet submarine." Referring to the probable *Foxtrot* contact reported by Admiral Anderson, now being pursued by the USS *Cony,* he went on to say that "there is a sub very close, we believe, and therefore it should be twenty to thirty miles from these [ships], and hence it is a very dangerous situation. The navy recognizes this [and] is fully prepared to meet it."

Kennedy asked what might happen if a U.S. destroyer was sunk by a Soviet submarine while trying to board and search a Russian merchant ship. Not receiving an acceptable answer, he went on to say, "I think we ought to wait on that [boarding] today. We don't want to have the first thing we attack [be] a Soviet sub. I'd much rather have a merchant ship."

WHEN WILLIAM J. REED HIT THREE days without sleep, Commander Kaye insisted that he head over to the officer's quarters at Fort George G. Meade, Maryland, and find an empty bunk for a few hours. Reed tried to argue but lost. He also tried to sleep, but his head kept spinning over the brewing crisis near Cuba. He knew there'd been hundreds of possible submarine sightings made by ships, planes, SOSUS, and Boresight stations, and so far they could not prove which ones were more accurate, though Reed harbored his own bias.

All hits were thoroughly analyzed by experienced ASW submariners, surface ship jockeys, or pilots at ASW Force Headquarters in Norfolk, Virginia, but those professionals were still limited by the mantra "Bad in, bad out." Reed's job, which kept him awake for these past few days, mandated turning that bad into at least something decent. When they heard that the Boresight HFDF station at Homestead got a solid hit on a *Foxtrot* near the Walnut line, Kaye congratulated Reed and his team for doing that job well by improving the bearing accuracy on Homestead's equipment.

Based on what Boresight hits they did have, NSA estimated that four *Foxtrots* now encircled Cuba, one near the Bahamas east of Florida about 100 miles off the Walnut line, one southwest from there about 500 miles east of Cuba, another 100 or so miles south of that sub, and the fourth one around 700 miles west of the others, south of Cuba down near Jamaica.

Reed forced his mind away from the threatening *Foxtrots* and finally dozed off. A few hours later, Kaye shook him awake.

"Sorry to wake you, BJ," Kaye said, "but we need to prepare for a high-level meeting."

Reed yawned and rubbed the sleep from his eyes. "A meeting? Where?"

"At the White House," Kaye said. "We've been summoned by the president."

ON BOARD B-130, ON STATION EAST of the Bahamas, Captain Second Rank Nikolai Shumkov's face heated as he issued a string of profanities. The temperature in Compartment Five, which housed three large engines that reeked of diesel fumes, increased to more than forty degrees centigrade after they entered tropical waters. Sweat gushed from every pore on Shumkov's body and stained his clothes. Senior Lieutenant Viktor Parshin, B-130's chief mechanic, stood near one of the engines and tried in vain to wipe the black oil smudges off his hands with an old rag.

"I'm sorry, Captain," Parshin said, his voice almost a yell against the throng of two engines. "One engine is still down, and a second could die at any time."

"I should have listened to you back at Sayda Bay," Shumkov said, shaking his head.

"No matter, Captain," Parshin said. "If you had, we would not even be on this mission."

Shumkov could tell by Parshin's bright red face that the man had been on duty far longer than the mandated thirty-minute rotation in the superheated compartment. He also knew that his chief mechanical engineer spoke the truth. They would not be here had they elected to make repairs after discovering the hairline fractures in the drives. They would have been left behind.

B-130 hit the water for the first time in September 1960. That made her an older sibling to the other three Project 641 boats on this mission. More experienced but not as healthy, B-130 was born with defects and suffered often from mechanical ailments. She came out of the yards with flaws in two of her diesel engines. When shipyard engineers discovered the hairline cracks, they insisted on immediate repairs, but the builders refused. Shumkov suspected a cover-up. No one wanted to admit to the mistake.

The diesels ran fine, but Shumkov feared the day when one or more would suffer a coronary and die, most likely in the middle of an important deployment that required extended use. Parshin expressed concerns about their engines, as well as their aging batteries. The geriatric two-volt cells, 448 of them located on the lower decks of Compartments Two and Four, were due for replacement. The electrolyte in these 650-kilo batteries ran hot during recharging, which could cause a fire or even an explosion. They also took longer to charge, sometimes more than twelve hours, which made B-130 a laggard behind the other boats.

Shumkov ignored his chief engineer's warnings and decided to live with the risks. Thriving on adventure, he could not imagine being left out of such an exciting mission. Now his decision came back to worry him at the most inopportune time. Here in the Caribbean Sea, surrounded by the enemy, they were unable to surface and make repairs.

While all three of B-130's propellers could be spun via the Kolomna diesel engines, they usually turned two props with one engine and used one to charge the batteries, with one resting. Due to the ignored drive cracks, they lost one of the engines, and a second now hung by a thread. Should they lose that one, they'd be down to one diesel that might fail at any time. They'd have to run only on the slow emergency motor while snorting to recharge batteries, and their ability to run from the Americans would be greatly diminished. If they lost that third engine, B-130 would be forced to head home under tow with her tail tucked between her legs. Shumkov sickened at the thought. Not only would he be excused from history, but he'd also be riddled with guilt for abandoning his duty to the other captains.

Shumkov glanced at the port engine gauges, their needles resting at zero. A maze of small round indicators filled the engine control panel, along with a shiny metal main fuel valve that looked like a rudder wheel on an old sailing ship. Lamenting his fate, as he turned to leave, he said, "Don't stay in here too long, Viktor. That's an order."

"Yes, sir," Parshin said.

Shumkov shot through the hatch into Compartment Four. The tantalizing aroma of lamb drifted by and reminded him that he hadn't eaten but a little borscht and goulash for dinner at midday. That was eleven hours ago. Now, at almost 11:00 P.M. boat time, the cooks were preparing tea and the snack meal. Hot piroshki. The tasty meat pie was Shumkov's favorite, especially since he really didn't care for the bread. Project 641–class boats did not have the luxury of a bakery. Bread arrived on board prebaked and was stored unrefrigerated in plastic bags that contained a small amount of alcohol. When the cooks warmed the bread, the alcohol evaporated and offered freshness, at least in theory. For Shumkov, however, the flavor paled in comparison to his wife's baking.

Although Shumkov normally ate his meals in the wardroom, he entered the galley and smiled at the chef. The rotund *michman*, who always produced grade-six quality meals, belied the running submarine joke that chefs should stay skinny in order to squeeze into their corner of the pressure hull. The small space held an assortment of bottles, boxes, and condensed milk cans, along with a large plate of fresh piroshkis sitting on the edge of the wooden counter. The chef returned Shumkov's smile as he held up the plate. Shumkov grabbed one and took a bite. He started to leave, stopped, turned, and grabbed another one.

Two sailors sat on the blue-cushioned bench in the galley nearby. Each displayed a five-digit number on his uniform that designated department, position, compartment, and shift numbers. One read 4-44-23, which meant Communications, Radio Room, Compartment Four, Shift Two, third in charge. The radio-room number reminded Shumkov that he needed to send out a burst transmission about the condition of his boat, but he decided Moscow could wait a few minutes. He sat on the bench next to his men and lost himself in the rich taste of lamb as he devoured the piroshki. Halfway through a second pie, his elation was interrupted by a distinct change in the vibrations running through the boat.

Parshin bolted through the Compartment Four hatch. The chief engineer spoke no words; his face told the story. The second diesel engine had just died. Shumkov sat his half-eaten piroshki on the table.

Two minutes later, he met his brown-haired electronics officer, Lieutenant Cheprakov, in the radio room. As the officer in charge of division Boyovoi Chesti (BCh) Four, Cheprakov worked closely with the five English-speaking OSNAZ operators. The wide-eyed young men, in between listening to jazz and news on Voice of America, were having the time of their lives monitoring radio intercepts between American ships, planes, and shore stations. Apparently the entire U.S. Navy now shared one mission in life: to find the four Project 641 boats. Shumkov hoped that the Americans would continue to fail, but now, with two engines down, he knew that the winds of luck were shifting.

Shumkov wrote out a message and handed it to Cheprakov. The electronics officer made a few modifications to ensure encryptability by the SBD and gave the edited message to the radio operator. The *michman* studied the message. When he looked up at Cheprakov, his fear-filled eyes seemed to say, "Is this true?"

Cheprakov returned an affirmative nod. "It's true. If we lose one more engine, we'll need a tugboat to take us home."

The radioman typed in the message on the SBD keyboard. The burst transmission electrified the tropical clouds, bounced across the ionosphere, and landed on a receiver dish in Moscow.

The split-second radio waves also tickled a GRD-6 antenna just over 500 miles away in Homestead, Florida. An hour later, led by a Boresight fix, an ASW plane visually detected B-59's snorkel mast. Operators issued the next sequential contact number. Someone in Admiral Anderson's Flag Plot room stuck a flag on the board, designating the submarine sighting as C-18. The USS *Cony*, patrolling an area near the Bahamas, received urgent orders to pursue the *Foxtrot* and force her to the surface.

As NIGHT DESCENDED ON WASHINGTON, D.C., the temperature chilled along with any prospects of a fast or easy resolution to the crisis. At 1:45 A.M. on October 25, President Kennedy responded to Khrushchev's earlier threat—delivered via Westinghouse's president, William Knox—that Soviet submarines would sink any U.S. destroyers that tried to stop Russian merchant ships. Kennedy stated that the United States took appropriate action after receiving repeated assurances that no offensive missiles were being placed in Cuba, and that when these assurances proved false, the deployment "required the responses I have announced. . . . I hope that your government will take necessary action to permit a restoration of the earlier situation."

Around 7:15 A.M., the USS *Essex* and USS *Gearing* (DD-710) steamed toward the Soviet freighter *Bucharest* with orders in hand to intercept and board if necessary. Prior to the boarding, however, the navy decided to let this Russian vessel pass after concluding that she was capable of carrying nothing more than a cargo of petroleum. Instead, the fleet received orders to observe the tanker *Graznyy*, as her deck might be loaded with missile field tanks.

SOMETIME THAT WEEK, ON OR ABOUT the morning of October 25, Briefcase in hand, sitting next to Commander Kaye, William J. Reed shifted nervously in the backseat of a black sedan. Gray clouds outside the windows descended on the nation's capital. Reed knew that almost every American wondered if they'd live to see another fog-filled day. Earlier he'd spent hours preparing overhead projector slides and a typed memorandum. The navy gave Kaye the honors of delivering these to President

Kennedy and a few select members of the ExComm group, and Kaye asked Reed to come along. The ensign hoped that his technical input would not be needed.

The vehicle pulled to a stop in front of the White House. Reed and Kaye stepped from the car and walked toward the entrance. The majestic six-story building shimmered in the sun as the iconic fountain splashed water into a cold October breeze. Commander Kaye stopped for a moment to gaze at the sight. Reed pulled his wool dress coat tight around his neck and smelled crisp air. With each breath, he drew in an equal measure of awe and admiration. He popped to attention and saluted the American flag as it flapped in the wind. Kaye did the same. Both lowered their arms and walked toward the entrance.

As they strolled, Kaye pointed at the White House complex. He explained that the group of buildings included the central executive residence, flanked by the East Wing and West Wing. He let out a chuckle and said that the place had 132 rooms, 35 bathrooms, 28 fireplaces, 8 staircases, 3 elevators, 5 full-time chefs, a tennis court, a bowling alley, a movie theater, a jogging track, a swimming pool, and a putting green.

"A bowling alley?" Reed said. "Hell, Joyce and I are hoping next year we can afford a place with four bedrooms and a little bigger kitchen."

"Be happy with what you have, BJ," Kaye said. "Most Russians live in tiny apartments and share bathrooms with a half-dozen neighbors. They've never even seen a bowling alley."

"Point taken," Reed said as the two approached the building.

Kaye pointed again. "That's where we're headed. The West Wing. In there is the president's Oval Office, senior staff offices, and room for about fifty employees. There's also the Cabinet Room, where the president does most of his business." He pronounced *business* as *bid-ness*.

Although Reed already knew quite a bit about the White House, he also knew that Kaye loved to show off to impressionable listeners, so he let his boss ramble on.

Kaye pointed out more as he led Reed into the large entrance hall. The commander explained that in 1806, President Thomas Jefferson transformed the hall into an exhibition area for artifacts from Lewis and Clark's famous expedition to the Western Territories. President Ulysses S. Grant started another tradition of hanging presidential portraits in the entrance hall and the perpendicular cross hall.

"I guess when you're the chief, you get to decorate your teepee anyway you want to," Kaye said.

Reed smiled, said nothing.

An aide greeted them at the entrance and directed the pair to a windowless office in the West Wing. The large conference room displayed a podium, a pull-down projection screen, and an overhead projector for transparent slides. As they stood alone near a rectangular conference table, Kaye said they called this the Fish Room. He said Teddy Roosevelt requested that the room be built in 1902, and he used it as his office. When they expanded the West Wing and built the Oval Office in 1909, they turned the room into a waiting area.

When Reed asked why they called it the Fish Room, Kaye glanced upward and said that Franklin Roosevelt put in the skylight in 1934, along with an aquarium and fishing mementos. He always called it the Fish Room, and the name stuck.

Reed pointed to a large sailfish mounted on the wall. "Roosevelt's?"

"Nope," Kaye said as he rested a hand on the back of a high-back leather chair. "Kennedy's. He caught the thing on vacation in Acapulco."

"Now I know why I like the guy," Reed said.

The door opened, and President John F. Kennedy entered, along with Defense Secretary McNamara, Deputy Secretary Roswell Gilpatric, National Security Adviser McGeorge Bundy, and the director of the National Security Agency, Air Force Lieutenant General Gordon Blake.

Reed had previously met General Blake and found him to be a "man's man" who combined a frequent Midwest smile with a sharp mind and hard-charging work ethic. Blake hailed from Iowa and graduated from the U.S. Military Academy in 1931; he earned the Distinguished Flying Cross as a communications officer flying on B-17s during the war. He also earned a Silver Star, Legion of Merit, Air Medal, and several campaign battle stars and now displayed a vast array of colorful ribbons on his uniform.

General Blake was Commander Jack Kaye's immediate boss. Reed didn't know it then, but in less than three years, Blake would award him a certificate of appreciation for his dedicated service to the NSA, specifically for his contributions to the Boresight program.

McNamara motioned for Reed and Commander Kaye to take a seat. Reed's throat tightened, and he longed for a drink of water. He and Kaye removed their covers and sat. Reed found a glass, filled it from a pitcher on the table, and gulped downed half.

McNamara explained to the president that he'd received minimal information from Admiral Anderson about a new naval technology called Boresight, that this new system appeared capable of locating Khrushchev's *Foxtrot* submarines with better accuracy than SOSUS or other means. McNamara felt that such a capability could provide an advantage in the current Cuban negotiations and so asked Anderson to set up a meeting to brief the president and select members of ExComm. He reminded the group of the highly classified nature of the information and turned the floor over to Kaye.

Commander Kaye took a sip of water, cleared his throat, and walked to the podium. He pulled a few slides from a manila folder and placed one on the overhead projector. He flicked on the switch, and the projector's fan hummed to life. He explained that, since the early days of World War II, the United States used high frequency direction finding systems to locate enemy submarines. After the war, in 1960, the Soviets switched to a new type of ultra-short burst signal that they adopted from recovered German technology. They phased over from standard HF transmissions, and by December of that year, they were using the burst exclusively. The navy could no longer find those subs.

Kaye pointed to Reed and said that the ensign discovered the burst and helped the NSA and navy design a new technology under Operation Boresight. Kaye then started to explain how the system worked. When McNamara drilled him for more details, Kaye asked Reed to step to the podium.

His mouth dry, Reed downed some more water, grabbed his folder, and approached the podium. Kaye gave him a pat on the shoulder as he walked past and returned to his seat. Reed's eyes locked with Kennedy's. The president offered a smile and a nod.

Reed recalled hearing stories about how Kennedy's torpedo boat, PT 109, had been rammed by the Japanese destroyer *Amagin* during the war and had sunk in the Pacific, and about how Kennedy had hurt his back in the collision but still managed to swim to shore while towing a badly burned sailor using a life-jacket strap clenched between his teeth. Kennedy found the rest of his crew on a nearby island, where he scrawled a rescue message on a coconut given to Solomon Islander scouts. That coconut wound up saving the lives of his men. Kennedy turned the thing into a paperweight that now sat on his desk in the Oval Office.

"Go ahead, Ensign," the president said softly. "Just tell us what you know."

In that brief moment, Reed understood how the man's infectious charisma earned him the presidency. Using more than a dozen slides, with McNamara grilling him for facts, Reed translated the technical details of Boresight into layman's terms. Questions were asked about bearing accuracy, frequency of transmissions, distance limitations, and the number of stations operational.

Reed said that the Soviet navy, reflecting the ways of its authoritarian government, did not trust its submarine captains, and so required that they send a coded update at least once a day. That message consisted of a short burst signal that could now be detected by Boresight intercept stations. Multiangulation could then be used to locate the source of the transmission. Several stations equipped with Boresight technology received hits on four *Foxtrot* submarines, as they neared Cuba, that were sending twenty or thirty transmissions at a time—probably due to reception verification difficulties with Moscow. U.S. stations cross-referenced bearing hits to direct ASW forces toward the *Foxtrots*. That was the good news.

Reed then delivered the bad news: given the nascence of Boresight technology, the limited number of operational sites, and the distance of those sites from the targets, as well as the inexperience of the operators, ballpark fixes of between forty and sixty nautical miles were the best they could accomplish today.

"Are we talking closer to forty or sixty?" Kennedy asked.

"Up until last week, it was closer to sixty," Kaye said. "Thanks to Ensign Reed and his team, we're now closer to forty for most stations."

Kennedy glanced at McNamara. "How good is that for our ASW boys, Bob?"

"Not much better than SOSUS, Mr. President," McNamara said. "We'd still need to throw too many planes and ships on those fixes to find the subs. I'd sure like to get that number down to thirty miles or less."

"There's something else, Mr. President," Kaye said.

"Go ahead, Commander," Kennedy said.

"It's just a speculation, sir, but the NSA reported that a *Foxtrot* submarine fired two nuclear torpedoes off Novaya Zemlya last year. If the Soviet subs heading to Cuba are carrying—"

"We know, Commander," McNamara said. "But we believe it's unlikely."

"Unlikely, but not impossible," Kennedy said, his face somber.

The room fell silent. Filigree danced in a ray of sunlight that beamed through the skylight in the Fish Room. Kennedy stared at the table for a long moment, then looked up. His deep brown eyes pierced Reed's social armor and reflected the hope of an entire nation. "Ensign Reed, with those subs still in the picture, I don't have a strong hand against Khrushchev. And if they are carrying nukes, God help us all. So I need you to get us better than thirty miles. Do you think you can do that?"

Reed's legs went numb. He glanced at Kaye. The commander's eyes opened wider than submarine hatches, but he said nothing. Reed stood up straight, looked back at Kennedy, and said, "Yes, Mr. President. I believe I can."

On the way out of the White House, Commander Kaye grabbed Reed by the arm and said, "How the hell are you going to get to thirty miles in a matter of days?"

"I have no idea," Reed said. "But if I don't, like the president said, God help us all."

As THE TWO WERE LEAVING THE building, a man stepped in front of Reed and blocked his exit. "What the hell are you doing here?"

Reed studied the man's face, then smiled. "Lieutenant Commander Quittner?"

"I'm retired now, so that's Mr. Quittner to you," Quittner said with a grin.

Arnold Quittner served as Reed's former executive officer on his first ship, the PCS-1380. After more than a decade, the man sported some gray and a few extra pounds, but his voice still grumbled like a Mack truck in low gear.

"Retired?" Reed said. "I thought you'd serve forever."

"I am serving," Quittner said, "just not in the navy. I'm one of Kennedy's legal advisers now."

Commander Kaye tilted his cover back and said, "I guess Kennedy could use all the legal beagles he can get right now."

Wearing a suit and tie, his hair a tad longer than navy regulation, Quittner said, "My plate's definitely full these days. I can't even begin to tell you how much legal maneuvering it takes to move dozens of political and military chess pieces around without ruffling lots of feathers."

"You'd think that under the circumstances," Reed said, "the president would have carte blanche."

Quittner shook his head no. "Until there's an official war proclamation, every special interest asshole on the planet wants his say. Even when we're staring down the barrel of a gun, there's always some guy that cares more about his personal pocketbook."

"Somebody needs to hogtie those bastards and brand them traitors," Kaye said with a scowl as a couple of congressmen walked past.

"I wish we could," Quittner said as he brought his wrists together. "But that's hard to do when you're handcuffed by the legal system."

"Maybe the sheriff needs to change a few rules," Kaye said.

"Maybe," Quittner said. "I do agree it's a pain in the ass sometimes, but I'd rather have the problems of a democracy than those of a dictatorship."

"Our forefathers never said that freedom would be easy," Reed said.

The three shook hands, and Reed and Kaye stepped back into the cold.

CHAPTER NINE

Anger is an acid that can do more harm to the vessel
in which it is stored than to anything on which it is poured.
—MARK TWAIN

EARLY IN THE EVENING OF OCTOBER 25, in a lab at Sanders Associates in New Hampshire, Reed forced his tired mind to stay focused. He had no idea how he was going to solve Kennedy's requirement to improve Boresight accuracy, but he hoped that he might find an answer here. Petty Officers Odell and Denofrio, the two stars on Reed's technical team, were huddled together with a group of engineers near the back of a large room filled with the evidence of technology. A dozen randomly placed tables sat nearby, covered with various devices, colorful wires, test equipment, and tools.

Reed removed his cover and coat and placed both on a rack near the door as a man in a lab coat approached. Light skinned, polished, and projecting a professor's demeanor, he introduced himself as Dr. Charles Skillas, the team lead for Antisubmarine Warfare DIFAR—whatever that meant. Reed and Skillas exchanged pleasantries and found seats near the group of engineers. Reed took those few seconds to try to recall a few details about Sanders Associates.

Eleven engineers and scientists from Raytheon founded the company in Waltham, Massachusetts, in July 1951. Royden Sanders Jr., one of the original eleven associates, became the company's namesake. Sanders Associates moved into a vacant textile building in Nashua in 1952 and focused on designing and building electronic systems, aircraft self-protection systems, and surveillance and intelligence systems, in-cluding submarine detection equipment. More than a thousand people worked at the facility, making Sanders the largest employer in New Hampshire.

Over the next several minutes, while Odell and Denofrio mingled with the engineering team, Dr. Skillas asked how they might help the navy. Reed divulged information about the Boresight program and ac-curacy issues without delving too far into sensitive areas. Sanders was a defense contractor, and everyone there had signed a stack of nondisclo-sure papers, but Reed still operated under strict orders not to provide civilians with more information than they absolutely needed to know.

Reed said that he researched Sanders Associates and knew they were working on directional hydrophone sensors for passive sonobuoys. He wanted to know more about the program. Skillas said that the navy awarded them a $600,000 contract to create DIFAR, an offshoot of the LOFAR technology used in the SOSUS system. DIFAR stood for Direc-tional Frequency and Ranging, and the key word was *directional*. The technology they'd developed used an internal compass to help deter-mine a more accurate direction to noise generated by Soviet submarines.

Skillas explained that dogs have the ability to screen incoming sounds and focus an ear on a noise to determine the exact location of the source. In essence, they can "turn on" unidirectional hearing, whereas human hearing is omnidirectional—meaning we can hear almost every-thing coming from everywhere. That works fine for people, but omnidi-rectional listening devices are not optimal for finding submarines. Sanders developed "dog hearing" unidirectional technology for airplane-dropped sonobuoys that could do a better job of figuring out an accurate bearing to a contact. They were also working on a technique for dropping those sonobuoys from ASW aircraft—as high up as 30,000 feet—using precise patterns they called CODAR.

Reed knew that sonobuoys used transducers and radio transmitters to record and transmit underwater sounds, and that there were three types: passive, active, and special-purpose buoys. Passive sonobuoys used hydrophones to listen for underwater sounds, and active buoys used transducers that "pinged" just like submarine sonar systems. Special-purpose buoys were used to capture environmental information, like water temperature, depth, and acoustic layers.

Skillas explained that the directional frequency and ranging DIFAR sonobuoy they were working on was passive, and the main component included a directional hydrophone that recorded accurate bearings to targets. The navy asked Sanders to build a prototype that could detect submarine noises in the low-frequency 5–2,400 Hz range and could operate for up to eight hours at depths down to 1,000 feet.

The brain behind this invention was an AQA-7 signal processor. This allowed for doglike directional hearing. The system processed incoming signals and output submarine position, speed, and direction information onto electrosensitive paper.

All music to Reed's ears. He told Skillas that they needed a way to increase bearing accuracy for the Boresight systems to gain better location fixes on submarines and wondered if any part of Sanders's technology could be used to help them. Skillas rubbed his chin and conferred with the others on his team. They asked dozens of questions. Reed and his team provided answers. The engineers drew on blackboards, fingered slide rules, and thumbed through technical diagrams and manuals.

After an hour, Skillas said, "We have good news and bad news."

Reed's heart sank. He'd been hoping for all good news.

"Give us the bad news first," Petty Officer Odell said.

"Why not the good news first?" Petty Officer Denofrio asked.

Odell snorted. "Maybe you don't understand, pretty boy, but when I go into a bar, I always hit on the second-ugliest woman first."

"You're right," Denofrio said. "I don't get that at all."

"I do that," Odell said, " 'cause if she says no, I still have one to go. So I always want the bad news first."

Skillas shook his head and flashed a smile. "Okay, the bad news is that we don't think our processor is adaptable to the systems you're using. Those burst transmissions are only seven-tenths of a second long, and they're coming from a recorder. Also, the AQA-7 is designed to work with underwater noise, not high-frequency radio transmissions. We don't think our signal processor will help much, given that scenario."

Reed lowered his head. "So, what's the good news?"

"The good news is that some of the math and other technology we used to design DIFAR might help you."

Reed lifted his head. "Math?"

"Math," said Skillas. "Like Gaussian on a plinth."

"Gesundheit to you, too," Petty Officer Denofrio said.

Reed held back a laugh. "What's that mean in English?"

"A bell curve on a baseline," Skillas said.

He went on to explain that the mathematical term *Gaussian* was named after Karl Gauss and was widely used in signal processing to define filters, which is what they did at Sanders. Skillas walked over to a blackboard. He drew a bell curve, which resembled a tall anthill with curved slopes. He placed several dots near the top of the hill. "These are normal, accurate bearings to a submarine as detected by your Boresight system." He then drew a few dots on either side, near the base of the hill. "These are wild bearings caused by inaccuracies and interference. With a standard high-frequency transmission, ruling out the obvious bad apples is not hard. With a burst transmission, you might get only a couple hits, all of which could be the bad apples. We need to help you find a way to get more good apples and better discern the bad from the good."

Reed asked, "How do we do that?"

"Well," Skillas said, "you need to cheat."

"Cheat?" Reed said.

"I don't like cheating," Denofrio grumbled.

"I do," Odell said, raising his hand. "I'm all for cheating if we get to win."

"How do we cheat?" Reed asked.

"You need more verifiably accurate bearings," Skillas said.

"Swell," Denofrio said like a New Yorker who'd just missed a taxi. "We'll just ask the Russkies to pretty please burst a bunch more times each day."

"Denofrio's got a point," Reed said. "We only get a limited number of hits."

"From the target, yes," Skillas said. "But now you need more check bearings."

Reed said, "Using known references to calibrate for bearing inaccuracies is something we've been doing since the early days of DFing."

"Not with burst signals," Skillas said. "Now you have to compensate for the dimension of time. Inaccuracies and interference will be different for an after-the-fact Soviet burst signal than a regular high-frequency transmission."

"No shit," Denofrio said.

Reed leveled a disapproving stare.

"Sorry," Denofrio said. "I meant, that's certainly an accurate assumption, Dr. Skillas."

"You also need to automate your bearing calculations instead of using that inaccurate manual string board compass rose you're using now," Skillas said.

"No shit," Odell said, throwing Reed an I-don't-care look. "But how do we do that?"

"Well," Skillas said, "that's where the software programs that we developed for our DIFAR project might help. Along with a new type of computer."

"New computer?" Odell said.

"The GYK-3," Skillas said. "It's in development right now at the Naval Research Laboratory, but it might work for your application."

"No shit?" Denofrio said, his face lighting up like Times Square.

"No shit," Skillas said.

Reed nodded. "Okay, I get it. What you're saying is that we'll have higher bearing accuracy if we set up a simulated burst signal on some of our ships and have them transmit immediately following a Soviet burst hit, and if we automate bearing locations using a computer like the GYK-3 instead of using manual string boards."

"Precisely," Skillas said.

Skillas and his team offered other tips they thought might improve things, such as using cesium clocks to better synchronize time accuracies between the stations and better ways to calculate interference and compensate for the effects. They also recommended reducing any man-made interference from electrical equipment or nearby power lines.

"You need to increase your good apples and reduce your bad ones," Skillas said. "Otherwise you'll never find those guys."

"No shit," Reed said.

An hour later Reed excused himself from the meeting to catch a flight to California. As he stood outside the brick structure nestled in the snow, he stifled a sneeze and glanced skyward. More snowflakes were beginning to fall. Down a curved sidewalk, a bundled figure trudged through the white and approached the building. The short man stopped near Reed and held out a hand. "I'm Dr. Ralph Baer."

Reed shook Baer's hand. "Ensign Bill Reed."

Thin, friendly, and bespectacled, Baer said, "Are you coming or going?"

"I'm not sure anymore," Reed said.

"Sounds like you need a vacation," Baer said empathetically.

"Definitely."

Baer gave a chuckle as he turned to enter the building. "Nice meeting you, Ensign Reed."

"Good meeting you, too, Dr. Baer," Reed said.

As Dr. Ralph Baer entered the foyer, Reed had no idea that he'd just met one of the most brilliant inventors in the world. This unassuming man was destined to create the first video game and spawn a multibillion-dollar industry that would change the world in ways unimagined.

That evening, Reed caught a military MAC flight headed to Northern California. After landing, held up by a half-dozen cups of French roast, he drove through the north gate of Skaggs Island and stepped back in time. Reed started his career as a communications technician at this facility after serving aboard the PCS-1380. Skaggs Island was a drained area of San Pablo Bay tidelands that sat about twenty-five miles northeast of San Francisco. In the early fifties, the Naval Security Group came to Skaggs to set up the HFDF facility and turn the island into a bonafide Huff Duff.

As Reed drove toward the ops building, he noticed that things hadn't changed much in the several years since he had brought his family there in 1957. The recreational buildings, theater, chapel, and bachelors' quarters still stood at attention like worn soldiers adorned in faded gray uniforms. Rows of single-story homes lined sad, narrow streets, where navy brats played with dogs destined for abandonment after the next military-ordered move.

Reed pulled to a stop in front of the small operations building near a massive elephant cage. Although Skaggs was one of the first stations to receive the new Wullenweber in 1962, most of the kinks were not yet ironed out. Reed had visited the station a couple of times over the past few months with his team to install and launch the Boresight equipment. Now he was back to deploy some of Dr. Skillas's recommenda-

From 1946 through 1954, diesel-powered submarines like the USS *Cubera* (SS-347, ABOVE) and USS *Blenny* (SS-324, BELOW) conducted top-secret espionage missions deep inside Soviet territorial waters. The Soviets sometimes caught these boats and harassed them. Diesel subs ran out of air after a few days submerged, and crews almost died on several missions. This problem eventually spurred the use of nuclear power in submarines. *U.S. Navy photographs, courtesy of ussubvetsofwwii.org*

ABOVE: The USS *Seawolf* (SSN-575) was America's second nuclear-powered submarine. Almost three decades after her launch, while conducting a top-secret mission in 1981, a storm trapped *Seawolf* in the sand off the coast of Russia for four days. Her crew of 190 came within a breath of not coming home.

U.S. Navy photograph

LEFT: The "father" of submarine nuclear power, Hyman G. Rickover, touring the USS *Nautilus*—the navy's first nuclear-powered submarine—in 1954.

U.S. Navy photograph, circa 1954

RIGHT: Officers and sailors in the control room of the USS *Nautilus* (SSN-571) while under the Arctic ice. Launched on January 21, 1954, the *Nautilus* was the first vessel to complete a submerged transit under the North Pole.

U.S. Navy photograph

"Scratchy" the bear, raised by the author's family while in Turkey, unwittingly helped William J. Reed find a Soviet "burst" signal that made a significant difference in the outcome of the submarine Cold War. *Author's collection*

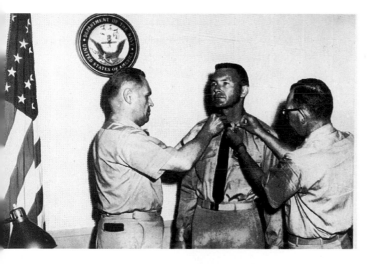

The author's father, William J. Reed, received a promotion from senior chief to ensign after finding the Soviet submarine "burst" signal while in Turkey in 1960.
Author's collection

Reed was also awarded a letter of commendation from his boss, Commander Frank Mason, for finding the signal.
Author's collection

In one of the most dramatic, untold episodes of the Cold War, four Soviet *Foxtrot*-class submarine commanders led their boats deep into the waters around Cuba in October 1962, which on several occasions led us to the brink of nuclear war. *Left to right:* Capt. Dubivko (B-36), Capt. Shumkov (B-130), Chief of Staff Arkhipov, and Capt. Ketov (B-4) left Russia on October 1, 1962. They arrived off the coast of Cuba three weeks later. *Courtesy of Ryurik Ketov*

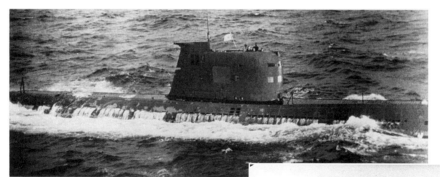

RIGHT: The U.S. Navy was primarily focused on blocking Soviet merchant ships from bringing more nuclear missiles to Cuba.
U.S. Government photograph

ABOVE: But they also forced three of the *Foxtrot* submarines, including Captain Savitsky's B-59 (*above*) to the surface during the Cuban Missile Crisis. The United States did not know at the time that each *Foxtrot* carried a nuclear-tipped torpedo capable of destroying everything within a ten-mile radius, and all four Soviet subs came within minutes of firing. *U.S. Navy photograph*

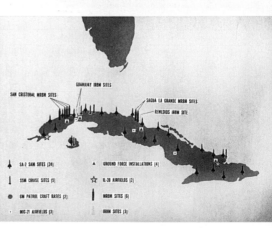

SOVIET MILITARY BUILD UP IN CUBA

GUANAJAY IRBM SITES
SAN CRISTOBAL MRBM SITES
SAGUA LA GRANDE MRBM SITES
REMEDIOS IRBM SITE

SA-2 SAM SITES (24)   GROUND FORCE INSTALLATIONS (4)
SSM CRUISE SITES (5)   IL-28 AIRFIELDS (2)
GM PATROL CRAFT BASES (2)   MRBM SITES (6)
MIG-21 AIRFIELDS (3)   IRBM SITES (3)

Living conditions aboard the Soviet *Foxtrot*-class submarines involved in the Cuban Missile Crisis were deplorable, and crews suffered from heat exhaustion. The captain's stateroom is pictured above, the hatch leading up to the conning tower and bridge is pictured to the right, and the navigation table in the command center is pictured below.

Author's collection, Maritime Museum of San Diego, California

During the Cuban Missile Crisis of October 1962, a nuclear missile strike from Cuba could have killed eighty million Americans within five minutes.

U.S. Government CIA photograph

President John F. Kennedy meets with Secretary of Defense Robert McNamara. They and other Executive Committee members planned for a quarantine of Cuban waters.

U.S. Government archive photograph

During the quarantine, destroyers like the USS *Barry* were ordered to fire at the rudders of Soviet merchant ships, like the Ansov, who refused inspections.

U.S. Navy photograph by Frank Cancellare

LEFT: Prior to the Cuban Missile Crisis, Ensign William J. Reed discovered the "burst" radio transmission used by Soviet *Foxtrot* submarines and then briefed President Kennedy and his advisers on the technology during the crisis. Reed worked to improve the systems to help locate the *Foxtrots* near Cuba. This effort allowed Kennedy to play hardball with Khrushchev, which forced the Soviet Premier to back down and return his ships to Cuba.
Author's collection

ABOVE: William J. Reed's family in Maryland in 1964. *Left to right:* Pamela, William, W. Craig (author), and Joyce. *Author's collection*

LEFT: After the Cuban Missile Crisis, William J. Reed traveled the globe to upgrade U.S. listening stations, so they could detect Soviet burst signals. One of those missions took him to Greece, where he was involved in a CIA firefight with Soviet spies. *Author's collection*

Top-secret listening station sites resembled large "elephant cages" and were positioned along the Pacific and Atlantic rims near Imperial Beach, California (ABOVE), Okinawa, Japan (BELOW), and more than a dozen other locations. *Author's collection*

During the early 1960s, William J. Reed helped deploy top-secret listening stations around the world in an effort to locate Soviet submarines.

GNU Free Documentation License, Version 1.3

ABOVE: The USS *Thresher* (SSN-593) was lost with all hands on April 10, 1963. CENTER RIGHT: Navy divers aboard the bathyscaphe USS *Trieste* (DSV-0) descended to 8,400 feet to search for her remains. *U.S. Navy photographs*

BELOW: Navy diver Nihil Smith—whose best friend, Joe Walski, was aboard the *Thresher* when she sank—brought up a pipe found by the *Trieste* in 1963, validating the sub's location on the bottom. BELOW RIGHT: The *Trieste* found other remains, including the crushed sonar dome. *U.S. Navy photographs*

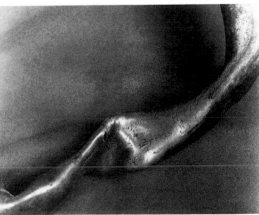

The North Koreans captured the spy ship USS *Pueblo* (AGER-2) on January 23, 1968, and found a working KW-7 encryption unit, which they sold to the Soviets. *U.S. Navy photograph*

ABOVE: The Soviet *Golf II*–class missile boat K-129 disappeared in the Pacific on March 8, 1968. The Soviets blamed the United States, based on transmissions they intercepted on the KW-7 using keylists provided by the spy, John Walker. *U.S. Navy photograph*

LEFT: The Soviets may have retaliated for the loss of K-129 by ordering an *Echo II*–class submarine to sink the USS *Scorpion* (SSN-589) with a torpedo on May 21, 1968. *U.S. Navy photograph*

LEFT: Navy deep-sea saturation divers breathing a helium mixture, deployed from the USS *Halibut* (SSGN-587, BOTTOM LEFT) to depths of 400 feet near Russia. The divers "tapped" signals from secret communications cables that were recorded and analyzed by "spook" technicians like Frank Turban. *U.S. Navy photograph by Senior Mass Communication Specialist Andrew McKaskle; U.S. Navy photograph*

Cold War submarine "spooks," circa 1976. *Left to right:* Bob Jordan, John Whitmire, Skot Beasley, Scott Hendren, Frank Turban, Dwight Anderson. *Courtesy of Frank Turban*

LEFT: During the latter half of the Cold War, U.S. Navy divers completed Ivy Bells cable-tapping missions using large "beast" pods. *Courtesy of Russian Ministry of Security's museum at the Lubyanka Prison*

In the latter two decades of the Cold War, the USS *Parche* (SSN-683, TOP) and USS *Richard B. Russell* (SSN-687, BELOW) delivered these divers to the Sea of Okhotsk and Barents Sea at depths down to 700 feet. *U.S. Navy photographs*

USS *Haddo* (SSN-604) entering New Zealand harbor in January 1979. As the boat's rescue diver, the author is standing on the back of the boat in the event a sailor falls overboard. *Author's collection*

The author on the *Haddo* entering Apra Harbor in Guam—where the author was born. *Author's collection*

Crew members in the control room of the USS *Haddo,* circa 1978. The author is at the far back right and Lt. Edwin L. Tomlin is on the far right. *Author's collection*

The *Haddo* and other Cold War fast-attack submarines hunted Soviet ballistic-missile boats like the infamous *Delta III* (*below*), often coming within a few dozen yards to tail these targets. *U.S. Navy photograph*

ABOVE: While conducting Holystone espionage operations, the USS *Haddo* and other Cold War fast-attack subs frequently photographed the *Typhoon*-class (BELOW) and other Soviet submarines, through periscopes or by deploying navy divers such as the author. *Author's collection; U.S. Navy photograph*

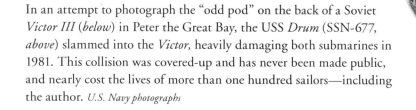

In an attempt to photograph the "odd pod" on the back of a Soviet *Victor III* (*below*) in Peter the Great Bay, the USS *Drum* (SSN-677, *above*) slammed into the *Victor,* heavily damaging both submarines in 1981. This collision was covered-up and has never been made public, and nearly cost the lives of more than one hundred sailors—including the author. *U.S. Navy photographs*

tions. He knew that the Fred Ten antenna array at Skaggs improved bearing accuracy by an order of magnitude over the old GRD-6 sites, but Skaggs was almost 2,700 miles away from Cuba. The Fred Ten elephant cage usually couldn't get a hit at better than 3,200 miles on a good day, and one ionospheric hop of 2,700 miles was a stretch. But if they used some of the suggestions from Sanders Associates, along with a bit more tweaking on the systems, they just might get down to that thirty-mile radius President Kennedy had requested.

Reed spent a sleepless night at Skaggs working with the technicians and implementing new procedures. Subsequent to his meeting at Sanders Associates, he'd contacted Commander Kaye, who called Admiral Alfred Ward, who talked to his boss, Admiral Anderson, and got permission to have a destroyer install a simulated burst transmitter and start transmitting when asked to by Net Control—after the Boresight stations got some burst signal hits—then change course a few times and retransmit. That way the Boresight stations could correct for inaccuracies and interference using a known target location. Reed didn't have time to find and install a computer to automate bearings, so he planned to connect later with some of the engineers at NRL to talk about using the GYK-3 to improve Boresight detection capabilities.

Exhausted, with nothing left to do but wait and pray that the new tricks would work, Reed boarded a plane and flew back to Maryland.

IN AN OCTOBER 26 MEETING, PRESIDENT Kennedy ordered the State Department to proceed with Cuba invasion preparations. SecDef McNamara reported that, in the event of such an invasion, CINCLANT estimated more than 18,000 casualties in the first ten days of fighting. Kennedy stated that despite the cost in lives, only an invasion could ensure the removal of all missiles from Cuba. Just prior to issuing the order to invade, a few of the conservative members of ExComm persuaded the president to delay the invasion and continue with military and diplomatic pressure.

At 1:00 P.M., John Scali of ABC News attended a lunch meeting with Soviet embassy official (and KGB station chief) Aleksandr Fomin, who stated that "war seems about to break out." Fomin asked Scali to use his influence to explore a diplomatic solution. The structure of such a deal, Fomin intimated, should include assurances that the Soviet Union would remove its weapons from Cuba, and the United States would state publicly that an invasion of Cuba would never occur.

The State Department received a message at 6:00 P.M., written personally by Nikita Khrushchev, which Robert Kennedy described as "very long and emotional." The contents outlined a deescalation plan similar to the one proposed earlier by Fomin.

Just before 9:00 A.M. on October 27, Radio Moscow broadcast a message from Khrushchev. In contrast to the letter of the night before, the message offered a new trade: that the missiles in Cuba might be removed in exchange for the removal of the U.S. Jupiter missiles from Turkey.

AT 10:25 A.M., A NEW INTELLIGENCE message arrived, and John McCone, the director of the Central Intelligence Agency, announced: "We have a preliminary report which seems to indicate that some of the Russian ships have stopped dead in the water."

Dean Rusk leaned over to McGeorge Bundy and said, "We're eyeball to eyeball, and I think the other fellow just blinked."

President Kennedy surmised that Khrushchev temporarily halted the advance of his cargo ships to avert an immediate confrontation, as well as to buy time to contemplate his next move. Following suit, Kennedy directed that no Soviet ship should be intercepted until the situation could be properly assessed. ExComm members issued a collective sigh of relief, but the sigh did not last long.

IN CUBA, AN ALARM SOUNDED AT 9:10 P.M., and the hairs on Major Grechenov's neck stood on end. He grabbed the radio and pressed the talk key, issuing a report that the SA-2 site near Banes had just detected a U-2 reconnaissance plane entering Cuban airspace. The controllers designated the contact as target thirty-three. A minute later, the antiaircraft division commander received authorization to destroy the target. Grechenov issued the order to Sergeant Varankin, the commander of the reconnaissance and targeting station. Varankin provided an affirmation. A few seconds passed before he reported a target lock. Grechenov asked for and received range, speed, azimuth, and altitude readings. He imagined the pilot sitting in the cockpit of the American spy plane. He wondered if the man had a wife, children, a happy home.

Grechenov abruptly ended such thoughts as he raised and lowered his arm. Sergeant Varankin gave a nod and pressed the firing key. He counted off the seconds until a tiny fireball appeared in the sky like a dying star gone supernova.

Major Grechenov did not know Major Rudolph Anderson, the pilot of the doomed U-2 and the first casualty of the Cuban Missile Crisis. He also harbored no concerns that his actions might have just started a war with the Americans. When President Kennedy and the ExComm group learned about the downed U-2, they worried that war was imminent. Two Soviet freighters, the *Gagarin* and *Komiles,* steamed to within a few miles of the Walnut line. The waiting U.S. fleet headed them off and maintained, for the moment, a Mexican standoff.

Admiral Anderson reported to Kennedy and McNamara that various SOSUS stations had verified seven possible contacts on Soviet conventional submarines near Cuba, but determining accurate bearings and submarine classifications was still problematic. Most of the contacts were labeled with a reliability one or two status out of a possible three, with location radius accuracy of greater than thirty miles in every case. One SOSUS report indicated difficulty in tracking contacts due to their distances from an underwater array, and another admitted to a misclassification of a *Foxtrot* submarine as a possible *Golf* class.

Anderson then reported the good news: Ensign Reed and his team had improved the bearing accuracy at the Skaggs Island Wullenweber Boresight station to around one half degree, and two other GRD-6 stations down to less than two degrees. The combined improvements were now capable of reliability-three fixes, with accurate submarine classifications within a radius of better than twenty-five miles. President Kennedy expressed his gratitude and authorized a letter of commendation for Reed and others involved in the Boresight program at NSA Section A-22.

AROUND 300 MILES SOUTH OF BERMUDA, just below the surface, Captain Savitsky sent a transmission to Moscow to acknowledge new orders for B-59. They were assuming a patrol area to the east of Dubivko's B-36 and 170 miles north of B-130's former patrol zone due east of Cuba. Savitsky was unaware that Boresight recorders, now upgraded with better location accuracy, had intercepted his burst signal.

Savitsky moved B-59 between the Soviet freighters and the U.S. fleet operating near the Bahamas southeast of Florida. He was now well inside the Walnut line. Moments later, Admiral Ward's ASW forces were on the move. Although Ward was provided with longitude and latitude coordinates for the *Foxtrot*—courtesy of Boresight—he was not authorized to know from where those fixes originated. Ward relayed the information to the ASW carrier USS *Randolph,* which was escorted by the destroyers *Bache* (DD-470), *Beale* (DD-471), *Cony* (DDE-508), *Eaton* (DD-510), and *Murray* (DDE-576). Given the volatility of the situation, nerves were stretched to the breaking point. One miscue could result in disaster.

A swarm of S2F Tracker aircraft and Sea King helicopters locked on to Savitsky's B-59, designated as a *Foxtrot*-class submarine, and the USS *Beale* dropped grenade-sized depth charges while pinging the area with active sonar. The USS *Cony* raced in and also splashed warning charges. Not only did the quantity of dropped charges exceed the "four or five" promised by the United States in their rules of engagement, but the devices from the *Cony* and ASW aircraft detonated in close succession to those dropped by the *Beale,* which created louder than normal explosions. Using dropped sonobuoys, ASW aircraft from the *Randolph* approximated B-59's location in front of the Soviet freighters and inside the Walnut line. The *Randolph* alerted CINCLANT. Five minutes later, a red telephone rang in the White House.

AFTER RECEIVING THE CALL, PRESIDENT KENNEDY raised a tired hand to his face and covered his mouth. He opened and closed his fist. His face drawn, his eyes pained and almost gray, he said, "Isn't there some way we can avoid having our first exchange be with a Russian submarine . . . almost anything but that?"

Secretary of Defense McNamara responded with a single word: "No," followed by a slow, careful explanation: "There is too much danger to our ships . . . our commanders have been instructed to avoid hostilities if at all possible, but this is what we must be prepared for, and this is what we must expect."

SENIOR LIEUTENANT PAVEL ORLOV CLENCHED HIS teeth as the dull thud of another explosion shook his submarine. Standing next to the navigation plot in the control center of B-59, on the port side of the boat, Orlov wondered why he'd volunteered for this suicide mission. He envisioned a transfer to a tropical land, that's why. But when Captain Savitsky told the crew that the Moscow main navy staff canceled the plan to establish a submarine base in Cuba, and instead directed them to run in circles amid dozens of U.S. warships near the Bahamas, the wind died in Orlov's sails.

As if that wasn't bad enough, the incessant heat and inability to shower brought irritating rashes and oozing, painful sores. A storm lashed at them for two days, causing several sailors to vomit on the deck when they tried to snort. And then there was the battle of the egos. Their submarine was cursed with the presence of Brigade Chief of Staff Captain Vasily Arkhipov, whose by-the-book viewpoint clashed with Captain Savitsky's bend-the-rules style of command.

Although Arkhipov outranked Savitsky, the captain of a ship or submarine always held the position of ultimate authority. Nevertheless, only a fool would ignore a superior officer's strong "recommendations," as this could ultimately have career-ending consequences. Diplomacy was not one of Savitsky's strong suits, however, and the constant disagreements with Arkhipov led to confusion and lower morale, if that was possible. Still, the crew continued to perform well, as to do otherwise taunted death.

For Orlov, not doing his best could also reflect dishonor on three generations of naval intelligence officers. Orlov's father received a transfer to the United States while working for the Main Intelligence Department (GRU) in 1945. Eight-year-old Orlov Jr. arrived with his family in Washington, D.C., and spent several years of his youth there gaining command of the English language. That served him well when he returned to Russia and became a naval intelligence officer assigned to the special OSNAZ group. There he learned about signals intelligence and how to operate ESM equipment designed to monitor U.S. radar and radio communications. He heard that his job was similar to that of American I-Branchers, or "spooks," as they were called. When his command gave him the opportunity to go to Cuba, he jumped at the chance.

When Orlov first reported aboard B-59, he and the other eight members of the OSNAZ team endured skepticism, criticism, and harsh treatment for being "nonquals." Many days went by before the OSNAZ group received a reprieve by producing reliable reports on NATO ASW movements. Then, slowly, attitudes started to change.

After B-59 passed just south of Bermuda, the Nakat ESM equipment picked up signals from American ships and planes in droves. Orlov and the other four English-speaking specialists spent long hours intercepting radio traffic. They determined that their pursuers were part of an ASW flotilla spearheaded by the aircraft carrier USS *Randolph*. Captain Savitsky thought he had avoided the hunter/killer group until that evening, a few hours after receiving new position orders and sending a receipt verification burst transmission to Moscow. That's when the floodgates opened and hell sent a swarm of locusts in the form of S2F Tracker aircraft and Sea King helicopters with dipping sonar. Destroyers operating in tandem with the *Randolph* soon followed, and it wasn't long before they surrounded B-59 and locked them into a tight cage. Then the explosions started.

American planes and ships now had them pinned down with active sonar and warning explosives. To Orlov, the grenade-sized depth charges sounded like the real things, and the quantity was far greater than the maximum of five promised by the U.S. Navy in their rules of engagement broadcasts. When Captain Savitsky had received the communication from the Americans outlining those rules—specifically about the requirement to surface and assume an easterly course—he snorted and said, "I will *never* surface." From Orlov's perspective, Savitsky and Brigade Chief of Staff Arkhipov had at least one thing in common: they were both stubborn mules. Neither wanted to show weakness in front of the other, so both held firm to their conviction not to surrender without a good fight.

Now here they were, more than 300 meters below the surface, batteries depleted, air fouled, and crawling along on the economy motor at three knots, still refusing to surface. Theoretically, with a full charge, they could endure a beating from the Americans for up to three days. The problem was, they had never been able to snorkel long enough to gain a full charge, and if they had to maintain three knots, they would never get away.

Acoustic estimated fourteen surface ships in pursuit, including the *Randolph* and a slew of destroyers. The dim emergency lights flickered, and body odor permeated the CC. Orlov forced his hand to remain steady as he assisted the navigator with the parallel contact plot, a skill he had learned at the naval academy. The plot lines showed that the Americans were following the canons of military art by surrounding B-59 and then tightening the noose. Another explosion rattled pipes. To Orlov, the ordeal felt as if someone had crammed him into a metal can and then started pounding on the thing with a sledgehammer. His chest heaved as his lungs fought to pull in sufficient air. The carbon dioxide level in the boat was becoming dangerous, and their store of lithium hydroxide was all but gone. They kept the dry powder in canisters and spread it across flat surfaces to absorb CO_2. They also carried chemical oxygen generator canisters that, when activated, heated up to create oxygen. Unfortunately, the excessive heat in the boat made using the canisters too much of a fire hazard.

Sweat stung Orlov's eyes as he laid a plot. He heard another explosion and then a loud thud. At first he thought something had come loose from the overhead. He glanced to his right. One of the duty officers had fainted from lack of oxygen and collapsed to the deck. Captain Savitsky summoned the boat's doctor. Then someone else fell down. Then another. Savitsky called for relief watchstanders as the fallen sailors were carried out of the CC. Orlov wondered if he'd be next, or if he'd ever see the snow-capped mountains of his youth again, taste his mother's cooking, or listen to his father's well-worn sea stories.

More loud explosions rocked the boat like a fishing bob on a stormy lake. The intensity of the blasts made them sound much louder than warning charges. Captain Savitsky came unglued. He started screaming that the Americans were now dropping real depth charges. Captain Arkhipov, who stood next to Savitsky near the conning tower ladder, countered that such a thing could not be true.

As the two argued, Orlov saw Valentin Grigorievich, the officer in charge of the nuclear torpedo, shoot through the forward hatch and enter the CC. "Captain, are we under attack? Should I ready the special weapon?"

"Affirmative," Savitsky said. "Ready tube number two."

"No!" Arkhipov yelled. "The conditions for firing have not been met."

"Sounds like the war has already started while we've been doing somersaults down here," Grigorievich said.

"I agree," Savitsky said.

"You'll kill us all," Arkhipov said.

"But not in vain."

Arkhipov grabbed Savitsky by the arm, "You can't do this; we don't have authorization."

Savitsky pulled his arm away. "I can, and I will. Our batteries are depleted, our air is gone, and I will not disgrace our navy." He sent Grigorievich forward to prepare the purple-nosed weapon. He then turned and yelled toward the fire control station, "Fire Control, distance to the carrier?"

A pink-faced *michman* called from around the periscope housing. "Twenty-two *cabletovs,* sir."

Two nautical miles, thought Orlov. He knew their nuclear torpedo could kill out to a radius of ten nautical miles, which meant that Arkhipov's words were true. They would die along with the Americans. Despite the heat, Orlov shivered, now certain that his next few breaths would be his last.

"Acoustic," Savitsky called, "bearing and speed on the carrier?"

"Control, target bearing two-one-nine," acoustic replied, "speed ten knots."

"Don't do this," Arkhipov pleaded. As if to mock his plea, another charge shattered overhead. The brigade chief grabbed a ladder handle to keep from falling.

"I am in command of this vessel," Savitsky said, "not you."

Orlov's head pounded from lack of oxygen as he studied the plot on the nav table. All fourteen warships were well within the nuke's kill zone.

"Watch Officer," Savitsky said, "make your depth sixty meters, course two-one-nine, speed six knots."

"Yes, sir."

"Acoustic, prepare first measurement."

Acoustic gave a bearing.

"Acoustic . . . zero," Savitsky said. "Regimen one minute."

His breathing labored, Orlov updated the plot to the carrier. Acoustic started feeding bearings to the torpedo attack crew—which included the navigator, electronics officer, and fire control team—every minute.

That crew, which Orlov now joined, plotted each change until the captain was satisfied that they had an adequate firing solution.

"Control, target bearing two-one-five, speed ten knots, distance nineteen *cabletovs*."

"Prepare to fire tube number two," Savitsky said.

Orlov's heart fluttered. He felt dizzy and thought he'd collapse like the others. He steadied himself on the edge of the plot table and fought to stay conscious. If he was going to die, he wanted to at least hear the explosion.

"Vitali," Arkhipov said to Savitsky, his voice almost a whisper, "please."

Another loud explosion threatened to shake loose the fillings in Orlov's teeth.

Savitsky narrowed his eyes and stared at the brigade chief. "Acoustic, prepare last measurement . . . zero."

The captain raised his arm. Orlov felt the hand of fear wrap about his throat like a giant snake. He closed his eyes and waited for the end.

Deputy Political Officer Ivan Maslennikov jumped through the aft hatch from Compartment Four. "Wait!"

Savitsky lowered his arm without issuing the final order to fire.

"If you fire now, Vitali, you will kill us all and start a war," Maslennikov said.

Savitsky pointed an angry finger skyward. "It's already started!"

"No," Maslennikov said. "Those are not full-strength depth charges. They have fourteen warships surrounding us. If they wanted to sink us, we'd already be dead."

Savitsky lowered his head and stared at the deck. His eyes darted back and forth, and Orlov could tell that the captain battled with his anger to make the right decision. A minute passed, then another.

Savitsky raised his head. "Attention in the CC. Cancel attack and prepare to surface."

Another explosion ruptured a seal and sent seawater shooting into the CC.

CHAPTER TEN

Survival is triumph enough.

—HARRY CREWS

IS FACE RED WITH ANGER, JOHN Scali of ABC News met again with Soviet embassy official Aleksandr Fomin at 4:15 P.M. on October 27. Scali demanded to know why the terms offered in the two letters from Premier Khrushchev were as different as night and day. Fomin stuttered and said that it must be due to "poor communications." Scali pointed an accusatory finger and shouted that the differing letters were obviously a "stinking double cross." He stated that an invasion of Cuba was only hours away. Fomin's eyes widened. He mumbled that a response was expected from Khrushchev at any moment and urged Scali to tell the State Department that they intended no treachery. Scali shook his head and said that no one would believe that to be true, but he'd deliver the message to the White House anyway.

At 8:05 P.M., Khrushchev sent a letter outlining the structure of a deal. President Kennedy responded that if the Soviet Union agreed to remove all weapons systems from Cuba, under U.N. observation, and halt further delivery of such weapons, the United States would remove all quarantine measures and refrain from invading Cuba. A secret memo accompanied the letter. Some experts and scholars speculate that President Kennedy confirmed his offer to remove the missiles from Turkey in that note. A confidential source that worked for the NSA on the Boresight and Bulls Eye programs believes that the president also laid down a trump card. This informed source speculated that the memo con-

tained day-by-day location coordinates for the four *Foxtrot* submarines, outlining their track since arriving in the Sargasso Sea. Unaware
of the existence of Boresight, Khrushchev could only guess as to how
the United States managed to obtain this information.

Until the receipt of that memo, the Soviet premier's demeanor exhibited confidence and defiance. He slammed his left hand on the table
when meeting with advisers and shouted assurances that Kennedy could
not invade Cuba. To do so invited suicide. His Project 641 submarines,
each carrying a nuclear-tipped torpedo, were his safety nets. With his
deadly *Foxtrots* now exposed by unknown means, he could no longer
deal from a position of strength, and so was forced to dictate a "surrender" letter to Kennedy. The president called the letter "an important and
constructive contribution to peace."

Others contend that a bartender helped end the crisis. Aleksandr
Fursenko and Timothy Naftali, in their book *One Hell of a Gamble*,
suggest that Khrushchev backed down due to the ears of a Russian im-
migrant from the Balkans named Johnny Prokov, a bartender at the
Tap Room in the National Press Club in Washington, D.C. They believe
that Prokov passed to Anatoly Gorski, a Soviet KGB officer, information overheard during a conversation at the bar between two celebrated
American journalists, Robert Donovan and Warren Rogers, both cor-
respondents with the *New York Herald Tribune*.

Fursenko and Naftali speculate that the KGB could have passed
this information immediately to Moscow, and Khrushchev have had it
on his desk within hours. Based on the hearsay overheard in a bar, the
stubborn Soviet premier finally became convinced that Kennedy was
serious about going to war over Cuba. To avoid that war, Khrushchev
infuriated Fidel Castro by pulling the plug on two massive operations
that spanned more than six months, involved hundreds of ships and
thousands of sailors, and cost billions of dollars. Certainly this conjec-
ture is feasible, but perhaps unlikely.

Despite Khrushchev's overture, however, the Cuban Missile Crisis
did not end. In fact, the conflict on the high seas escalated, and neither
side knew then just how close they came to Armageddon.

AT 10:52 P.M. ON BOARD THE destroyer USS *Cony*, Ensign Gary
Slaughter, the phone talker on the bridge, held his breath at the sight of

a *Foxtrot* submarine breaking the surface in a swirl of white foam. The *Cony* had chased the *Foxtrot* for more than twelve hours. Although the sight filled Slaughter with awe and pride, he did not know how close he'd just come to being vaporized by Captain Savitsky's B-59. The Soviet submarine, her sleek hull dripping with salt water, turned northeast and started to snorkel. Several men appeared on the bridge, all wearing Russian uniforms and covers.

Trained in Cyrillic transliteration tables, Slaughter ordered Signalman First Class Jessie to flash a communiqué to the men on the sub's bridge.

"What ship?" Jessie flashed with the *Cony*'s signal light.

"Ship X," came the reply.

"What is your status?" Jessie flashed back.

"On the surface, operating normally."

"Do you need assistance?"

"No, thank you."

Hours passed with no further communication as the *Foxtrot* slowly steamed northeast. The following morning, the *Cony* received another flash. The submarine asked for bread and American cigarettes. Slaughter smiled as the *Cony* maneuvered to within eighty feet of the Soviet vessel and prepared for a line transfer. One of *Cony*'s bosun's mates set up the shotgun-looking device designed to thrust a weighted rubber-tipped projectile attached to a long line. The bosun's mate fired the shot line toward the *Foxtrot*'s bridge.

When they saw the spearlike projectile flying toward them, the Russians yelled, pointed skyward, and ducked down below the bridge fairing. Slaughter let out a laugh. He recalled that Soviets didn't use shot lines, but instead employed bolo lines flung by a strong seaman. After they sent bread and cigarettes to the *Foxtrot,* the *Cony* shadowed the Soviet sub for several more hours, staying 500 yards off her port beam.

That evening a P2V Neptune ASW plane streaked down from the darkened sky and dropped a scattering of incendiary devices to activate the aircraft's photoelectric camera lenses. Again the men in the *Foxtrot*'s bridge yelled, pointed, and scrambled for cover. Inside B-59, Captain Savitsky interpreted the miniature explosions as a possible attack. He wondered if the Americans had tricked him into surfacing so they could easily sink his boat. Posturing to defend his craft, he armed his nuclear torpedo and opened his outer torpedo tube doors.

On the *Cony,* watchstanders shouted as the Soviet sub changed course. Sonar operators reported the sound of torpedo tube doors opening. Ensign Slaughter heard the *Cony's* captain hail the aircraft carrier *Randolph* and demand that the Neptune's pilot cease his photo runs. The CO then ordered Slaughter to have a signalman flash an apology to the Soviet submarine. Nervous minutes passed. Finally, the *Foxtrot* closed her torpedo tube doors and turned back toward the northeast.

ON OCTOBER 28, 400 MILES NORTH of San Juan, Puerto Rico, aboard the destroyer *Charles P. Cecil* (DD-835), Lieutenant John Hunter, the officer of the deck, saw something that bothered him. Hours earlier they'd been ordered by CINCLANT to search within a twenty-five-nautical-mile radius of a specific coordinate. Hunter did not know why they'd received that order, but he suspected that SOSUS or some other system detected a submarine. He had no knowledge of Boresight.

Hunter glanced at his watch. Just after 7:00 P.M. local time. He studied the green phosphor glow on the radarscope. A blip, one that resembled the irregular echo of a snorkeling submarine, or "snork" in ship jargon. Hunter contacted the Combat Information Center. Radarman Third Class Russ Napier had also detected the contact in the CIC. Hunter picked up the phone and called the ship's captain, Commander Charles Rozier. The captain granted permission to prosecute. Hunter then made a huge mistake.

Hoping that he might fool the Soviet submarine, he ordered a high-speed run toward the contact. He also called down to the sonar shack and asked them to turn off the active stack. No more pings. By the time the *Cecil* hit twenty knots, the radarscope went blank due to the ship's churn. When the *Cecil* neared the sub's last known location, Hunter contacted sonar and ordered them to turn on the active stack. Just before he hung up the phone, he heard a loud pop in the background. Sonar called back. The stack had fried itself. The *Cecil* just lost an ear.

Hunter stared at the radar screen. No more blip. Obviously, the submarine went deep. He swore under his breath. Not only did his trick not work, but his decision to switch off the active sonar backfired. They'd just lost the Soviet sub.

CAPTAIN DUBIVKO SURFACED B-36 EARLY IN the evening on October 28. His orders had not changed in days: to maintain a combat-readiness posture on station south of B-59 and B-130 just outside the center of the quarantine line. That position, of course, placed him square in the middle of a phalanx of American ships and planes. Avoiding those hunter/killers kept the crew on a tight edge while they sweltered in the heat.

Dubivko pulled himself up the ladder through the conning tower and edged onto the bridge. A setting sun splashed streaks of purple-orange across the horizon. His lungs savored the fresh air. The past few weeks had taken their toll. Due to excessive heat and lack of sleep, he had dropped almost fifty pounds. Now he could no longer climb a ladder without exhaustion.

Three of the crew stepped onto the bridge platform and leaned their heads back to smell the salt air. They appeared gaunt and weak with deep circles under their tired eyes. Here in enemy waters, Dubivko should be snorting versus surfacing to snorkel. But he was worried about his crew. Given the deplorable conditions, he didn't know how much longer they could endure. He decided to take the chance and surface, and allow the crew some fresh air and saltwater showers in the sail.

Earlier in the day, the OSNAZ group intercepted a radio report that several Russian merchant ships had reversed course and were headed back to the Soviet Union. Dubivko wondered what that could mean. Had Moscow abandoned Operations Anadyr and Kama, thus ending Khrushchev's ambitious plan to place missiles and submarines in Cuba? Were they preparing for war? Having received almost no information from Moscow, he could not even wager a guess. Assuming the worst, he ordered Zhukov not to use surface search radar to minimize detection. That decision came with a downside. An approaching enemy ship could surprise them, and with the additional men topside, diving would be delayed.

In the event of that scenario, given that they were in calm seas,

Dubivko ordered a depth that placed B-36's bridge lower in the water than normal while they snorkeled. He loved cruising on the surface at night with a light breeze ruffling his hair and caressing his cheeks. Despite the hardships they'd endured on this mission, standing high in the sail of his boat made everything seem right again. This was his little slice of heaven, and nothing could take that away.

Almost nothing.

Just after 7:00 P.M., Zhukov shouted up through the open hatch, "Contact bearing zero-five-zero!"

Dubivko called down, "Range? Contact type?"

Silence. Then, "Two thousand meters, probable American destroyer!"

Dubivko peered through his binoculars and scanned the horizon. Nothing but black. The destroyer was running without lights.

"Srochnoiya pogruganye!"

The bridge area cleared, and Dubivko slammed the hatch closed above his head. In the CC, he called over the communication line to Pankov, "Acoustics, bearing to contact?"

"Control, we are now detecting two contacts on bearing zero-five-five. Second contact has high-speed screws."

Torpedo?

Dubivko wondered if they were now at war, and B-36 was destined to become the first vessel sunk. "Watch Officer, all ahead full, right full rudder, depth seventy meters."

Dubivko cursed the unseen demons. He did not have a full charge on the batteries. His men were worn down to threads, and worst of all, the damage to the VIPS lid limited their depth to seventy meters. He was a David with one arm trying to fight dozens of Goliaths. At least he had rocks. One big rock in particular. But to fling that weapon from his sling meant ending their mission forever.

ON OCTOBER 29, AS THE CLOUDS split to reveal a red sky at morning, Commander Rozier strolled onto the bridge of the *Charles P. Cecil* and relieved John Hunter of the conn. After the sonarmen repaired the active stack, they managed to regain contact on the Soviet sub. Their prey tried a cadre of tactics to evade but failed. Rozier held on like a cowboy on a bucking stallion, one arm waving a hat. He knew the enemy was blessed with excellent thermal conditions, with the isothermal layer extending down to more than 300 feet, but he suspected fate had intervened to take that advantage off the table.

When Rozier received the flash from CINCLANT, providing coordinates and a defined search radius, he was more than surprised. He knew SOSUS was good, but not that good. When he read that the contact type was a probable *Foxtrot,* he let out a long whistle. Just what kind of new technology could be that certain of the platform type? Now, suspecting that they were prosecuting a *Foxtrot,* he wondered if something was wrong with the boat.

Soviets used three ratings for submarine diving depth characteristics. Working depth, referred to a window within which a boat could operate under normal conditions for extended periods and engage in combat. Limiting depth documented how deep a submarine could dive before she started hacking and wheezing. Most boats were restricted to a maximum of 300 excursions to this depth for their entire service life. Lastly, design collapse depth was where a submarine turned into Cream of Wheat.

Rozier knew that a *Foxtrot*'s limiting depth was 985 feet, which afforded a working depth of 854 feet. Yet their contact, based on active ping estimates, had not descended below 250 feet. Why? That boat's captain must know that the thermal layer started at 300 feet, and if they dropped below that threshold, they'd have a much better chance of avoiding detection. Yet they stayed above the layer. That could only mean an ignorant sub captain or some kind of damage that kept them shallow, and judging by the evasion tactics used so far, Rozier didn't think he was chasing an idiot.

The *Foxtrot*'s captain used a clever mix of tactics, including trimming the tanks, adjusting the bow planes, speeding up and slowing down, frequently changing course, and hiding in the *Cecil*'s wake. The wake tactic created a baffle that masked the destroyer's active sonar pings, given this at more than thirty knots, the *Cecil* churned the ocean to a depth of seventy feet. When everything else failed, the *Foxtrot* tried to inhibit the *Cecil*'s active sonar pings by shooting out chemicals that formed a cloud of hydrogen bubbles when mixed with seawater. That didn't work either.

The *Foxtrot* then tried something new. From the sonar shack, Sonarman Third Class Elroy Nelson reported that the sub's screws had gone silent. All he could hear now was the chatter of snapping shrimp in his headset. Rozier figured that the Soviet sub had stopped dead in the water. He circled the last known contact location, pinged, and waited.

A half hour later, the *Foxtrot* surged back to five knots and ran toward them in an attempt to get behind the *Cecil* and hide in the ship's wake again. Rozier refused to let that happen. At 500 yards distance, he ordered turns for fifteen knots and right full rudder to keep the *Foxtrot* pointed toward their bow.

The Soviet sub then released a noisemaker torpedo, revved up to full speed, and darted off to starboard. Rozier knew to watch for the tactic, given that Soviets were taught to do this during a close-in attack. The aft torpedo room on a *Foxtrot* housed four torpedo tubes—two filled with warshots and two with noisemakers. In a combat situation, Soviet captains were trained to fire off a noisemaker, followed by a warshot, then make a wide turn and sprint; slow again and go silent. Hopefully, any pursuers would follow the noisemaker and not the submarine. Rozier did not intend to fall for the ruse. Still, he couldn't seem to corner the submarine into a tight enough box to drop any warning charges.

Rozier decided to enlist some help by contacting a P2V Neptune ASW plane patrolling the area out of Jacksonville, Florida. The P2V, with the call sign Pollyboy 5, had just relieved another aircraft on station and came roaring toward the *Cecil*. The plane circled overhead as the *Cecil* directed the runs over the radio, helping the plane correct its lead angles and line up correctly over the target. Several "no joy" runs later, Rozier wondered if they'd ever nail this guy. Then the P2V reported "Madman," and Rozier smiled. The ASW plane's magnetic anomaly detection (MAD) equipment had picked up a fluctuation in the earth's magnetic field when it flew over the *Foxtrot*'s metal hull. Contact nailed.

Rozier stayed on top of his quarry throughout the night. More P2Vs arrived in the morning and lit up the ocean with active sonobuoys. They eventually validated contact C-20 as a *Foxtrot*. Rozier suspected sooner or later that the Soviet sub would run low on battery power and have to surface, but his patience for this game was running thin. Now that he'd cornered the *Foxtrot,* he decided it was time to drop a few loud incentives into the water. Little did he know that his decision was about to place his ship at ground zero.

ALMOST 300 MILES NORTHEAST OF CAICOS Passage, the USS *Blandy* and several other U.S. destroyers stared at a ghost. They had pursued the illusive *Foxtrot* contact for fourteen straight hours after she dove to avoid detection by ASW aircraft. Below the surface, Captain Shumkov on B-130 used up a quiver of tactical arrows to dodge the destroyers. In a last-ditch attempt, he shot a noisemaker torpedo from an aft tube. The noisemaker emitted the same frequency as a Project 641 submarine to fool enemy sonar operators. Normally, Shumkov would trail off in a preset direction at ten knots after firing one, but with batteries running low, that option no longer existed.

On top of the waves, *Blandy*'s sonar operators not only did not fall for the trick, but misinterpreted B-130's noisemaker as "high-speed screws in the water." *Blandy*'s skipper, Captain Edward Kelley, took evasive maneuvers and readied his weapons. Seconds before firing, sonar reported a "false alarm." The contact had only released a decoy that sounded like high-speed screws. Captain Kelley took his finger off the trigger.

Several hundred feet below the *Blandy*, Captain Shumkov ran out of options. Depleted batteries and thin air left him with no choice but to surface or die. Cornered, exhausted, and at his wits end after days of undersea conflict, with explosive charges raining down, Shumkov decided to end the ordeal by firing the special weapon.

His decision was motivated partly by pride and partly by the desire to save face in front of his political officer, Saparov. He ordered the special weapons security officer to make preparations to fire the nuclear torpedo. The security officer refused, citing that they did not have authorization. Fortunately, fate intervened. Due to the excessive heat and deplorable conditions on board, Saparov took ill and passed out. Shumkov no longer had an audience to bolster his ego. After considerable deliberation, he finally gave the order to stand down. Later he informed the crew that he had never intended to use the special weapon, as B-130 would not have survived the blast.

Captain Shumkov surfaced B-130 into the fading light of a setting sun on October 30. Not long after, while devouring another meat-filled piroshki, he heard his last engine die. Up until then, he'd clung to a desperate hope that his crew could repair at least one of the other two dead engines. Devastated, he now had no choice but to call Moscow and request a tow home. The bite of piroshki in his mouth turned bitter as he cursed the gods for forcing him to abandon his duty and leave his friends behind in hostile waters.

CAPTAIN DUBIVKO ON B-36 IMAGINED A miniature demon sitting on his shoulder and laughing. The gremlin had succeeded in taking away one of the most important tactical advantages available to any submarine: the ability to go deep. The thermal layer started at one hundred meters, but since they'd damaged the VIPS lid, they could not go below seventy meters, so evading enemy sonar became almost impossible. Inside B-36, the conditions became even more horrific. With temperatures exceeding sixty degrees centigrade in the engine room, and forty-five in other compartments, Dubivko ordered frequent bed rest for most off-duty personnel. Some of the men still passed out due to heat stroke not twenty minutes into their watch.

Lack of water exacerbated the problem. Reserves allowed for no more than 250 grams per day per man. Profuse sweating and dehydration increased skin rashes and decreased appetites. Many crew members lost more than thirty percent of their body weight. Fortunately, they'd stocked plenty of fruit compote, which the crew drank in place of water.

After they had detected the destroyer, they quick-dove to seventy meters. At thirty meters, they heard the loud chug of the destroyer's engines pass overhead. What they thought were high-speed screws in the water—indicative of a torpedo—turned out to be a towed sonar array.

The American destroyer continued to harass them by pounding the ocean with incessant active pings. Acoustics reported yet another contact in the port baffles. Given the noise interference generated by the destroyer and circling aircraft, discerning what type of platform this contact might be was not possible. Acoustics did report that the hydrophone effects appeared to be coming from a depth of twenty meters, but they could not be certain. Dubivko wondered if the contact might be an American submarine or one of his sister boats. If the former, Dubivko was a sitting duck. Unable to dive deep, with the batteries nearly depleted, the odds of dodging a speeding torpedo were dismal at best.

He walked to the nav table and asked Naumov to lay out the plot showing the patrol areas of the four Project 641 submarines. He traced the bearing line to the eastern edge of B-130's assigned patrol area. As the main navy staff still included the call signs of all four boats in their transmissions, Dubivko was unaware that Captain Shumkov had surfaced earlier, lost his last engine, and was preparing to be towed home. He also did not know whether hostilities were escalating or if the two superpowers were engaged in a shooting war. Given that the destroyer circled at 2,000 meters, he suspected that gun barrels remained cold, but he needed to know for certain.

Despite the risk, Dubivko ordered his boat to periscope depth. They needed to snort and assess the tactical situation. Also, they were overdue for a burst transmission from Moscow, and Dubivko wanted to determine if their orders had changed regarding the use of the special nuclear weapon.

Acoustics reported that the P2V Neptune had just made another pass. Dubivko glanced at his watch. Caution dictated timing the passes and lowering the periscope each time the aircraft came near, especially since nightfall remained an hour away. A sailor cracked the conning tower hatch open at twenty meters depth, and Dubivko hurried up the ladder, followed by Political Officer Saparov. He waited for the P2V to make another pass, then raised the periscope. Salt water dripped onto his shoulders, but the small space felt cool compared to the overheated CC. He made a quick visual sweep. Flashes of crimson accented ethereal clouds on the horizon and reminded him of Polyarny.

Dubivko spun the scope and marked off the bearings to each of three gray destroyers. Running without lights, the dark silhouettes bristled with guns and antennas. On a fourth bearing, Dubivko spied a single blinking light. He squinted. The shape of a helicopter came into view. Below the whirling blades, a long cable ran into the ocean.

Dipping sonar.

Dubivko lamented the good weather. Calm seas afforded no hiding places. He lowered the scope and called Potapov on the phone. The chief mechanical engineer reported that the electrolyte in the batteries was now almost pure water. That meant they needed to snort within fifteen minutes. Acoustics reported the sound of the P2V swooping overhead again. Dubivko counted off several seconds in his head, then raised the scope.

He focused on the stern of the closest destroyer and spotted a large reel of black cable on the fantail. One end of the cable ran over the side and disappeared into the water. A sonar transducer. Dubivko's pulse quickened. Transducers were towed by destroyers and lowered to depths beneath the thermocline to find hiding submarines. At the end of the cable hung a large teardrop-shaped device that improved the destroyer's ability to find submerged targets. For Dubivko, that posed a huge threat, not just from detection, but from collision. If B-36 came too close to that transducer, the heavy object could rip a hole in their side and send them to the bottom.

Dubivko checked his watch again. "*Opustit periscope.*"

The scope lowered back into the housing. He called Acoustics and asked for a depth update on the thermal layer. Unchanged at one hundred meters. The contact reported earlier, which Dubivko thought might be B-130, remained at twenty meters distant. If this was indeed Shumkov, he knew where the layer started, and unless B-130 also suffered from a depth-limiting problem, Shumkov should have descended below that layer. Still, Dubivko had to assume that the contact might be the B-130 and that the Americans had forced them to come shallow.

Dubivko struggled with his next decision. He turned over the conning tower to Saparov and slid down the ladder into the CC. He picked up the phone and summoned Alexander Pomilyev, the officer in charge of the special weapon. He met the young man outside his stateroom. They entered, and Dubivko sat on his bunk, while Pomilyev remained standing.

"I want you to ready the weapon in tube number two."

Pomilyev's eyes widened. "Sir?"

"We need to surface. If the enemy turns hostile, I intend to fire at the closest target."

"But, sir, that will destroy us as well."

Dubivko remained silent for a long moment, then said, "Prepare the weapon."

Pomilyev lowered his chin. "Yes, sir."

Dubivko returned to the conning tower, timed the P2V pass, and raised the scope. He swallowed his guilt and ordered a heading of due east, according to the surrender directions they'd received repeatedly from the Americans—rebroadcast from Russia. With their batteries gone, the least he could do was give B-130 a fighting chance to get away if she was nearby. If B-36 surfaced now, that distraction just might afford Captain Shumkov enough window to sneak away.

"*Vspletye!*"

ON OCTOBER 31, AT 4:00 A.M. aboard the USS *Charles P. Cecil,* Sonarman Elroy Nelson finally got relieved from the stack by Sonarman Allen Tuell. The twenty-something Tuell from Santa Cruz, California, yawned as he settled in for what he hoped wouldn't be a boring six-hour watch. Less than two hours later, at 5:53 A.M., Tuell heard a familiar sound in his headset. His trained ears discerned the shush of high-pressure air. The *Foxtrot* just blew her ballast.

Lieutenant John Hunter heard the 1MC loudspeaker crackle. "Submarine on the surface!" All about him, excited sailors in skivvies and shower shoes lined the railing to get a glimpse of the surfacing sub. The stubby nose of a Soviet *Foxtrot* broached the surface, followed by the long black hull and windowed sail. Here and there the black deck was mottled with weatherworn dull red patches. Hunter wondered if that might be rust. On the sail, three numbers shimmered white in the morning sun. They read 911. Hunter knew this was not the boat's real designation, but only a decoy to fool the enemy.

Enemy.

Hunter thought about that word for a moment. Were they really the enemy, or just seventy-eight men caught up in a high-stakes game of cat and mouse started by narrow-sighted men in power? Hunter squashed his political thoughts as several figures appeared on the bridge of the *Foxtrot.* One raised a flag atop the main antenna, which flapped in the breeze from the back of the sail. Hunter recognized the emblem as the main cruiser ensign of the USSR Naval Fleet. He knew this was a big moment, that he was now a part of history and witnessing the forced surfacing of a Soviet submarine in the Caribbean during the Cuban Missile Crisis.

His smile died as he saw the *Foxtrot* make a sudden turn away from the *Cecil.*

ON THE BRIDGE OF B-36, CAPTAIN Dubivko adjusted his binoculars and focused on the boxy destroyer still shadowing them. After intercepting a radio transmission sent to their American escort, the OSNAZ team told him the name of the ship was the *Charles P. Cecil*. Now less than one hundred meters away, the dark shape of the *Cecil* loomed over his boat like a massive guard marching him to the gallows.

Dubivko figured that the ship's captain maintained that proximity because he knew the "dead zone" for Soviet torpedoes—the distance traveled after leaving the tube before the weapon armed—was right at 150 meters. This safety measure, programmed into all torpedoes, prevented weapons from detonating too close to their firing submarine. If Dubivko shot a torpedo at this close range, the projectile would bounce harmlessly off the metal side of the destroyer.

The American ship flashed a message from their bridge. "Do you need help?"

Dubivko ordered an OSNAZ signalman to reply. "We do not need any help. Asking you not to interfere with our actions."

Dubivko scanned the destroyer. Several gun turrets pointed in their direction. The long barrels looked ominous in the fading light.

Were they preparing to fire?

Convinced that they were, Dubivko lowered his binoculars and called down to the CC. "*Boyevoya trevoga!*" He heard the crew scramble to battle stations. He then ordered a turn to port, away from the enemy ship. He needed to open the distance to more than 150 meters so the nuclear torpedo could arm itself. He called Pomilyev in the forward torpedo room and asked the special weapons officer to prepare to fire tube number two on his order. His hands shook as the boat's screws slapped the ocean and edged B-36 farther away from the destroyer. They had fifty meters to go before they were 150 meters away from the ship.

Dubivko thought about his home, his wife, and the life he would never see again. A part of him fought against his decision and tried in vain to talk the other part out of it. But pride and honor made him deaf to the voice of reason. He could not sit still while the Americans blasted his boat into shards of steel. He'd go down fighting with the last breath in his tired and worn body.

Dubivko glanced again at the radar repeater as the boat passed 150 meters. He started to call down to Pomilyev but hesitated for a brief moment. Zhukov called up from radio. "Sir, we just intercepted a radio transmission to the *Charles P. Cecil*—the destroyer that is shadowing us."

"Transmission?" Dubivko said.

"From President Kennedy."

"President Kennedy?"

"Yes, sir. The message to the *Cecil* reads, 'Thank you for your work. Keep the surfaced submarine here by all means.'"

Dubivko pondered the message and picked up his binoculars. He scanned back and forth along the destroyer's hull. The gun turrets were now turned away. He breathed a sigh of relief, changed course, and told Pomilyev to disarm the nuclear torpedo. Glad to still be alive, Dubivko descended back into the belly of his boat and ordered Zhukov to transmit a burst message to Moscow informing them of B-36's situation.

They sent the same message forty-eight times to the main HQ of the naval fleet until they finally received a response. The *Cecil* continued to shadow B-36 as the crew repaired the damaged upper lid of the VIPS, which earlier had prevented them from being able to deep dive. The diesels ran nonstop for the next two days to charge the batteries. Normally, the charge took only ten to twelve hours, but due to the long periods of high temperatures in the boat, the electrolyte in the batteries soared past sixty-five degrees centigrade. That required them to ventilate extensively to reduce the temperature back down to sixty so as not to cause an explosion.

Not blasted to pieces, not nuked by their own torpedo, not blown up by hot batteries, no longer debilitated by a broken VIPS lid, Dubivko swallowed a bite of piroshki, lit a cigarette, and contemplated an escape plan. A crazy idea popped into his head, and he called Zhukov and Pankov to his cabin.

"We'll need to distract them," Dubivko said as he sat on his bunk.

"We could try sending a signal in the circular regime using the Sviyaga hydroacoustic station," Pankov said, standing bent over in the cabin.

Zhukov, also standing bent, scratched his chin. "That might work. We'd have to rig our acoustics to send out a strong signal on the same frequency as their active sonar beam."

"Is that feasible?" Dubivko asked.

"I think so," Pankov said. "We know the frequencies they use, and we're now close enough, but we will need to send several strong transmissions."

"This will work for only a short time," Zhukov said. "Once we're too far away, their sonar will not be disrupted."

"How long?" Dubivko asked.

"Minutes," Pankov said.

"That's long enough," Dubivko said. "Make it happen."

The following day, just after lunch, with no helicopters or ASW aircraft in sight, Pankov flipped a switch and flooded the *Cecil*'s underwater ears with a high-pitched whine. Dubivko slammed shut the upper hatch and quick dove. The boat passed underneath the *Cecil*, and he was certain they missed the ship's hull by mere centimeters. He submerged B-36 to 200 meters, reversed course, and zigzagged away at a fast clip. Pankov lit off a stream of six-second transmissions to suppress the *Cecil*'s active sonar beams. The trick worked.

Hours later, Dubivko brought B-36 shallow to transmit a series of burst updates to the main navy HQ. This time they received an immediate reply. Unbeknownst to Dubivko, B-36's transmissions provided solid hits to three Boresight stations.

American ASW planes reappeared, but this time, with a full battery charge and the ability to go deep, B-36 managed to avoid detection. Dubivko attributed this to skill and luck, unaware that the Americans were using Boresight fixes to plot his homeward track. As long as he kept his submarine pointed toward Russia, they wouldn't have to sink his boat.

Now ONE HUNDRED MILES SOUTH OF Kingston, Jamaica, Captain Ketov on B-4 continued to evade American ASW forces. He harbored no illusions that his good fortune might be attributed to an upgraded sonar system, which afforded B-4 longer-range hearing. That, and a large dose of luck. While B-4 snorkeled throughout the night of November 2, Ketov inhaled the sweet night air, watched silver moonlight dance on the wave tops, and prayed that his good fortune would hold.

For certain, lady luck had played a role, but Ketov also gave credit to his crew. They'd done an excellent job of combining skill and intelligence to avoid the enemy. While doing that, Ketov noticed that ASW planes stalked them the most when they started to snorkel or snort at night, and aircraft numbers diminished during the day when B-4 ran on batteries. The planes appeared to come around the most in the early evening, not long after B-4 lit off her diesel engines. Based on this pattern, Ketov surmised this was no coincidence and might be due to detection by the American hydroacoustic stations embedded in the seafloor. Of course, he had no knowledge of Boresight, and so did not assume that B-4's late afternoon burst transmissions triggered an ASW response that placed aircraft nearby when the submarine came shallow again to snort.

Ketov stood on the bridge and listened to the muted hum of B-4's diesels. He thanked whatever god there might be that he had not fallen to the fate of his friend, Captain Shumkov, on B-130. Ketov could not even imagine losing his engines and having to be towed home. When he heard about the incident on the daily radio update, he wanted to call his comrade and express his condolences. He could not, of course.

Ketov glanced at his watch. An additional communications session with Moscow was scheduled for that night. He figured when the session started in twenty minutes, new orders would be issued. During the next five minutes, he let his mind go on vacation. He thought about his wife, his children, and all the things he had missed while living most of his life far away from home. Five minutes came and went before a report from radio jolted Ketov from his daydreaming. Radio detected a signal on the Nakat. The hit was weak but growing stronger. Ketov had observed earlier that some of the P2Vs and Tracker aircraft took to truncating their radar signals in an effort to fool the Soviet subs by making the planes appear farther away than they actually were. He did not intend to fall for this trick. He ordered the bridge cleared and quick dove to a depth of ninety meters.

Brigade Commander Captain Agafonov cornered Ketov in the CC and demanded to know why they went deep fifteen minutes before the scheduled session with Moscow. He appeared anxious to receive an update, as interference had blocked reception of the last one. Ketov explained his reason for diving: that he knew the Americans attenuated their radar signals to ensnare unsuspecting submarines. Agafonov refused to listen. He insisted on surfacing to complete the battery charge and to ensure a clear signal for the radio session. Ketov stood his ground, stating that the OSNAZ group earlier determined that the USS *Inde-*

pendence was operating in the area, and they detected dozens of ASW aircraft, including one likely headed their way at this moment. He told Agafonov that as long as he was in command, they were not going to surface.

"Fine," Agafonov said. "You're relieved of command."

Shocked, Ketov implored Agafonov to reconsider and to not place B-4 in jeopardy by surfacing. Agafonov waved him off and gave the order to come shallow. Angry, but having no choice, Ketov turned over command and retired to his cabin.

Agafonov brought the boat shallow and extended a periscope. The Nakat barked out a series of loud beeps as American radar got a hit on B-4's masts. Agafonov took the boat deep again, but by then, it was too late. A Tracker airplane from the *Independence* gained a solid lock and refused to let go. Agafonov tried to dodge as the plane dropped warning charges that rattled dishes and nerves.

Hours went by as Ketov remained in his cabin while a less experienced officer tried without success to avoid the enemy. Finally, after three hours of explosions and pinging, Agafonov became frustrated and angry. He called the special weapons officer in the torpedo room and ordered him to ready tube number two. Then he turned the boat toward the USS *Independence*.

The officer in charge of the nuclear weapon did as ordered, then called Ketov in his cabin. He informed his commander that Agafonov intended to fire the weapon. Ketov jumped from his bunk and ran toward the CC. He shot through the hatch and stood face-to-face with Agafonov.

They argued for several long minutes, then Ketov said, "Do you really want to go down in history as the man who started World War III?"

The brigade commander lowered his eyes. With his jaw clinched tight, he looked like a boiling kettle about to blow. He raised his chin and glared at Captain Ketov. The steam evaporated, and Agafonov relinquished command. Ketov relieved the commander and ordered the special weapons officer to stand down.

After taking back control of his boat, Ketov recalled the directives issued by Kennedy and the Americans that submarines near Cuba should surface and head east at a slow speed. He did not intend to surface, but he instead instructed the OSNAZ group in radio to broadcast a message to the Americans that B-4 would comply with the instructions. The ASW aircraft stopped dropping explosive charges.

An hour later, Ketov cheated. With ninety minutes remaining until dark, he made a radical course change, took the boat beneath the thermal layer, and increased speed. As night unfolded in the Caribbean, Ketov slowly brought the boat shallow. Acoustics listened but heard nothing. Ketov popped up a periscope but saw no planes. Nakat ESM detected no nearby radar. Satisfied that no Americans stood poised to harass them, Ketov brought B-4 to snorting depth and recharged the batteries.

He spent the next ten days dodging the Americans as they made low passes across the ocean's surface to snag him via magnetic or diesel exhaust detection. They never found him. On November 12, he received new orders to head 500 nautical miles north to an area known as the main U.S. Navy ASW line south of Newfoundland. Over the next few days, Ketov noticed that the American ASW forces seemed to follow him north along his track. He saw long-range reconnaissance aircraft flying at high altitudes and P2Vs thrumming along at low or medium heights. Sometimes the Nakat picked up three pairs of P2Vs circling around B-4's previous course. Ketov remained puzzled. Although they still snorted at night, they went silent during the day. So how was it that the Americans could follow him so accurately?

THE CREW OF B-36 CELEBRATED THE Soviet October revolution on November 7 by taking ocean water showers in the conning tower. Captain Dubivko decided to join them when First Officer Kopeikin wrinkled his nose in the CC as Dubivko walked past. When another submariner can no longer stand your stench, it's time for soap. He drenched himself in the cool, refreshing waters of the Sargasso Sea pouring through the showerhead in the sail and smiled as the water covered his tired body. Finally, he thought, the demons have stopped harassing me. Dubivko's smile vanished when a sailor interrupted his shower to deliver an urgent message.

Engineer Captain Potapov reported that the Compartment Five group commander, Senior Engineer Lieutenant Kobyakov, having become exhausted after enduring high engine-room temperatures for the past several weeks, failed to properly purge the air intakes on the two running diesels. Both engines seized up. That left B-36 with one working engine. Dubivko was now aware of what happened to Shumkov on B-130, and he harbored embarrassing visions of being towed back to port, or worse, of being forced to endure another babysitting escort by an American destroyer.

He dried off and clambered down to Compartment Five, where Potapov and Kobyakov updated him on the two dead engines. The two engineers spent the next few days cannibalizing one broken engine to try to get the other back online. Running with one engine made Dubivko's job of avoiding detection an almost impossible task. Still harassed by ASW aircraft, this debilitation forced him to keep B-36 stationary in the water while charging the batteries, then run slow on the electric engine during the day.

The OSNAZ group discovered that the helicopter carrier USS *Thetis Bay* (LPH-6) occupied the center of the deployment area, and her helicopters also exhibited an uncanny ability to follow B-36's track. Dubivko decided to forgo permission from the main navy HQ and moved B-36 200 miles south. Soon after, he received orders to return home. Hamstrung by two dead engines, B-36 could go no faster than six or seven knots, and Dubivko figured he would not see his wife again until late January, if at all. Then, as B-36 entered the Norway Sea, Potapov reported that he and Kobyakov had repaired one engine, so they now had two working. Dubivko sat on his bunk and let his shoulders relax after he heard the news, certain that the demons were finally departing. Unfortunately, they were only taking a short sabbatical.

When B-36 neared Iceland, bad luck prevailed again. B-36 ran out of fuel. Dubivko threw his arms skyward in frustration when Potapov gave him the dismal news. Potapov then suggested that they could mix oil with water to keep at least one engine limping along. That worked for a time, and Dubivko's spirits soared when a fuel tanker came to the rescue near the Lofoten Islands just off Norway. As luck would have it, black swells prevented the tanker from passing a refueling hose to the submarine. Disheartened once again, Dubivko dove B-36 and ran slow on the electric motor. Just when he thought he'd finally lost the battle with lady luck, his main chemist, Yuri Klimov, devised a way to use seawater to charge the batteries enough to temporarily power the electric motor. Now, if they could only make it home before the motor gave out.

BY MID-NOVEMBER, PRESIDENT KENNEDY AND PREMIER Khrushchev hammered out a deal that specified removal of all Soviet offensive weapons from Cuba. These included the nuclear missiles as well as the nest of IL-28 bombers. Days passed, but nothing happened. Members of ExComm met again to discuss options to deal with Khrushchev's obvious noncompliance with the agreement, including more blockade pressure and sending a strike force to take out the IL-28s.

Khrushchev responded to Kennedy's queries by saying that Russia intended to remove the IL-28s, but not right away. He stated that "it can be done in two to three months." Upset by this, Kennedy told his advisers that "we might say the whole deal is off and withdraw our no invasion pledge and harass them generally."

On November 16, the largest amphibious landing since World War II unfolded at Onslow Beach, North Carolina. The two-day exercise was a full-scale dress rehearsal for an invasion of Cuba. Six marine battalion landing teams, four assault boats, and two helicopter assault carriers were involved. Some 100,000 army soldiers, 40,000 marines, and 14,500 paratroopers supported by 550 aircraft and 180 ships stood ready to invade Cuba.

On November 20, Kennedy held a press conference to announce that Castro had finally agreed to remove the IL-28 bombers within thirty days. The following day, the president issued a proclamation to end the naval quarantine of Cuba.

Although tensions eased by mid-December between the two superpowers, the events that transpired during the Cuban Missile Crisis plunged the world into a Cold War that escalated to the brink of nuclear destruction several more times over the next two decades. U.S. and Soviet submarines were the common element each time.

IN LATE DECEMBER, WHEN B-36 DOCKED at the pier in Sayda Bay, only Brigade Chief of Staff Second Captain Arkhipov came to meet the boat. Dubivko didn't care. He was just glad that the mission was over and that the evil spirits that had tortured him while under way could no longer make his life a living hell. What happened next, however, made him wish he'd never returned home. He was expecting a hero's welcome. After all, he and his crew survived the longest excursion into enemy territorial waters since World War II. They endured horrendous conditions, impossible odds, torturous weather, and a host of debilitating mechanical failures. What he and the other three captains involved in Operation Kama received turned out to be just the opposite of what they deserved and is best described in Captain Dubivko's own words, edited for readability by the author:

> *On the next day after we returned to Gadzhiyevo, I was summoned to the Commission of the main Navy Headquarters to analyze the trip. Rear Admiral P. K. Ivanov—head of the Department of Combat Preparedness—headed the Commission. Unfortunately, the work of the Commission on analyzing the actions of the submarines in extraordinary conditions, according to practice established at that time, was aimed exclusively at uncovering violations of orders, documents, or instructions by the commander or by personnel. . . . We were accused of violating secrecy, failing to abide by NIS-58 instructions for submarine forces while trying to avoid U.S. anti-submarine aircraft and ships. They did not take into account the fact that if we abided by NIS-58 . . . then our submarines would never have arrived at our final destination.*
>
> *. . . The conditions in which our submarine brigade had to work were so difficult in the tactical, moral and physiological sense, that we were happy to have returned from that trip alive and healthy. This feeling was confirmed by Military Council of the Northern Fleet Rear Admiral F. Ya. Sizov when he said, "We did not expect you to come back alive. . . ."*
>
> *Some time later, all the submarine commanders who took part in the trip were summoned to Moscow for personal reports to the USSR Defense Minister. . . . Marshal [Andrei] Grechko refused to listen to my report on the problems and difficulties of the trip. . . . The only thing he understood was that we violated the secrecy requirements, were discovered by the Americans, and that for some time the enemy maintained contact on our submarines. . . . He offered*

*his opinion in the statement, "I would have better sunk than
come to the surface. . . ."*

*[Our] experience, gained through hardships, should
have been culled from all the submarines, analyzed, and
shared with other Fleets. To my regret, the Operations Department of the Navy General Staff failed in that endeavor.*

*I would like to once again express my appreciation and
gratitude to the entire crew of B-36 submarine for exhibiting
high standards of resistance and valor throughout the trip,
and for helping me, as their commander, to affirm our dedication to the Soviet Navy and to our Motherland.*

—CAPTAIN FIRST RANK, RETIRED, ALEKSEI DUBIVKO

ALTHOUGH PRESIDENT KENNEDY AND DEFENSE SECRETARY McNamara retained suspicions that the four *Foxtrot* submarines that converged on Cuba in October 1962 might be carrying nuclear torpedoes,
the truth was not revealed to the world until 1995, more than three
years after the Soviet Union collapsed. Not until then did we learn that
the Soviets shipped 161 nuclear warheads to Cuba. Ninety of those
were tactical, which would have killed tens of thousands of U.S. soldiers had Kennedy given the order to invade.

I spent my entire adulthood believing that President Kennedy
saved the world from nuclear winter during the Cuban Missile Crisis.
Certainly, he and the members of ExComm were instrumental in forcing
Khrushchev to back down at a crucial time during the brewing conflict,
but the other heroes of this undeclared war were the commanders of the
four *Foxtrot* submarines. Each had the opportunity, when faced with disaster, to pull the trigger and start a war. Each could have taken a dozen
or more American ships with them into the smoldering center of a mushroom cloud, and each could have propelled the world toward a devas-
tating October that would have changed the course of human history.
However, each made the right decision in the end, and those of us who
survived that time owe them our gratitude and perhaps our lives.

The events that transpired in the fall of 1962 brought the world to
the edge of nuclear winter not once, but at least a half-dozen times. The
Cuban Missile Crisis, perhaps better stated the Cuban Submarine Cri-

sis, spurred the creation of the Moscow–Washington hotline to ensure immediate, direct communications between the superpowers in the event of future potential conflicts. Although critics claim that Kennedy's actions prior to the crisis—particularly those related to the Bay of Pigs incident—likely caused the escalation in the first place, most agree that the outcome propelled the United States to a more confident stance as an international superpower. Unfortunately, the crisis also fueled fear in military and political circles that chilled the Cold War further, leading to an eventual "keeping up with the Joneses" competition between the United States and the Soviet Union. Both sides markedly increased spending on offensive weaponry, military technology, and counterintelligence.

As for Boresight, my father commented many years later that perhaps the success of this program was a double-edged sword. On one side was the technology that enabled U.S. ASW ships and planes to locate the *Foxtrots* and compel all but one to the surface. On the other side was the fact that, in helping our boys do a better job of forcing the Soviet captains into a corner, we also pushed those submarine commanders to the cliff's edge. While staring at that precipice, with nuclear weapons armed and ready, officers on all four *Foxtrots* nearly annihilated dozens of U.S. ships. Such an action would have escalated the crisis into a full-scale nuclear war that neither side could win. Only calm heads and perhaps divine intervention kept dozens of American cities from becoming radioactive wastelands.

CHAPTER ELEVEN

Even castles made of sand, fall into the sea, eventually.
— Jimi Hendrix

T HE EARLY SIXTIES USHERED IN THE next evolution in underwater warfare with the advent of a revolutionary new type of nuclear fast-attack submarine. Named after the USS *Thresher* (SSN-593), boats of this class brought to life almost every item on a navy wish list for the perfect stealth boat. From sonar to fire control to weapons to electronic-snooping "spook" gear, *Thresher*'s integrated design allowed sailors to operate as a synchronized team—like an ant colony with an attitude. Sub drivers were ecstatic. Faster, deeper diving, more quiet and maneuverable, these boats were the Ferraris of the fleet, which spurred fierce competition between skippers to snag a set of keys to one of the navy's hottest new vehicles.

Lieutenant Commander John Wesley Harvey did just that. Born into the Bronx backstreets of New York, he entered the U.S. Naval Academy in 1946. Harvey first served in the skimmer fleet aboard the aircraft carrier *Coral Sea* (CVB-43) before switching to submarines in 1952. He earned his submarine dolphins on the diesel boat *Sea Robin* (SS-407) before transferring to the USS *Nautilus* in July 1955. After a few runs on the *Seadragon* (SS-194), Harvey took command of the USS *Thresher* in January 1963. Four months later, he went to sea for the last time.

BUNDLED IN THICK NEOPRENE, SITTING ON the hard deck of the USS *Preserver* (ARS-8), Sonarman First Class Nihil Smith could no longer feel his cheeks. A harsh Atlantic wind swept its cold anger across the deck of the 200-foot-long rescue ship and swirled above an ocean full of death. Smith turned his frozen face to the right. The head diver flashed a signal. Smith stood and stepped toward the back of the *Preserver,* a double-tank "twin-ninety" scuba rig mounted on his back. As he walked, his navy diver booties scraped across the rust on the ship's deck, and the dense smell of the *Preserver*'s engine exhaust hurried him toward the edge.

Metal chains clanked and winches groaned as the *Preserver*'s crew went about their daily tasks. The *Preserver* first saw the light of day in San Francisco in 1944 and served as a support ship in the Pacific—a platform for navy divers conducting salvage and rescue operations in such exotic places as Saipan, New Guinea, Guam, Okinawa, and Bikini Atoll. In May 1962, navy divers on the *Preserver* helped recover Mercury astronaut Scott Carpenter after his *Aurora 7* capsule splashed down in the Atlantic. A year later, in early May 1963, *Preserver*'s captain got the call he hoped would never come.

The *Preserver* caught up with the deep-diving submersible USS *Trieste* (DSV-0) at the Boston Naval Shipyard, and the two vessels headed toward a location some 200 nautical miles east of Cape Cod, Massachusetts. The Swiss-designed *Trieste* "deep boat" bathyscaphe held a crew of two and could descend to the darkest part of any ocean on earth—nearly seven miles deep. Auguste Piccard, her Swiss designer, envisioned grandiose missions of scientific discovery for the sixty-foot-long submarine and named his 1953 creation after a quaint Italian seaport. Like any father, he probably knew his daughter might someday be called upon to search for the remains of a disaster.

That day arrived when, for the first time in the *Trieste*'s short ten-year life, she became a reluctant detective hunting for bodies. These cadavers, however, were not hidden in back alleys or buried under New Jersey dirt. They lay scattered across the bottom of a dark ocean more than 8,000 feet deep. In such a foreboding place, the odds of finding even the slightest clue were somewhere between zero and "worse than Vegas."

Nihil Smith stepped off the *Preserver*'s fantail and climbed onto the small skiff—the Boston Whaler with an outboard motor. He took a seat next to another diver as the *Preserver*'s crew lowered the dive boat into the water. A sailor on the skiff started the outboard, and the Boston

Whaler moved slowly away from the ship. Smith glanced back at the *Preserver.* Not a large vessel, the ARS-8 had two masts, a small superstructure, and a thirty-nine-foot-wide flat deck at the stern. She carried just two 40 mm AA gun mounts and four .50 caliber machine guns, but nobody expected her to fight any battles.

The skiff reached the dive spot, and Smith stood up. He watched his dive buddy step over the gunwales and splash into the blue. Following suit, Smith seated his scuba regulator, planted his right palm flat against his face mask and mouthpiece and jumped into the ocean.

Despite the thickness of his wet suit, the cold still launched his testicles into his throat. He sucked in several short breaths, forced his mind back to calm, and swam toward his dive buddy. The two met, traded "okay" hand signals, then looked down. Air bubbles circulated around Smith's fins, but they did not come from his twin-nineties. These bubbles ascended from a white submarine 100 feet below, on its way up from 8,000 feet deep. He bled air from his buoyancy vest and descended toward the mini-sub. As he cleared his ears to equalize with the ocean pressure, his eyes focused between his knees. Two vents mounted to the deck of the *Trieste* stared back.

To Smith, the strange twelve-foot-wide craft below looked nothing like a typical submarine. Instead, the *Trieste* resembled something from *Popular Science* magazine with her stump of a conning tower, flattened nose and heel, and round observation gondola bulging from her belly. The vessel's operators climbed down into the gondola through an entrance tunnel connecting to a hatch in the stubby sail.

The *Trieste*'s pressure sphere offered only enough space for two people, keeping them alive with independent life support provided by a closed-circuit rebreather system not too unlike the ones used by astronauts. Two thirty-eight-cubic-foot oxygen cylinders pumped in air, and canisters of soda lime scrubbed carbon dioxide from the air. No diesel engines ran inside the *Trieste*'s hull. Power came from two electric propulsion motors mounted externally and pressure-compensated to withstand the intense forces of the deep. Twelve-volt electric car batteries powered these motors from large boxes located on the after walking deck of the *Trieste*. Each box housed a dozen batteries. After every dive, all of the *Trieste*'s batteries, including the silver cell ones used inside the submersible, needed recharging.

Smith neared the curved white hull and grabbed onto an eyehook. His dive buddy did likewise. Together they rounded the side and swam toward the "grappling" mechanical arms underneath, near the observation sphere. A pair of eyes, inside the *Trieste,* stared through the single tapered cone-shaped block of Plexiglas mounted on the sphere. Smith gave the operator inside a signal, and an instant later, bright light from a set of quartz arc bulbs lit up the black ocean.

Smith could now see the bottom portion of the forward pellet ballast hopper protruding from the hull just in front of the sphere. He knew that inside the silo—a round cylinder about three feet in diameter and seven feet tall—nine tons of magnetic iron pellets were used as ballast. An identical hopper sat a few feet behind the observation gondola. Using pellets versus water helped increase descent and ascent speeds, given that intense water pressures down deep did not allow for air-filled ballast tanks.

The pellets, which filled each hopper, resembled oversized BBs. Their weight pulled the bathyscaphe down toward the bottom. When the crew cut off power to large electromagnets attached to the hoppers, the pellets dropped to the ocean floor, and the *Trieste* shot toward the surface. The crew refilled the ballast "shot tubs" from stacks of twenty-five-pound pellet bags lashed to the deck of the *Preserver.*

Smith kicked his fins and swam toward the mechanical arms. His dive buddy moved alongside and pointed. Smith followed the diver's finger and shifted his eyes downward. The *Trieste*'s bright arc lights reflected off something shiny. A short pipe. Smith's heart skipped a beat, and he wondered if this was the proof they had been searching for. He pulled the pipe free and brought it to the surface. Back on board the *Preserver,* experts examined the bent piece of metal. In a small compartment filled with navy personnel and scientific experts, Smith watched as one expert took out a magnifying glass and focused on the side of the pipe. The gaunt man with wire-rim glasses wrote down a part number, then flipped through a set of blueprints. The expert thumbed to a page, looked at the part number on the paper, focused back on the blueprint, and raised a stubby chin. Without speaking, the man nodded his head up and down.

Smith stepped outside the room and walked to the fantail. A cruel Atlantic gale swatted at the large ship as she bobbed up and down in the roiling swells. On the horizon, a bruised and swollen sky swallowed the last ounce of sunlight, leaving behind purple cotton clouds.

NIHIL SMITH WAS NO STRANGER TO the dangers faced by submariners. He joined the navy at the age of seventeen after striding into the recruiter's office in Azusa, California, in 1956. He had four uncles who had served their country during World War II; two of them went navy, and one became a submariner. That uncle often talked about the high morale and esprit de corps that could only be found on the boats. Inspired by his uncle's stories, Smith volunteered for subs and reported aboard the diesel boat USS *Trigger* (SS-564) in Norfolk, Virginia, in 1958. He earned another stripe when he passed the test for sonarman second class and a navy diver pin after graduating from the rigorous training course in Hawaii that same year. While qualifying in submarines during 1959, Smith came to relish the boat life and learned firsthand about the camaraderie that his uncle had talked about.

"Everybody depends upon everybody," his uncle said. "You're on the boats because you want to be, not because you're forced to be. Where else can you have that kind of togetherness? Where else can you depend on your mates because they all take pride in what they do?"

On the *Trigger,* Smith met just such a mate, a radioman named Joe Walski. Smith and "Ski" became close over the next year as they helped each other qualify in submarines. They earned their dolphins on the very same day. By the time Smith transferred off the *Trigger,* he and Walski were best friends. As Smith walked across the gangplank of his first boat for the last time, he felt like he was leaving a brother behind.

Smith reported to the diesel submarine USS *Barbel* (SS-580) in Kittery, Maine, and soon discovered that his "brother" Walski received a transfer to a nuclear-powered boat going through an overhaul, also in Maine. Smith was ecstatic. He raced over to the drydock and met Walski near the brow. They hugged, and Walski invited Smith for a tour of his boat. Smith slid down the ladder and whistled. Compared to the *Barbel,* Walski's new submarine seemed like a luxury hotel: wood paneling, shiny pipes, new Naugahyde-covered benches, and all the latest equipment. Walski spent the next hour dangling a carrot in front of Smith, using every angle to try to convince him to leave the *Barbel* and request a transfer to this newer SSN, known as the USS *Thresher.* Smith almost conceded but eventually declined, quoting the saying "Diesel boats forever!"

Sad but certain that he'd made the right decision, Smith sauntered back to the *Barbel*. Walski went to sea on his youthful nuke boat, while Smith rode the *Barbel* down to New London, Connecticut. Their respective schedules made visiting each other impossible. Smith then rode with the *Barbel* through the Panama Canal to her new home port in San Diego. There he received orders to the Naval Electronics Laboratory and was assigned to the research vessel *Trieste*, which had just returned from making the world's deepest dive in the Marianas Trench.

When Smith completed second class diver training in March 1963, he called to tell his quasi brother about the accomplishment. Walski was excited about the news but couldn't talk long, as the *Thresher* was preparing for upcoming sea trials. They vowed to make time to visit each other again, perhaps sometime later that year. Neither knew that this was a promise that could not be kept.

The lead ship of her class and the pride of Portsmouth Naval Shipyard, the USS *Thresher* launched on July 9, 1960. She underwent sea trials throughout '61 and '62, testing a host of modern and complex systems and weapons. After she arrived in San Juan, Puerto Rico, on November 2, 1961, the *Thresher*'s nuclear-trained "nukes" shut down the reactor. Equipment failures and mistakes combined to create serious overheating in the engineering spaces that resulted in a partial evacuation of the crew. This event became the first omen signaling the *Thresher*'s eventual demise.

The second portent occurred while the submarine sat moored at Port Canaveral, Florida. A tugboat accidentally slammed into the *Thresher*'s side and damaged a ballast tank. After repairs at Groton, Connecticut, the *Thresher* headed south to conduct more tests and trials near Key West, Florida. Thereafter she cruised up to Kittery, Maine, and stayed in the dockyard for refurbishment until the spring of 1963.

After her refit, with Radioman First Class Joe Walski on board, on April 9, 1963, the *Thresher* pulled away from the pier. Lieutenant Commander Harvey informed his crew that they were about to conduct rigorous postoverhaul sea trials. The submarine rescue ship USS *Skylark* (ASR-20) accompanied the *Thresher* to the trial area off Cape Cod. On the morning of April 10, 1963, *Thresher* started the first of her deep-diving tests. When the submarine neared her test depth of 1,300 feet, Lieutenant Commander Harvey called the *Skylark* on the Gertrude underwater communicator. The message came through garbled, but what could be deciphered indicated trouble. The words "minor difficulties, have positive up-angle, attempting to blow" were the last heard from the ill-fated submarine. She died that day along with 129 officers, crewmen, and military/civilian technicians.

No one knew for certain what happened to the *Thresher* until after Nihil Smith wrestled that small pipe, brought up from 8,400 feet deep, away from the grappling arms of the *Trieste*. The two-man crew of the bathyscaphe had previously spent weeks searching with no luck. Large grids were set up using color-coded markers and numbers for each search section. Every morning, the *Trieste* sank to the bottom and illuminated the ocean floor. She spent all day covering the grid coordinates, her crew hoping to find the remnants of the *Thresher*. They found nothing save a yellow boot that resembled part of a "canary suit" worn by nuclear personnel during reactor emergencies—not enough to validate *Thresher* as the source.

Each day Smith and the other divers prepped the 35 mm cameras, filled the ballast tubs with pellets, tested the radios, cleared debris from the *Trieste*'s hull, washed her windows, and "checked her oil." Then they waited. Hours later, usually near sunset, they suited up and prepared to meet the deep-diving vessel when she came back up. Those monitoring the *Trieste*'s ascent called out her depth, having tracked the submersible via hydrophones. When she neared one hundred feet, Smith and another diver climbed into the Boston Whaler and went out to meet her.

Their work didn't end there. At the completion of each day's dive, crews on the *Trieste* and *Preserver* prepared for another dawn and descent into the abyss. They hauled power cables attached to floats over to the *Trieste* to recharge the lead acid batteries. They wrestled with a high-pressure air line to charge up the antichamber blow system, which expelled air from an area near the sphere to allow the *Trieste*'s crew to exit the submersible after surfacing. While some of the crew processed the film taken during the day's search, others replaced oxygen cylinders in the sphere, along with CO_2 canisters and silver cell batteries. Finally, they refilled the pellet ballast hoppers from the twenty-five-pound bags found on deck and checked and rechecked that everything was ready to go again.

Preparations usually concluded around midnight, giving everyone less than six hours of sleep before the 6:00 A.M. predive check. That's when Smith suited up and stood ready to hit the water by 7:00 A.M. The *Trieste* then disappeared again under the waves while the support crew waited. Days and weeks went by. The *Trieste* occasionally spotted what looked like a debris field but never brought up any solid evidence, until they found the pipe.

After that, over the course of many months, the *Trieste* located parts of the *Thresher* strewn across a 160,000-square-yard area. Every day, Smith watched as more pieces were salvaged, all the while painfully aware that his best friend's body lay somewhere in the debris. First they found the stern planes, followed by the fairwater planes. These append-ages once looked and functioned much like wings on an airplane to move the boat up and down in the water with the grace of a dolphin. Now they were bent and twisted backward in grotesque shapes that resembled the branches of a dead tree.

As an orange-red sun sank into the sea and scattered tentacles of dying light across the wave tops, Nihil Smith stood in silence on the deck of the *Preserver* and thought about those final minutes. He wiped away a tear and imagined the look of shock on the faces of the *Thresher*'s crew as a wall of water filled the boat. He pictured the sub-marine imploding at 2,000 feet deep and saw the terror-filled eyes of his best friend, Joe Walski, as frigid salt water filled his lungs. To this day, Smith is still haunted by that nightmare.

PHOTOGRAPHS OF THE FALLEN *THRESHER,* ALONG with recovered parts, allowed a special court of inquiry to conclude that the submarine suffered a joint failure in the saltwater piping system. A shipyard worker connected the joint using silver brazing instead of a standard weld. The *Thresher*'s loss prompted the SubSafe program, which dictated a higher degree of quality control for all U.S. submarines. This decreased the quantity of submarines originally planned for, but increased safety and reliability.

On the other side of the world, still stung by the Cuban Missile

Crisis, Admiral Sergei Gorshkov pushed his navy in the opposite direction. The Soviets sacrificed quality for quantity, producing an average of seven nuclear and six diesel boats per year, which resulted in a spate of accidents. K-8 had a reactor coolant leak in October 1960. In July 1961, eight men died on K-19, a *Hotel*-class missile boat, when her reactor overheated. B-37 endured an internal torpedo explosion in January 1962, and intelligence reports indicated that the *November*-class K-3, the Soviet's first nuclear sub, continued to have leaks in her steam generators, along with other persistent reactor issues for years. Despite the problems, Secretary of Defense Robert McNamara expressed constant concerns about Soviet submarine escalation and the need to accurately locate those boats.

That concern fueled McNamara's desire to insert himself into the middle of the Boresight program. He contacted Jack Kaye at the NSA and informed the commander that A22 Desk now reported directly to him. McNamara showed up at staff meetings and asked a million questions. Some liked his inquisitive, "need to know" manner, and some didn't. When William J. Reed first met McNamara at the White House during the Boresight briefing with President Kennedy, he leaned more toward the "didn't like" camp. McNamara's hard-charging style seemed confrontational, especially when the secretary questioned him to death over minute technical details about Boresight. After McNamara got involved with the Boresight program, and Reed started working with him on a more frequent basis, his opinion of the man changed.

Those who knew Robert Strange McNamara sometimes said that his middle name fit like a glove. Possessing the number-crunching mind of a computer, the bull-nosed personality of General George Patton, and the diplomacy of Henry Kissinger, McNamara knew what he wanted and "damned the torpedoes" until he achieved his goals. One of those quests entailed the worldwide proliferation of Boresight. McNamara had witnessed firsthand the effectiveness of the system and its ability to diffuse high-seas confrontations by exposing the Soviet's undersea predators. As an insurance package, he insisted on equipping several more GRD-6 stations with Boresight technology while building more Wullenweber sites under Project Bulls Eye to improve accuracy.

Having been battle tested during the Cuban Missile Crisis, and in light of Soviet submarine advances, Boresight quickly became a hot property. McNamara propelled the program to a lofty number-two status, subordinate only to the Polaris missile program, which allowed ballistic nuclear submarines, like the USS *George Washington*, to lob radioactive warheads from more than 1,000 miles away.

Employing Doberman pinscher diplomacy, McNamara pressed NATO allies to allow inspections of select real estate abroad to ascertain the best locations for Wullenweber elephant cages. In addition to already operating installations at Edzell, Scotland, and Hanza, Japan, projected foreign locations included Galeta Island, Panama; Rota, Spain; and Sabana Seca, Puerto Rico. Skaggs Island, California, remained the only functional North American facility with construction planned for Adak, Alaska; Homestead, Florida (replacing the GRD-6); Imperial Beach, California; Marietta, Washington; Northwest, Virginia (also replacing the GRD-6); Wahiawa, Hawaii; Winter Harbor, Maine; and the Pacific island of Guam.

Locations were not picked at random. The geometry involved in optimizing direction finding required math degrees to calculate prime locations. These equations also dictated that, ideally, stations should be spaced 120 degrees apart—or about one-third of a compass circle away from one another. This spacing allowed for more accurate bearing hits on targets.

Not all of the "Fred Ten" Wullenweber arrays built by the navy were used for HFDF operations, but most were. The Air Force built similar Wullenweber "Flare Nine" antennas at other locations, including Karamürsel, Turkey, but none were intended for HFDF use. Navy personnel borrowed some of these arrays for Boresight operations where those locations were suitable, or where signal intelligence requirements, such as transmitter fingerprinting, could best be met.

For Reed and the A22 team at NSA, the remainder of 1962 and most of 1963 entailed Boresight system refinement and expansion. Although McNamara pushed for rapid deployment, he did so with an eye toward efficiency and economy. Still, for the most part, Boresight equipment requisitions could override virtually any other project save Polaris. Backups to backups were ordered and installed. Contractors and subcontractors, such as Sanders Associates and ITT Federal Systems, had near-blank checks. Components used had to be properly interfaced with one another and with existing equipment. Incompatibilities, glitches, and kinks had to be ironed out in time for McNamara's QA inspections. As was his modus operandi, McNamara frequently micromanaged many of the processes until he was satisfied that his subordinates could live up to his idea of perfection. Some found this annoying, but Reed wished that every public servant kept such a close eye on efficiency and accountability. He felt that America's national debt wouldn't be so high if that were the case.

A Northern California native and son of a shoe sales manager, Robert McNamara graduated from UC Berkeley in 1937 and earned an MBA from Harvard two years later. He gained his hard-nosed accounting skills by working at Price Waterhouse in San Francisco until late 1940, whereupon he returned to Harvard to work as an assistant professor. He served as a captain in the Army Air Force during World War II, then joined Ford Motor Company, where he earned the title of "whiz kid" by helping implement modern management control systems. Hard work and brilliance earned him the presidency of Ford in November 1960, which proved short-lived after President Kennedy recruited him as secretary of defense in 1961.

More than two years later, in the fall of 1963, after Reed received a promotion to lieutenant junior grade (LTJG) and became the primary technical liaison for the A22 Desk at NSA, McNamara requested a meeting at his office in the Pentagon. They spent hours peering through magnifying glasses at potential map locations for Boresight stations.

Halfway through the meeting, McNamara leaned over Reed's shoulder, pointed a long finger at the toe of Italy on the map, and said, "There. Build one there."

Although the location looked quite suitable from a geometric standpoint, Reed said, "Mr. Secretary, I don't believe there's a decent way to get to and from that location. That's all nearly inaccessible mountain country."

McNamara waved a dismissive hand. "Planes, trains, buses, or horses. There's always a way in or out. Go take a look and report back."

Reed laughed. McNamara did not. Reed's face turned serious. He cleared his throat and said, "Yes, sir, Mr. Secretary. I'll get right on that."

Weeks later, after traversing political land mines with the Italian government, Reed rented a burro and a guide and rode for two days across the worst terrain ever constructed by God. He returned from the trip with a sore ass, a grumpy demeanor, and a no-go report for McNa-mara. Getting supplies and construction materials in and out of that location would be impossible at worst and a nightmare at best. McNa-mara finally conceded and handed Reed a bottle of aspirin. Reed fig-ured that was about as close to an apology as he'd ever get.

Engineers at the Naval Research Laboratory made their own con-tributions to the Boresight and Bulls Eye programs via a new digital computer with a multiprocessor and magnetic core memory called the AN/GYK-3. Originally used in the NORAD (North American Aero-space Defense Command) early warning project, the GYK represented the edge of the edge in computing technology. NRL contracted with Bur-roughs Corporation in 1961 to develop a prototype that turned out to be every bit as impressive as touted by the designers. Burroughs delivered the computer in 1962, which offered the ability to better track moving targets as compared to other systems that were more suitable for man-aging production inventories.

Judged by today's standards, the computer boasted primitive com-puting power, capable of handling no more than 625 bytes and around one hundred records per minute. Outputs were sent to CRT displays and high-speed printers. Today we take for granted that our mobile phones can manage millions of bytes of data and breeze through thou-sands of tasks in a second. Back then such staggering numbers were considered science fiction. These early-generation processors could, how-ever, work on up to twenty tasks at the same time. Despite being slower than a geriatric snail, the GYK's multitasking capabilities created beam-ing grins on the faces of design engineers.

Based on recommendations he'd received from the engineers at Sanders Associates, Reed contacted Bruce Wald and others at NRL to discuss how they might use the GYK computer for Boresight. A com-puter could help automate the previously manual and tedious task of plotting fixes to targets by pulling strings of yarn across a compass rose. With the GYK computer, that process could be completed faster and with far greater accuracy. Later that year, the Wullenweber Bulls Eye and GYK computer-infused Boresight projects eventually blended into a single program called Clarinet Bulls Eye, which later changed to Clas-sic Bulls Eye to conform to standard naming conventions.

With the Cuban Missile Crisis now over, and with the leaves of autumn turning fields of green into swirls of auburn and gold, Reed at first thought he might be able to find time to relax with his family in their two-story Maryland home. But across the ocean, a smitten enemy escalated plans to leapfrog their adversaries by building an arsenal of underwater weapons that could plummet the world back into the Dark Ages. That fact eventually led Reed to the pinnacle of his career, followed by an end that no one expected, least of all him.

THE BEGINNING OF THAT END STARTED when Captain Kaye, at Mc-Namara's bidding, promoted Reed to one of the most prestigious positions in A22: head of field operations. In short, he became NSA's head Boresight/Bulls Eye troubleshooter. The job entailed not only frequent trips to all stations in the Atlantic and Pacific, but also additional travel to such locations at Cheltenham, England, for meetings with the heads of NATO intelligence departments and with various international R&D laboratories. Reed admitted years later that pride and a strong sense of duty prevented him from admitting the truth to his boss—that such an assignment required three people, not one. The responsibility and relentless travel eventually took its toll, but in the meantime, Reed carried a pocketful of government "paper money" travel vouchers. He also carried a gun.

Any information about Boresight technology, from concept to equipment to operational procedures, received top-secret "eyes only" classifications. Sending documents, training materials, or tapes with burst signal examples by mail was out of the question. Top-secret codebooks, plans, instruction manuals, operating guidelines, and protocols needed to be hand-delivered by armed courier to dozens of HFDF stations worldwide. Carrying a briefcase handcuffed to his wrist, Reed and others like him traveled the globe with nervous fingers touching the trigger of concealed weapons. Every shadow hid a potential KGB agent, and every friendly smile seemed dubious.

Wearing civilian clothes, unencumbered by his rank as compared to others more senior, accepted into the intelligence community alongside CIA spies and NSA operatives, Reed flew to Norway to assist with Boresight installation and training for an antenna system located in Vadsø. He departed for London, thence to Norway in late November 1963. When he arrived in Oslo, several Norwegian intelligence officials met him at the airport. On the way to his hotel, they discussed the strategic importance of having direction-finding systems at this station, given that, if someone threw a rock about a hundred miles from Vadsø, they'd hit Murmansk, one of the Soviet Union's most important and widely used submarine naval bases.

The navy's main concern centered around the older *Golf-* and *Hotel*-class ballistic missile boats, and the new *Echo I, Echo II,* and *Juliet* guided missile submarines, not to mention the nuclear-powered *November*-class subs. While SOSUS still had a chance of finding the *Golf, Hotel,* and *Juliet* diesel subs when they snorkeled, *Echos* and *Novembers* sported nuclear reactors and sound levels just above the "quiet sub" classification of less than 150 dB. That decibel level actually rivaled the USS *George Washington:* not exactly quiet, but not real easy to find with current sonar systems. Moreover, the *Echo II* could now carry newer *Shaddock* SS-N-3 antiship cruise missiles with ranges of up to 245 nautical miles.

The *November*-class actually trumped the USS *Nautilus* and *Seawolf* on the capability front, and to prove the point, the Soviets sent the *Leninsky Komsomol* to the North Pole. She arrived in September 1963. With the sting of the Cuban Missile Crisis not yet dissipated, all these Soviet submarine advancements kept the U.S Navy awake at night. Finding and tracking these boats, especially the *Hotels, Echos,* and *Novembers,* referred to as the "HENs" by NATO, remained paramount and prompted Reed's visit to Norway.

Starving and tired, Reed arrived at his hotel in the evening, showered, shaved, and found the restaurant downstairs. There he met a collection of drunken Norwegian army officers who'd just survived a secret ski-troop exercise. Since a broad patch of Norway's border carved a line down Russia's left flank, Reed figured the "exercise" probably took place near Hesseng or Kirkenes. While the soldiers hooted and howled in a language Reed had learned only sparsely, a waitress appeared. Through Reed's partial Norwegian and her broken English, he managed to order a steak. Thirty minutes later, his food had yet to appear, and a sudden hush fell over the room. The Norwegian soldiers stopped laughing and yelling. Their voices descended to whispers, and their faces registered shock and concern. Other patrons stopped eating. Their looks and tones mirrored the army officers. Confused, Reed flagged his waitress.

Tears streamed down the girl's face as she approached Reed's table. In her broken English she said, "They shoot him."

"Shoot who?" Reed said.

Having overheard Reed speaking English, an American couple darted over from another table. A rotund man, clutching his wife's hand, said, "I speak a little Norwegian. Have you heard the news?"

"What news?" Reed said.

The man looked at his wife. His eyes misted. She started sobbing. He held her in his arms and said something unintelligible. As Reed watched them, others in the room started crying.

"What's going on?" Reed asked the man. "Why is everyone so sad?"

The man wiped his cheeks and said, "President Kennedy's been shot. He died a few minutes ago."

Reed's stomach knotted, and his hunger vanished. A million questions ran through his mind. Had the Soviets finally retaliated in response to the crisis? Was the United States now at war with Russia? Were Joyce and the kids in danger? He had no answers and would not until morning, when he could contact the U.S. embassy. For now, he could think of only one appropriate thing to do. He grabbed a glass and walked over to the table occupied by the Norwegian army officers. He held up his empty glass, and one of the soldiers poured from a bottle. When Reed held his glass high, the officers stood from their table and did likewise. No words were spoken as the half-dozen military men offered a toast to a fallen hero.

REED RETURNED TO THE STATES AFTER helping the station in Norway optimize their Boresight systems. The following year, on July 5, 1964, carrying a concealed Webley revolver, he left JFK Airport bound for London. He couldn't recall the last time he'd actually fired his pistol at the range. Making time to do that hadn't been a priority. He'd never been shot at and figured the odds of that were just about nil—that is, until he got a call from the CIA, who cordially compelled him to report to Greece on the double.

When he arrived in Athens, a CIA operative met him at the airport and said he needed Reed for a sensitive mission. He said his name was Fred. No last name was given. When Reed inquired as to the nature of the operation, Fred said that the NSA was concerned that Soviet spies might have obtained several Boresight technical documents. Once the word *spies* fell into the sentence, the CIA took jurisdiction. As such, the mission entailed validating that Soviet or Soviet-friendly contacts had indeed absconded with top-secret documents, and if so, their orders were to prosecute the said suspects with appropriate action. When Reed asked for the definition of "appropriate action," Fred shrugged and said, "Dunno. We usually make that shit up on the fly."

Reed shook his head and slouched into the seat of the Mercedes. Fred pulled away from the curb at the airport and sped into downtown Athens. Night blanketed the city, and shimmering lights accented the city's marbled works of beauty. The Mercedes had that new car smell, overpowered now and then by Fred's aftershave.

"Why me?" Reed asked.

"We need your technical knowledge," Fred said.

"My technical knowledge? What, to catch a spy?"

"No," Fred said. "To validate that said contacts have obtained said documents."

"But if said contacts have indeed pilfered said documents," Reed said, "then who's to say that said contacts haven't already said something to someone or photographed said documents and have, say, already delivered the information?"

"Not my problem," Fred said. "I've been ordered to retrieve said documents from said contacts, have you validate their validity, and take appropriate action."

"Which appropriate action you'll be making up on the fly."

"Precisely," Fred said as he careened around a corner. Reed held onto the door handle and wondered if he'd live long enough to meet said contacts or examine the documents.

"Do we know where said contacts are located?" Reed asked.

"Yes." Fred said.

"Are you going to tell me?"

"They're in a Greek village some distance from here."

"Does this village have a name?"

"Yes."

"You don't say."

"No, I don't say."

"Fine," Reed said. "Enough said."

The CIA put Reed up in a local hotel for the night, and the next morning, Fred poured him into a black van and sped toward the Greek village. Nothing was said.

When they arrived, another operative named Bob, who'd arrived earlier in the Mercedes, took charge. He said they were waiting for the Greek authorities, as they were under orders to cooperate with the locals. Once the police officers arrived, his orders were to observe the unsubs, verify that they fit the description of said suspects, then take appropriate action. Like Fred, Bob did not define "appropriate action." Bob told Reed to stay in the Mercedes, and Reed offered no argument.

Thirty minutes later, a truck full of badge-brandishing, rifle-toting, macho Greeks rolled in. They stormed out of the truck, took up cover positions, and pointed guns. The tallest one approached Bob. The two spoke while pointing at a gin-joint restaurant a half-mile away. It was a dusty town with a scattering of cars, a few peasants milling about, and some thin dogs and chickens here and there. A stench hung in the air that reminded Reed of an Oklahoma barn.

When the conversation between the head guys ended, the wait began. An hour went by, then another. Nothing happened. Starving, Reed wondered how long the operation might last. A dark Volvo pulled in front of the restaurant. A half-dozen men slammed car doors and walked inside. The Greek cops raised their heads and ears. Reed noticed fingers twitching. While Bob peered through a pair of binoculars, the head Greek guy waved a hand in the air. His team of six ran screaming toward the restaurant with guns blazing. Bob and Fred dropped jaws, looked at each other, and went running after them, yelling at Reed to stay put in the Mercedes. Again, Reed offered no rebuttal.

The bad guys in the gin-joint restaurant started firing back through the windows as the Greeks approached. Reed grabbed a pair of binoculars from the front seat and pointed them through the window of the Mercedes. Two of the Greeks dropped onto the cracked road in pools of blood. Not believing his eyes, Reed watched one of the Greeks throw something through the restaurant window. If that's a grenade, he thought, there won't be any documents left to validate. The projectile spewed out smoke instead of shrapnel as the Greeks knocked down the door and ran inside. Bob and Fred went in after them, albeit with a little more caution. Binoculars glued to his eyes, Reed heard muffled gunshots and saw bright flashes through the windows. Then silence.

Long minutes passed. Reed scanned back and forth between the windows and the door. Nothing but smoke. No movement of any kind. Then a hand. Bob's hand. He stepped through the door and signaled for Reed to come over. With trepidation, Reed stepped from the car and walked the half-mile to the restaurant. A wave of nausea hit as he went past the fallen Greeks, eyes open and flies dancing about the blood. He stepped through the door of the restaurant, and his knees buckled. A half-dozen more bullet-torn bodies lay soaked in crimson. Smoke billowed around the room. Bob motioned Reed to a table. He opened a worn leather satchel and removed a few dozen pieces of paper, then handed them to Reed and asked him to validate the documents as Boresight-related or not. Reed examined the stack. He read each page, then reread them to be sure he hadn't missed anything.

"Well?" Bob asked.

"Goniometer," Reed said.

"Gonorrhea?" Bob repeated. Fred laughed. So did one of the Greeks.

"Not quite," Reed said. "These papers are technical specifications for a German goniometer—the same one we use in our Wullenweber Bulls Eye sites."

"So said contacts did have secret documents?" Bob asked.

"No," Reed said. "The goniometer design is two decades old. We use it in our arrays, but it's not related to the Boresight equipment. In fact, it's not even classified."

"Said document is not even classified?" Fred said.

"Not classified," Reed said.

"No shit," Bob said. He grabbed the stack of papers from Reed's hand, scanned his eyes across the dead bodies on the floor, and said, "It is now."

AFTER THE GREEK INCIDENT, SOMETHING CHANGED in Reed. The excitement and enthusiasm he once had in his work dimmed. Still, he maintained his competency and helped the Canadians enter the Boresight fold at Masset, Gander, and Newfoundland as part of the Canadian-U.S. Atlantic HFDF network. These sites were later upgraded to Wullenweber or similar arrays. By 1967, more than a dozen Wullenweber elephant cages were built or under construction around the world, and the navy enjoyed a continuous stream of tip-offs, flashes, and fixes on Soviet submarines. As for Reed, he decided to leave the program and take a less stressful post in San Diego.

Reflecting on this decision years later, he said that had he stayed in the Classic Bulls Eye program, he might have made full-bird captain one day, which is rare for an up-through-the-ranks Mustang officer. On the other hand, he probably would have sacrificed his sanity and maybe even his soul to gain the world. In hindsight, he did not regret walking away from his James Bond life and was proud to have served as he did, for in doing so, he paved the way for others to take the Boresight/Bulls Eye programs to new heights of stardom.

CHAPTER TWELVE

*If you reveal your secrets to the wind, you should not
blame the wind for revealing them to the trees.*
                                        —KAHLIL GIBRAN

O N DECEMBER 29, 1964, THE U.S. Navy made a huge
blunder by granting Petty Officer John Anthony Walker
a top-secret security clearance. While his background
appeared clean, a more thorough psychological screen-
ing may have revealed cracks in Walker's armor. Nevertheless, the navy
welcomed him as an R-Brancher and opened the proverbial intelligence
kimono. Walker now had access to the navy's most sensitive cryptographic material.

By April 1967, the navy assigned Walker to the Naval Communications Area Master Station in Norfolk, Virginia, a drab, windowless,
two-story building off Terminal Boulevard. At first glance, no passersby
would suspect that inside the facility, operators sent and received thousands of encoded messages every day to and from ships and submarines
operating in the Atlantic. Walker's responsibilities as a watch officer in
the message center included handling classified communications with
U.S. submarines, specifically reading and dispatching top-secret traffic
bouncing between the message center and on-station subs.

Sophisticated cryptographic machines and Teletypes lined the walls
of a small room in the message center, a few feet away from Walker's
desk. The machines clicked and hummed twenty-four hours a day as
messages were scrambled and unscrambled, traveling ship to shore and

shore to ship. The navy code-named its crypto system Orestes, which employed a device called the KW-7, an online, send/receive message encryption unit installed in shore stations and aboard ships and submarines. To send messages over a secure UHF Teletype circuit, a model 28 Teletype forwarded the prepared message to the KW-7. The unit keyed a UHF transmitter to send out the message. Given its wide use, the navy viewed the KW-7 as one of its most critical encryption devices, used for more than eighty percent of the messages sent and received by the entire U.S. fleet, and of utmost importance to submarine operations.

The KW-7 came in a small gray box with a panel covering three-fourths of the upper surface area bordered by a set of switches, dials, and lights across a black panel on the lower section. Original versions used wire cords to set the daily encryption key. A newer version came with a small square bulge on the gray cover panel to make room for rows of plugblock modules underneath. Various combinations of the plugblocks allowed for encryption key changes. R-Branchers had to set up the plugboard manually, which everyone hated. Daily keylists were provided to R-Branchers, who in turn changed plugs or block modules to alter the encryption. As such, keylists were considered more than just top secret; they were literally the keys to the hen house.

In December 1967, when John Walker began stealing keylists for the KW-7 and delivering them to the Soviets, he harbored no concerns that this action might result in the deaths of ninety-eight submariners within three months, and another ninety-nine submariners three months after that. Walker suppressed any guilt he might have felt by assuring himself that as valuable as the keylists might be, the documents he stole were only half of the puzzle needed by his Soviet handlers. Without an actual working KW-7 device, the keylists offered little value. Walker didn't account for Soviet cunning and skill, and therefore never imagined that America's largest adversary would obtain a working KW-7 device less than a month later.

On January, 11 1968, the spy ship USS *Pueblo* (AGER-2) sailed out of Sasebo, Japan, with orders to monitor Soviet naval activity around the Tsushima Strait. Their SIGINT mission entailed using Russian-speaking I-Branchers listening via ESM masts to gather intelligence from North Korean ship and shore platforms. For weeks prior to their departure, intercept operators at the CIA's Foreign Broadcast Information Service picked up threats coming from North Korea warning U.S. "espionage boats" to stay out of their territorial waters or suffer severe consequences. These transmission reports were never forwarded to the NSA.

While the *Pueblo* sat off the coast of North Korea in early January, not more than a whisper outside the international twelve-mile limit, the CIA intercepted more transmitted warnings, some pointed directly at the USS *Pueblo*. Again, the NSA was not informed, and the *Pueblo* remained on station, her deck coated with ice and her crew shivering in the bitter cold. In the distance, North Korean mountains, their black slopes covered with white, jutted toward a bleak sky. Inside the spy ship's thin metal bulkheads, a team of twenty-eight spooks rotated shifts in the tight SIGINT space. Racks of equipment hummed and blinked, including a KW-7 encryption device, typewriters, Teletypes, a WLR-1 intercept receiver, and 600 pounds of top-secret documents.

Although the *Pueblo* carried twenty-two weighted bags to toss these documents over the side, a small incinerator to turn them to ash, and two paper shredders to grind them into oblivion, the combination was not capable of destroying more than a small percentage of the secret material in the event of an emergency. As for the destruction of sensitive equipment, the ship carried no explosives—only sledgehammers and axes, which were no match for hundreds of pounds of steel-encased crypto gear. No U.S. destroyers patrolled nearby. No aircraft carriers and no submarines stood at the ready. No jet fighters sat fueled on runways in the event of an incident with the North Koreans. The NSA and the navy sent the *Pueblo* into the den of the lion alone, adorned with nothing more than a small knife and a loincloth.

On January 21, the lion stirred. A Soviet-made SO-I-class submarine chaser came within two miles of the U.S. ship. The following day, two *Lenta*-class DPRK fishing trawlers cruised within twenty-five yards—close enough for *Pueblo*'s sailors to stare into the eyes of the trawlers' crews. Commander Lloyd Bucher broke radio silence and tried to send a situation report, but the signal did not go through.

The next morning, on January 23, the North Korean island of Ung-do hid behind a dense mist some sixteen miles away. Inside the *Pueblo*'s SIGINT area, an R-Brancher jolted to attention as he detected radar signals coming from two North Korean SO-1-class subchasers, one of them using the call sign SC-35. On the bridge, Bucher scanned the horizon with binoculars. His eyes widened as the subchaser SC-35 shot out of the mist and raced toward the *Pueblo*. The small gunboat, brandishing a 3-inch cannon and two 57 mm guns, pulled close and demanded that the *Pueblo* announce her nationality. Commander Bucher told his crew to raise the American flag and hydrographic signal—the ship's cover as a research vessel.

The subchaser closed further, pointed her guns, and ordered the *Pueblo* to surrender. Bucher had no intention of handing over his ship, so he went to all ahead two-thirds and attempted to run. *Pueblo*'s speed paled in comparison to the subchaser's, and to make matters worse, three North Korean torpedo boats joined the chase. SC-35 made a third swing around the *Pueblo* and hoisted the signal, "Heave to or we will open fire."

Now four miles outside North Korean waters, Bucher tried to maneuver away until two MIG-21 fighter planes roared overhead and yet another torpedo boat and subchaser entered the fray. He contemplated putting up a fight, but *Pueblo*'s ammunition had been stored below decks, and the machine guns were covered in cold-weather tarpaulins. Bucher knew he had to raise his arms in surrender, but not until his crew could destroy the cryptographic equipment on board. He gave the order to do so and asked his engineering officer, Gene Lacy, if they had time to scuttle the ship. Lacy shook his head no. Even flooding the *Pueblo* would take forty-five minutes, and if the North Koreans shot holes in the life rafts, the crew would die within minutes in the freezing ocean.

Bucher continued evasive maneuvers for another two hours to buy time while the crew tried to shred documents and smash the crypto gear into unusable pieces. Given the volume of sensitive material on board, and the hardened steel cases used to house the cryptographic systems, the crew failed to destroy much before the North Koreans sent 57 mm cannon shells toward the *Pueblo* and opened fire with machine guns. Most of the rounds missed, but a few landed, slicing Bucher's ankle and buttocks.

Communications Technician Don Bailey had been frantically typing away on the KW-7 unit for more than an hour, sending and receiving messages to and from Japan. R-Branchers in Kami Seya implored Bailey to destroy the cryptographic equipment, but the sledgehammers and axes proved ineffective. Kami Seya promised air cover, but no U.S. planes were within range, save a squadron of seventy-eight fighters in Japan, which were forbidden from flying combat missions due to international agreements. The South Koreans requested permission to save the *Pueblo* with their 210 armed jets, but General Charles H. Bonesteel refused, citing a potential unwarranted escalation. A dozen F-105s were finally given authorization to fly from a nonregulated location in Japan. They requested clearance to sink the North Korean ships and then refuel in South Korea, but they received the opposite instructions, making their arrival in time impossible. When someone shook President Lyndon Johnson awake that morning to brief him on the incident, the North Koreans had already bagged their prize.

As ordered, Bucher followed the North Korean vessels until the *Pueblo* neared the coast of North Korea. Around 2:00 P.M. he ordered all stop to check on the destruction of the papers and equipment in the SIGINT area. SC-35 closed to less than a mile and opened fire. More than 2,000 rounds ripped through the *Pueblo*'s thin steel and slammed into the wardroom, laundry room, and passageways. Near Bucher's stateroom, Fireman Duane Hodges slumped to the deck as a projectile tore off most of his leg and sliced open his gut. Blood spilled from his intestines and coated the deck in red. I-Brancher Marine Sergeant Robert Chicca watched in horror as blood oozed from his thigh, and Radioman Charles Crandal screamed in pain as hot metal shards impaled his leg. Fireman Steve Woelk reeled backward as sharp pieces of shrapnel burned into his groin and chest.

On the bridge, Bucher immediately ordered full ahead to appease the North Koreans and get them to stop firing. He gave the conn to Lacy and ran down to the SIGINT area. On the way he saw the mangled body of Duane Hodges in the blood-soaked passageway. In the SIGINT space, spooks were still trying to cram secret papers into mattress covers and pound crypto equipment into oblivion. Neither effort was going well. Most of the sensitive material, including top-secret code lists, along with the KW-7 and other crypto gear, remained intact.

Out of options and fearing for the safety of his crew, Commander Bucher did the unthinkable. For the second time in history, and not since Commodore James Barron turned over the USS *Chesapeake* to the Brits in 1807, he allowed a foreign power to seize control of a U.S.

Navy ship during peacetime. Once in port, *Pueblo*'s crew members were blindfolded, kicked, beaten, spat on, and marched off the *Pueblo* at bayonet point. A team of North Koreans hurried aboard and grinned at the sight of a fully operational, undamaged KW-7.

The North Koreans motored the *Pueblo* to Wonsan, while her former crew were tortured, starved, and held captive in a POW camp. During his captivity, Commander Bucher faced a mock firing squad, enacted to force a confession. He did not relent until the North Koreans threatened to execute his crew. Given that none of the Koreans spoke enough English to write Bucher's confession, they had him pen the document himself. As a tongue-in-cheek retribution for killing one of his crewmen, Bucher wrote the words: "We paean the North Korean state. We paean their great leader Kim Il-Sung." The North Koreans never caught on.

While the *Pueblo*'s crew endured horrendous conditions and unthinkable cruelty, the North Koreans made a deal with the devil. They turned over the captured KW-7 to their Soviet comrades in exchange for money, weapons, perhaps even a few cases of Stolichnaya. That same month, John Walker's current Soviet handler—the vodka-drinking, chain-smoking General Boris A. Solomatin—convinced him to steal a KW-7 operating manual in exchange for a big bonus. Walker gladly complied.

Over the next two months, Soviet communications experts used the manual, their cache of stolen keylists, and the operational KW-7 to gain access to the U.S. Navy's most sensitive communications traffic. Overnight, almost every message sent between American shore facilities to ships and submarines of the fleet became an open book. General Solomatin, the Soviet's KGB chief in Washington at the time, commented many years later that "Walker showed us monthly keylists for one of your military cipher machines. [He] enabled your enemies to read your most sensitive military secrets. We knew everything."

Oleg Kalugin, former KGB chief at the Soviet embassy in Washington and John Walker's first handler, validated Solomatin's comments after the Cold War. In an unaired CBS interview, he stated that "John Walker's information, on top of *Pueblo*, definitely provided the Soviets with the final solutions to whatever technical problems they may have had at the time. . . . We certainly made use of the equipment from the *Pueblo*." Kalugin also confessed in his memoirs that before March 1968, the Soviets were intercepting and deciphering encrypted navy messages as a result of the Walker–*Pueblo* intelligence windfall. This fact played a pivotal role in the downing of two submarines—one Soviet and one American—and brought the world once again to the brink of war.

Although the NSA knew that the North Koreans captured the *Pueblo,* they were not able to communicate with her crew during their eleven-month imprisonment. As a result, the NSA assumed incorrectly that Pueblo's crew destroyed the KW-7 prior to the ship's capture. Worst case, they thought, even if the KW-7 remained intact, and the North Koreans turned the device over to the Soviets, without the daily keylists, the encryption unit had the usefulness of a boat anchor. Of course, the NSA did not know that John Walker had been delivering the Soviets keylists for months, and had also given them an operations manual.

Cautious voices at the NSA called for a replacement of the KW-7, just to be safe, but budget-minded objectors overruled. Replacing that many units in the field would be far too costly, and besides, did we mention that the Soviets need the keylists to decipher our messages? A series of charged incidents followed the Walker-*Pueblo* debacles, beginning with the deployment of a Soviet *Golf II* ballistic missile submarine in early 1968. K-129, a Project 629A diesel-powered boat with pendant number 722, joined five similar vessels of the Fifteenth Submarine Squadron based at Rybachiy Naval Base on the Kamchatka Peninsula. The *Golf IIs* represented the Twenty-ninth Ballistic Missile Division at Rybachiy, under the command of Admiral Viktor A. Dygalo.

K-129's skipper, Captain First Rank Vladimir Kobzar, had just completed two back-to-back seventy-day combat patrols in 1967 and was looking forward to some R&R. He held back his anger when new orders demanded that K-129 undertake yet a third patrol. While preparing for that patrol, eleven "strangers" walked across the gangplank and descended through the hatch. To this day, aside from a few Soviets in command at the time, no one knows who these strangers were or why they were on board.

On February 24, 1968, powered by three diesel engines, K-129 twisted away from the pier. Veiled by darkness, the boat did not use running lights. Captain Kobzar stood on the bridge, bundled in a furlined *ushanka,* and watched waves crash into the bow of his submarine. Distant lightning flashed and mingled jolts of white with threads of morning indigo on the horizon. K-129 shuddered as Northern Pacific waves pounded her sides. Kobzar captured his final memories of day-light, cleared the bridge, and dove his boat under the roiling waves.

Armed with three SS-N-4 Sark submarine-launched ballistic mis-siles (SLBMs), each fitted with a one-megaton warhead, K-129 con-ducted a deep-dive fitness test with ventilation valves closed and ballast tanks sealed. The submarine leveled off at a depth of 200 feet and as-sumed her patrol course—due east to clear the shoreline shallows.

Fifteen miles outside the bay, Kobzar brought K-129 to periscope depth to transmit the first of a series of mandatory mission reports to the naval main staff at fleet headquarters in Vladivostok. Confident that any listening Americans could not decipher their codes, Radio Officer Senior Lieutenant Alexander Zarnakov sent a millisecond burst signal on the SBD radio reporting that K-129 had entered deep waters to start her mis-sion. Miles away, a Soviet radio dish grabbed the burst signal and pushed charged electrons down a wire into a receiver/decoder.

Hundreds of miles away, at Classic Bulls Eye stations in Japan, Guam, and Alaska, R-Branchers recorded the microsecond bursts, ana-lyzed the recordings, and sent tip-offs to Net Control in Hawaii. NC sent out a flash and later received back more bearings from other sites. Operators utilized newer GYK-3 Boresight computers to determine an approximate fix, along with other parameters to determine that they'd found a *Golf II*-class submarine.

NC informed COMSUBPAC (Commander, Submarine Force, U.S. Pacific Fleet) of their findings, who informed the commander of Subma-rine Squadron One in Pearl Harbor, Hawaii. After noting which subma-rines were operating near enough to Kamchatka, the commander of Squadron One composed a flash message to the appropriate attack sub-marine. They encrypted and sent that message using a KW-7.

While the Soviets had no clue that their burst transmissions were being multiangulated by U.S. Bulls Eye stations, the Americans had no idea that their encrypted communications to and from submarines were as clear and open as ordinary phone calls. The Soviets now knew that an American attack submarine had been sent to trail K-129. They probably suspected that the Americans found K-129's location via the $1.5 billion, 1,300-mile SOSUS acoustic detection system surrounding the Pacific.

Six days out from port, on February 29, K-129 failed to send a routine report that she had crossed the International Date Line. While some suspect rogue activities as the cause for such a breach, Soviet radiomen concur that if a boat's CO knows the enemy may be nearby, he'll forgo a radio transmission—this is especially true for ballistic missile submarines with primary missions to stay hidden. K-129's captain likely knew that an American submarine was operating in the area, in that the naval main staff intercepted the American submarine's position report via the stolen KW-7. Armed with that knowledge, K-129 would have gone deep, snuck under a sound layer, and slipped silently away in another direction. Reports show that K-129 also left her normal mission area on that date.

One could again argue that K-129 went outside her assigned box because someone on board intended to go rogue. Given that the officers on K-129 had families and loved ones in Russia, this scenario seems unlikely. SOSUS stations reported that once K-129 switched to batteries or ventured too far from hydrophone locations, tracking her became difficult, if not impossible. Could it be that the Soviets knew this fact, having gained that information via intercepted KW-7 transmissions, and ordered K-129 to go outside her box to test SOSUS detection capabilities and range limitations?

Over a twelve-day period, K-129 failed to transmit position reports twice. While this may lead some to believe rogue sailors were the cause, such a conclusion again seems suspect. The fact that K-129 transmitted as prescribed on other days and then skipped a few days is probably related more to evasion than mutiny. Had Captain Kobzar received a warning that a U.S. sub was closing the gap, he would have avoided the risk of coming shallow to transmit, instead remaining below a thermal layer to hide.

The most reasonable scenario has the *Swordfish* or another American boat stalking K-129 as directed by Bulls Eye and SOSUS fixes, then K-129 receiving KW-7 intercepted warnings from the naval main staff and avoiding the U.S. submarine like the plague. This game of cat and mouse ensued until K-129's life came to an abrupt end.

What happened to the Soviet *Golf II* on March 8, 1968, just fifteen days away from home and 1,560 miles from Honolulu? Several SOSUS arrays recorded "an isolated, single sound of an explosion or implosion, a good-sized bang." Some speculate that, despite this submarine's

ability to fire her ballistic missiles while submerged, for some unknown
reason, rogue individuals risked surfacing K-129 to commence an unwarranted attack on Hawaii. While doing so, in the nick of time, the
submarine suffered a casualty and sank. This conclusion is based primarily on the fact that K-129 got tapped for her mission six months
early, had eleven "strangers" on board, missed a few radio updates, and
traveled outside her assigned patrol grid.

When one considers the Soviet's newfound ability to decipher KW-7
transmissions, perhaps these seemingly odd circumstances begin to make
sense. The Soviets may have sent the noisy diesel submarine to sea six
months early to test their ability to intercept KW-7 transmissions. They
might have assumed that K-129 would be detected snorkeling by
SOSUS, which would trigger those transmissions. As for the eleven
strangers, one could conclude that these sailors were part of an OSNAZ
group of signal intelligence operatives—similar to "spooks" on U.S. submarines. They would have been tasked with observing American radio
transmissions and analyzing responses, transmission frequencies, and so
on. We've already explored the reasons why K-129 may have missed
radio transmission times and steamed outside her grid.

Even if Hawaii was not more than 500 miles beyond the range of
K-129's missiles, one could argue for the rogue theory if not for the fact
that forensic evidence confirms that she did indeed surface. This type of
submarine did not need to surface to fire her missiles, snorkel, or go
rogue; in fact, such an act makes surfacing all the more unlikely. The only
reason why K-129 might come up from the deep while on patrol is because she had no choice. This might be the case if she suffered a catastrophic failure or endured an accident that damaged something badly
enough that underwater repairs were not feasible, such as a collision
with a U.S. submarine.

Given the advanced sonar systems available on the USS *Swordfish*
or similar subs, had a U.S. boat been directed to the football field
wherein K-129 played, the American attack boat would have eventually found its prey. Having done so, the directives for all sub skippers in
1968 pushed them into close prowling range behind their targets. But
which boat received the orders to tail the *Golf II*?

Under the command of Captain John Rigsbee, the *Swordfish* had
been deployed in the area on WestPac since February 3. On March 17,

nine days after K-129 sank to the bottom of the ocean, the *Swordfish* pulled into Yokosuka, Japan, under the cover of night, seeking repairs on her conning tower, periscope, and ECM mast. Official reports attributed the damage to ice impact while conducting classified operations in the Sea of Japan on March 2. The navy told the Soviets that the *Swordfish* was 2,000 miles away from K-129 at the time of the incident. Most points in the Sea of Japan are over 4,000 miles away from Hawaii, but less than 1,000 miles from Yokosuka. If the *Swordfish* was indeed in the Sea of Japan (that is, near Vladivostok), and whacked her periscope and ECM mast there, why did she not arrive in Yokosuka until two weeks later?

Instead, the *Swordfish* limped into the harbor at Yokosuka while a Japanese/Soviet spy, stationed covertly on Honshu Island, observed the event. His Soviet handlers wrote the report off as routine and all but yawned. When retired admiral Ivan Amelko heard the news, his reaction was a bit more dramatic. He accused the United States of causing the demise of K-129 and killing ninety-eight men. The United States, of course, denied the accusation. Perhaps eyewitness accounts can help us clear up this mystery.

In late March 1968, excited about the prospect of another mission into dangerous waters, Communications Technician Second Class Frank Turban strutted through the door of his bunker in Kami Seya, Japan. Most of the working areas at Kami Seya were housed in underground tunnels to protect equipment and personnel from earthquakes. As a T-Brancher, his job revolved around unusual signals, and whenever one came along, his ears perked. While the Bulls Eye section at the Kami Seya Naval Security Group Activity, commanded by Captain J. W. Pearson, worried over high-frequency direction-finding tip-offs and flash reports from Net Control, Turban focused on special signals.

Nothing could be more special than something new and unusual transmitted from a Soviet submarine. BRD-6 ESM masts on board U.S. submarines, designed by Sanders Associates in New Hampshire, collected these signals. The BRD-6, which stands for Boat Radio Direction Finder, contained a miniature HF direction finder and burst signal receiver/recorder, similar to the one used in Bulls Eye stations. In fact, the engineers at Sanders gained much of their initial knowledge for this design from the meeting they had with William J. Reed and his team in 1962.

While at Kami Seya, Turban and ten other spooks received orders to report to the USS *Swordfish* for a top-secret SpecOp. Spooks were not official members of any submarine crew, did not claim any particular boat as home, but instead "snuck aboard" just prior to departure and usually stood watches behind a curtain in the radio room. They did not often mingle with the crew to ensure no secrets were told by accident. After a months-long mission, the CT team disappeared with equal stealth, such behavior conjuring the nickname "spook" by submariners.

While preparing for his *Swordfish* adventure, Turban heard from R-Branchers involved in burst signal HFDF operations at Kami Seya that they'd gotten several Bulls Eye hits on a *Golf II* submarine near Hawaii before the boat went completely dark. She hadn't transmitted again since March 8. Turban thought nothing of this report at the time as he caught a flight to Hawaii to assemble with the other spooks and prepare for their upcoming mission.

In late April 1968, Turban and his team met the *Swordfish* in Okinawa, Japan. He noticed that the front of the submarine's sail seemed a bit shinier than the rest of the boat but passed this detail off as routine maintenance. Using a towed underwater camera inside a mini-sub called the *Underdog,* the *Swordfish* deployed near Vladivostok to prowl for lost Soviet missile parts. Tethered to the boat like one of the heads of Hydra, the *Underdog*'s twelve-foot-long, two-ton aluminum body contained high-resolution cameras and bright battery-powered strobe lights. She also came with whiskerlike towed sonar and shark fin rudders and bow planes for maneuverability. A small propeller pushed the *Underdog* through the water, while the tether allowed for remote operation from the submarine.

Operators sat in front of monitors and examined the ocean floor as the *Underdog* darted across search patterns and relayed images back to the *Swordfish*. While this type of mission was considered quite dangerous, as it required penetration deep into Soviet waters, Turban noticed that their skipper often ran from his own shadow. When predators came stalking, the boat's executive officer wanted to uncollar and chase the bastards. The CO refused, instead content to stay as far away from potential encounters as possible, earning him the nickname of Charlie Tuna in reference to the Chicken of the Sea television commercials. Turban found the CO's gun-shy attitude strange for an attack-boat driver but asked no questions. After all, he was not officially part of the crew, only a temporary ranch hand with an assignment to find, record, analyze, and fingerprint interesting new signals.

The *Swordfish* remained on station in the Sea of Japan for sixty-eight days, while Turban's team, consisting of an officer in charge (OIC) and his assistant (AOIC), two linguist I-Branchers, two R-Branchers, and another special signals T-Brancher like Turban, did their thing. There were also three R-Brancher specialists who worked with the WLR-6 ECM radio equipment. Not once did Turban hear anything from the crew about a collision on March 2, not a word about repairs in Yokosuka or damage of any kind.

In light of the above, the pink elephant question lingers. The collision on March 2 (or perhaps March 8) took out *Swordfish*'s periscope and ECM masts, placed in her jeopardy, and lowered her mission capabilities to almost zero. If she had been in the Sea of Japan, why did the *Swordfish* not arrive in Yokosuka until two weeks later? On the other hand, if she indeed had been near K-129, a speed of fifteen knots allowed the *Swordfish* to pull into Japan on March 17 with ease.

Still, let's assume that the navy's statement is true, and that the *Swordfish* did not collide with K-129. We know that U.S. Bulls Eye stations intercepted K-129's transmissions, and SOSUS heard her snorkeling more than once. Armed with the knowledge about K-129's whereabouts, the navy certainly would have sent a U.S. submarine to investigate. But which boat?

On January, 13 1968, Commander Hugh Benton assumed command of the *Guardfish* (SSN-612) in Pearl Harbor, Hawaii. Shortly thereafter, his boat deployed for a six-month WestPac special operations mission. The *Guardfish* was on station near the Sea of Japan in March 1968. The USS *Barb* (SSN-596) was also on station near Kamchatka, but reports indicate that she was "surprised" by Soviet naval activity pursuant to K-129's loss, making her a less likely candidate. Perhaps the *Swordfish* or the *Guardfish* pursued K-129, and maybe one of them came a bit too close. Until sailors on either boat are willing to talk, or the navy finally releases the classified logs, the truth won't come to light.

Regardless of what really happened, given access to many of the facts we've just examined, the Soviets reached the verdict that the *Swordfish* rammed K-129 and caused her to die. They could not, of course, admit to the world that they suspected this because they possessed the means to intercept and decode KW-7 transmissions. Hands tied, they could do nothing more than hurl accusations. The fact that they had a fly on America's wall, and that it eventually led to the first premeditated sinking of a U.S. submarine since World War II, remained a secret.

BY THE THIRD WEEK OF MARCH, Soviet naval headquarters declared K-129 missing and organized a massive air, surface, and subsurface search-and-rescue effort into the North Pacific from Kamchatka and Vladivostok. U.S. intelligence experts examined SOSUS logs and determined an approximate location for the recorded "bang." The Soviets did not have the advantage of SOSUS, and so searched for K-129 in the wrong location. They finally gave up and went home, and soon after officially announced the loss of the *Golf II* and her crew.

Six weeks later, on Friday May 15, 1968, on station in the western Mediterranean, the USS *Haddo* (SSN-604) received the coordinates for a nearby Soviet *Echo II*–class nuclear submarine entering the Atlantic near the Straits of Gibraltar. The *Haddo*'s skipper did not know that the fix came from Bulls Eye stations in the Atlantic. The *Echo II*'s track put her on a course bound for the Canary Islands and a potential rendezvous with Soviet surface ships operating in the area. The *Haddo* went to work and slipped in close behind the *Echo*.

On May 20, Vice Admiral Arnold F. Schade, commander of the Atlantic Submarine Force, ordered the *Haddo* to hand off her tail of the *Echo II* to the USS *Scorpion* (SSN-589). On the *Scorpion*, Commander Francis Atwood Slattery picked up the trail of the enemy submarine as she made her way toward the Soviet flotilla. Photographed by U.S. satellites, the flotilla consisted of an unusual collection of vessels not normally known to operate together. The strange formation was conducting exercises in waters outside typical Soviet operational areas, in the vicinity of a U.S. nuclear test site, and lofting large balloons that housed surveillance equipment.

Commander Slattery brought the *Scorpion* to periscope depth around midnight on May 20. Radio operators started transmitting to the naval station in Rota, Spain, that *Scorpion* had arrived on station, but interference prevented validation of reception. A day later, they finally connected with a navy communications station in Nea Makri,

Greece, which relayed *Scorpion*'s message to SUBLANT (Submarine Forces, Atlantic). NAVCAMS (Naval Communications Area Master Station) in Norfolk received the relay from Greece on a KW-7. The message read that Slattery estimated interception of the Soviet surface group in about six hours. Slattery did not know that Soviet ears were listening to his report.

On May 22, the *Scorpion* arrived on station. Several hours later, without warning, the boat shook violently as an explosion rocketed through the sub's compartments. The crew scrambled to battle stations as years of experience and training kicked into high gear. Hatches were dogged, valves shut, and switches thrown, but to no avail. In just over a minute and a half, the submarine shot toward the bottom.

At 6:59 P.M. Greenwich Mean Time, the secret listening stations in the Canary Islands, Newfoundland, and Argentina began a 190-second recording of fifteen eerie acoustic signals as they reflected from the Plato Sea Mount. The signals bounced off convergence zones and found their way to an array of hydrophones mounted on the seafloor, placed there by the Air Force to monitor Soviet nuclear tests. Spinning reels of high-speed recording tapes documented the tragic sound that overpowered the background noise of whistling biologics. The high-energy release was rich in low frequencies with no discernible harmonic structure—a bubble pulsation from an underwater explosion. Operators recognized this as the frightening sound of a dying boat.

In the months following the death of the *Scorpion,* the theories as to her demise were many and varied. Initially, the most prevalent theory pointed to a "hot-running" torpedo that caused a catastrophic explosion in the torpedo room. But the experts at Ordnance Systems Command insisted that such a detonation was virtually impossible. An explosion could occur only if the torpedo hit an object while running at full speed.

What really happened to the *Scorpion?* Theories and speculations abound and have been addressed ad nauseam in several books, but most concur that the evidence points to an external explosion. One smoking gun all but validated this conclusion. In 1982, at a navy SOSUS training class in Norfolk, Virginia, students observed a LOFARgram printout of a top-secret recording made in May 1968. The scrawled lines on the paper depicted a hostile encounter between two submerged contacts. After twenty minutes, a third contact appeared. This submersible was launched from one of the submarines, and the scribbles verified a high-speed screw. The targeted submarine's signature shifted in width and size, indicating evasive actions as the torpedo neared. Seconds later, the paper filled with black ink as the high-speed screws caught up with the evading submarine and ended her life.

The students were told that the recording was made during *Scorpion*'s encounter with an *Echo II* submarine. Authors and experts bring additional facts to bear to substantiate this claim, but to what end? Until governments are willing to expose the truth, the official U.S. Navy position as to *Scorpion*'s demise remains as "cause unknown." If the conspiracy theorists are correct, then the lessons learned from allowing the greed of John Walker to manipulate entire nations will stay buried, and the cause-and-effect results of this act remain as a black stain on the pages of history.

CHAPTER THIRTEEN

Any sufficiently advanced technology is indistinguishable from magic.

—Arthur C. Clarke

THE SOVIETS FINALLY GAVE UP THE search for their lost K-129 and went home in May 1968, not long after the sinking of the *Scorpion*. The U.S. Navy let another month go by and then called in the cavalry. Captain James F. Bradley Jr., from the Office of Undersea Warfare, had already spearheaded the creation of the perfect platform to undertake the mission of finding the missing *Golf II*–class boat. The forty-six-year-old Bradley served as the navy's chief underwater espionage planner, often conducting meetings in a soundproof suite in the E ring of the Pentagon. Quiet, efficient, and whip smart, Bradley's official title of "Naval Operations, Navy Department" alluded to nothing covert. Only those close to the captain knew that he wielded the power to craft clandestine intelligence missions for all of the navy's attack submarines.

Under Bradley's orders, the nuclear submarine USS *Halibut* (SSGN-587), commissioned in January 1960, received a $70 million upgrade that included the mysterious *Fish*. Just like the *Underdog* tethered to the USS *Swordfish*, the *Fish* mini-subs contained deep-water cameras and sun-bright strobe lights to find interesting objects on seafloors. *Halibut*'s two Westinghouse Electric–built *Fish* cost $5 million each and resided inside a hump-shaped dome on the submarine's bow dubbed the "bat cave."

The dome once housed Regulus missiles back when *Halibut* was a teenager and resembled a miniature torpedo room complete with metal racks to house the *Fish*. The first *Fish* suffered from a host of issues, some caused by the intense pressures placed on cameras and equipment operating at 20,000 feet deep. These problems took months to correct. With kinks finally ironed out, the refurbished *Halibut,* commanded by C. Edward Moore under the highest "Brick Bat 01" authority, used her *Fish* to skim ocean floors on scavenger hunts for spent Soviet toys—like intercontinental ballistic missile (ICBM) warheads that might contain guidance systems for intelligence perusal. Bradley's program, dubbed Operation Sand Dollar, did not pan out as planned, as the boat was plagued by mechanical and computer failures. However, the concept of using the *Fish* to find and photograph things in dark places proved invaluable.

Paul Nitze, secretary of the navy, caught wind of the debacle and blasted Bradley, Moore, and others involved in the project. Sand Dollar's failures were soon forgiven when Nitze found out about K-129 and the proposal from Bradley to use the *Halibut*'s *Fish* to locate the downed submarine. The SOSUS team had provided location fixes based on the "bang" recordings that documented K-129's last gasps for life. Nitze approved the mission, and Captain Moore set to sea in the *Halibut* with do-or-die orders to locate K-129, take detailed photographs via the *Fish,* and place salvage markers. After months of frustration, snagged cables, and near-disasters, the *Halibut*'s mysterious *Fish* delivered success, and the precise position of the *Golf II* submarine was no longer a mystery.

Armed with *Halibut*'s K-129 photos and location coordinates, Carl Duckett, then CIA deputy director for science and technology, spearheaded the effort to recover the *Golf II* submarine intact. Captain Bradley dissented, telling Duckett that "you can't pick up the goddamn submarine, or it will fall apart." Bradley countered with a plan to send in minisubs, cut open the hull, and retrieve only the tidbits desired at a fraction of the cost of full retrieval. Duckett dismissed the plan and remained determined to haul up K-129 in all her glory, claiming that inside the *Golf II* there should be Soviet encryption equipment—including a burst radio transmitter and codebooks—and three nuclear SS-N-5 SERB missiles.

A detailed examination of Soviet radio, missile, submarine, and torpedo technology could be invaluable, offering the navy an opportunity to design effective countermeasures and improve ASW and Bulls Eye detection equipment. The Soviet codebooks alone might be worth the expense of the salvage—almost $200 million—and would allow analysts to decode hundreds of hours of burst signals and other transmissions recorded over the years at various Bulls Eye stations. Besides, Duckett insisted, he didn't want to chance leaving anything behind.

Could it be that Duckett overruled Bradley's far less costly plan to ensure that no part of K-129 could be salvaged by the Soviets, especially that part that contained metal fragments from the hull of the U.S. submarine that rammed the Soviet missile boat?

Dissenters quieted, the project to recover the K-129 intact moved ahead. The CIA established the Special Projects Staff to oversee the program, which they code-named Azorian. Duckett tapped John Parangosky, a senior official in the CIA's Directorate of Science and Technology, to run the show. A former Army Air Force lieutenant in World War II, Parangosky joined the CIA in 1948, where he put his expertise in photographic techniques and project management to use for the U-2 and A-12 spy plane programs. Parangosky recruited U.S. Naval Academy graduate and former diesel submarine officer Ernest "Zeke" Zellmer to run the day-to-day operations. President Nixon personally approved the creation of the Azorian task force in August 1969, and CIA director Richard Helms compartmentalized all top-secret project work under the heading of "Jennifer." Only a select handful of individuals at the White House and within intelligence communities knew anything about Azorian.

In early 1970, Joseph Houston, Itek Corporation's chief optical engineer for underwater systems, had just returned from a test run aboard the USS *Dace* (SSN-607). Houston spearheaded the development of the new Type 18 periscope, recently mounted on the *Dace* for sea trials. After a week under way, he wanted to enjoy a few days off. He had joined Itek in 1964 and over the next few years compiled an impressive résumé in optics that included the inventions of various testing equipment, advanced optical lenses, and high-resolution cameras, including ones used in several spy satellites under Project Corona. Within an hour after stepping onto the pier from the *Dace,* Houston received a summons from Itek's vice president of special programs, John Wolfe. The two met over lunch in Wolfe's office.

"Have you ever been to Mars?" Wolfe asked as Houston found a seat.

"Mars as in the planet?" Houston asked, wondering if this had something to do with his previous work on the camera systems for *Viking 1,* the first American Mars lander.

Wolfe gave a quick nod. "Yeah, Mars."

"Not lately," Houston said. He knew that the broad-shouldered and bushy-headed Wolfe had a flair for the dramatic and wondered if his boss really meant Mars or was just laying the foundation for a metaphoric bombshell.

Wolfe examined the contents of the box lunch on his desk, brought by his secretary, and wrinkled his nose. "I haven't either. Then again, neither has anyone from NASA."

"Is this about *Viking 1?*"

Wolfe shook his head no. "It's about solving the problem of taking detailed, high-resolution metrology photos of a hostile environment that we've never been to or seen before."

"I'm not sure I'm following you," Houston said, opening a bag of chips.

Wolfe said, "We need your expertise in submarine optics and deep-ocean camera systems to help one of our government clients. They need to take thousands of pictures with no measurement distortion so we can build a virtual model for a rig that can mine manganese nodules."

"Mars has manganese nodules?"

Wolfe waved a hand in the air as he swallowed a bite. "No, but it might as well be Mars. We need accurate metric data to set up a drilling rig more than three miles deep in the Pacific Ocean."

Houston lowered his sandwich. His eyelids flung open as the metaphoric bombshell landed. "Three miles? No large body camera can withstand that kind of deep-sea pressure, let alone record high-quality metric data. And even if you could, proper lighting would be a major challenge."

"That's why I called you," Wolfe said with a smile. A small piece of bread clung to his front teeth.

"Who's the client?" Houston asked.

"The CIA."

Houston almost choked on a gulp of water. *The CIA?* That definitely conjured some curiosity. Houston knew that Frank Lindsay, Itek's chairman of the board, had met Howard Hughes during World War II, back when Lindsay was a gun-blazing, parachute-popping OSS operative. Houston heard that Hughes kept in touch with Lindsay over the years and had recently contacted him about some hush-hush CIA-sponsored project involving deep-sea mining.

Houston said, "So why is the CIA interested in mining manganese nodules off the ocean floor?"

Wolfe said, "I could tell you, but then—"

"Yeah, I know," Houston said, "you'd have to send me on a permanent vacation to New Jersey. So what are the requirements for this camera and lighting system?"

"We need to . . . drill an oil well 16,500 feet deep, and we need accurately measurable photographs of the area so we can ensure we're designing the equipment to the right dimensions."

"How large an area are we talking about?" Houston queried.

Wolfe's voice dropped another few decibels. "Three hundred fifty feet long by fifty feet wide."

Houston almost gagged again on his water. "What kind of wellhead has those dimensions? That sounds more like a submarine, not a wellhead."

Wolfe's face turned white. He rose from his chair and grabbed a bunch of papers off his desk. He walked over to Houston and handed him the stack. "Read this agreement, sign it, and then we'll talk."

Project Azorian, of course, had nothing to do with building a wellhead or mining for manganese nodules on Mars. Instead, Houston was joining a massive team, spearheaded by Howard Hughes, tasked with salvaging an entire 2,700-ton submarine sitting 16,500 feet deep in the Pacific. Houston had no doubt that this top-secret mission would go down in history as one of the most ambitious, expensive, and massive ever undertaken. He suspected that the consequences of failure were huge, but they probably paled in comparison to the backlash certain to come from the Soviets should they find out.

After some further discussion with Wolfe regarding how the project would be organized and managed, Houston departed with visions churning in his head of a rig capable of supporting multiple cameras and lights. The enormity of the ocean area to be surveyed complicated the task by a few orders of magnitude. Also, the lighting had to be perfectly uniform, and the camera lenses needed to survive enormous pressures.

To make matters worse, Houston discovered that his project head was the mysterious John Parangosky, who insisted on being called Mr. P. to maintain his clandestine anonymity. Although Houston found Mr. P. highly intelligent, a great listener, warm with people, and "the calmest SOB on the planet when the *fit* hit the *shan*," he could also be a slave driver when it came to meeting design parameters and deadlines. "Parangosky had the demeanor of Alfred Hitchcock in the 1942 movie *The Man Who Came to Dinner*," said Houston of his former boss. "He was a heavyset engineer with a receding hairline and a wry sense of humor. He rarely displayed emotion and could smooth out the rough edges of a meeting within fifteen minutes."

Mr. P. assembled a small team of engineers and technicians in the winter of 1969. Included in the group were Joe Houston, John Wolfe, Floyd Alvarez, Ernest Zellmer, and a dozen others. Mr. P. also leaned heavily on Alex Holser as his "right-hand man" to assist with project management tasks.

The team often met in a conference room at Itek's corporate headquarters in Lexington, Massachusetts. "Itek's building had a gleaming glass 'swinging sixties' architectural feel," said Houston. "The main lobby jutted out over the parking lot and you entered via a ramp that resembled a vintage airplane staircase. A smattering of Pieter Mondrian artwork covered the walls in the foyer, which resided on the second floor along with the executive offices, cafeteria, and conference room where the Azorian team met."

Over the next year, Houston cycled in and out of dozens of team meetings where the group struggled to formulate solutions for a seemingly impossible task. No country in the world had ever succeeded in salvaging anything this large or this heavy from this deep. The Azorian team debated over four potential solutions: The "Brute Force" plan entailed using a Rosenberg winch to hoist the entire 2,700-ton sub up from the bottom. The "Trade and Ballast/Buoyancy" concept suggested the use of buoyant material to float the boat to the surface. An "At-Depth Generation of Buoyancy" idea proffered employing gas at depth to "balloon" the K-129 upward via chemically created hydrogen or nitrogen cryonic gases. Finally, despite Duckett's previous nixing of the downscaled plan proposed by Captain Bradley to use mini-sub robots to scalpel their way into the submarine's hull, Mr. P.'s team explored this possibility as a backup plan.

In late July 1970, the group opted for the Brute Force concept using a sling crafted from pipe-strings wrapped about the K-129 and hoisted upward via winches mounted on a yet-to-be-built 565-foot-long ship. By October of that year, an executive committee, headed by Deputy Secretary of Defense David Packard and formed to scrutinize Azorian project progress and cost overruns, estimated the program's odds of success at less than ten percent. Only the promise of an intelligence bonanza saved Azorian from cancellation in August 1971, and the CIA got the green light to proceed on October 4, 1971.

With Mr. P. breathing down his neck, Houston set about solving an almost insurmountable optics and lighting problem. While others on Project Azorian were involved in building the HMB-1 barge, massive grappling claws, and winch system for hoisting K-129 from the bottom, they all relied on Houston's success. They needed to transfer real life into bits and bytes to perform analysis and create what-if scenarios. One can't control the real world of turbulence, lighting, shadows, and whatever else Mother Nature fashions. These parameters need to be manipulated in a make-believe virtual world to better understand the true nature of the environment. That required precise, clear photographs free of shadows or blur that might distort measurements and assumptions.

Houston explained the difficulty of creating such a uniform background to Mr. P. by using the example of driving down a road on a foggy night. By glancing out the window, you'd notice that the streetlight illumination is brightest in the center of the beam but diminishes toward the edges of the road. This phenomenon is prevalent with single-beam lights due to dust and particulates in the air that scatter the light. In water deeper than 10,000 feet, few such particulates exist, so the clear water doesn't spread out the light. More lights cover more area, but then you get shadows in the areas in between each beam. Those shadows can distort dimensions and details. Mr. P. was impressed by Houston's explanation but still tapped his watch impatiently.

Houston spent long days designing a four-legged stool to hold an array of lights built by Hydro Products in Southern California, as well as a way to allow the array to hover over specific objects and light up areas eight times the size of the beams. He also designed the means to gain complete uniformity and eliminate shadows across the entire dimensions of the field. In this way, every protuberance, recess, and mechanical detail on the sunken sub could be accurately measured and transferred to blueprints and models used by the rest of the team.

Houston knew that if his lighting or camera caused designers to measure one detail wrong, a $200 million project could fail. Worse, if something catastrophic happened, the Soviets might find out about Project Azorian, which could provoke an international incident. Even more concerning, a massive failure could get people hurt or killed. With the weight of the project's success resting on his shoulders, Houston delivered sample photos to Mr. P. The pictures were of a pond in Houston's backyard, taken by hanging a camera from a tree while Houston's son, Brant, positioned objects to photograph.

Mr. P. grabbed the photos from Houston's hand. He pulled out a magnifying glass and glared. He ran the glass across the images, wrinkled his nose, and said, "Where'd all those catfish come from?"

Houston copped a sheep grin and shrugged. "The pond in my backyard."

"Nice," Mr. P. said. "Very nice."

Houston nodded, found a seat, and let his aching shoulders slump. Thereafter the team referred to his innovative strobe-light rig as "the catfish solution."

The remaining design and build phases for Project Azorian involved revolutionary concepts for underwater photography that pioneered the development of a wide-angle lens with zero distortion. This included a means of reducing the pressure on the outer lens by a factor of seven, which allowed conventional lens designs to achieve distortion-free performance. Houston published a technical paper on the topic in the Marine Technology Society's 6th Annual Conference and Exhibition in July 1970. Sometime later, Houston met Carl Duckett, the CIA's deputy director for science and technology and Mr. P.'s boss. Always smiling, fiercely patriotic, handsome, and smart, Duckett possessed an encyclopedic memory and razor-sharp wit. He often used both to fight bureaucrats when they threatened to squelch technological advancements that he believed were critical to national defense. An engineer by training, Duckett almost never thought *inside* the box.

Like any good engineer, Duckett knew the project's HMB-1 barge—designed to grapple and retrieve a downed submarine three miles deep—had to undergo a series of quality assurance tests before she could resurrect K-129 from the dead. To do this, he towed the barge to a remote location off Long Beach. As a warm California sun soaked beaches and blondes in ultraviolet radiation, Duckett pulled the plug and watched the unmanned HMB-1 barge sink into a 6,000-foot-deep trench. Unbeknownst to him or others on the team, fishermen on a trawler observed the entire operation. From their distant vantage point, they swore that a ship had just sunk with all hands on board. Being good citizens, they called the Coast Guard, which sent out a cutter to investigate. Seeing no signs of a craft or survivors, the Coast Guard assumed a false alarm had been called and went home.

Hours later, the barge miraculously reappeared, just as ordered by Carl Duckett. Rejoicing, the fishermen called the Coast Guard and said that God had saved the sunken ship, miracles do happen, and as changed men, they had sworn off liquor and women for life, or at least the weekend. Duckett knew nothing of the affair until Mr. P. read the story in the paper the following day. When informed about the incident, Duckett smiled and quoted Arthur C. Clarke, who once said that "Any sufficiently advanced technology is indistinguishable from magic."

Armed with new "magical" deep-diving cameras designed by Joe Houston, the research vessel *Sea Scope* steamed to the K-129 site. *Sea Scope* once sailed the seas as the USS *Harrier* (AM-366), a U.S. Navy Admirable Class Minesweeper commissioned in 1945. The *Harrier* received a name and designation change not long after World War II. Although reclassified as an oceanographic and research vessel, the *Sea Scope* came chockful of deep-sea reconnaissance toys, including sonar, ESM, underwater television and photographic equipment, and a magnetic and seabed exploration grappling system. The *Sea Scope* arrived over the *Halibut*-marked site of K-129's remains and used the deep-diving cameras and sophisticated equipment to provide a detailed map in support of the *Glomar*'s mission.

Photos taken by Houston's camera rig revealed that K-129 remained intact, save for a ten-foot-wide hole blown in her side just aft of the conning tower—probably caused by the catastrophe that sank her. Judging by the photographs, the navy estimated the *Golf II* accelerated to 200 knots on her way down before she collided with the bottom. She was intact but most likely quite fragile. Given that K-129 might be broken beneath her steel outer plating, bringing her up from the bottom would be a long, slow, difficult task, not too unlike those arcade games that challenge you to grab a toy with an imperfect crane and lift it out of the tank. Only in this case, the toys are 16,000 feet down in a tank of dark seawater.

On November 4, 1972, the *Glomar Explorer* put to sea under the guise of a mining vessel in search of manganese nodules. Forty hand-picked CIA agents made up nearly one-fourth of the crew. Nineteen months later, after a battery of test runs, the *Glomar* headed toward Hawaii. Upon arriving at the site on July 4, 1974, using a sophisticated guidance computer and bottom-placed transducer, the *Glomar* team went about the delicate task of maneuvering the HMB-1 into position over K-129. Computer readouts flashed with information in the control room, while operators chewed on fingernails.

In the meantime, a Soviet SB-10 salvage tug passed within 200 feet of the *Glomar Explorer*. The tug's crew of forty-three flashed smiles and cameras and unnerved the CIA operatives. The SB-10 finally moved off and *Glomar*'s crew commenced with the salvage.

The five grappling claws were attached to a length of pipe on the ship, resembling six interconnected tongs suspended from a long platform. The entire affair, nicknamed "Clementine," looked like a giant squid clinging to a pipe. The pipe was fed through a hole, making its way down 16,500 feet to the Soviet sub. Glomar's crew, which consisted mostly of roughnecks who once worked on oil rigs, constructed the tether from sixty-foot-long sections of pipe they linked together like sectioned tent poles. Television cameras equipped with strobe lights monitored the progress, which took several long days to complete. Finally, Clementine reached the bottom.

Several more days passed as the crew, using sophisticated computers, made careful calculations to allow Clementine to drape a steel net over the boat's conning tower. Slowly, they wrapped one of the claws around the sub, then another, followed by a third. Silt and sand shot out from beneath K-129 as the claws reached underneath her delicate sides. On board the *Glomar*, nervous fingers twitched as they gripped joysticks, typed in keyboard commands, and flipped switches. Three and a half miles down, K-129 waited patiently for her ride to the surface. More hours passed as readings were checked and double-checked. Camera images were studied and debated. Finally, convinced that they were ready to hoist, someone gave the signal to proceed with only three claws wrapped about the boat versus all five.

K-129 shuddered as she rose from the ocean floor. Sections of pipe reappeared from the water as Clementine struggled to bring her prize home at six feet per minute. Workers dismantled and stowed the sixty-foot pipe sections, while cameras below the surface kept a watchful eye on the claws. One wrong miscalculation, one incorrect move at this point, could end a mission that had spanned years and cost hundreds of millions of dollars. Unfortunately, that miscalculation happened, and K-129 rocketed back to the seabed. Clementine's claws still gripped the hull, and the massive arms were pulled down to the bottom along with the submarine. Unbeknownst to project engineers at the time, K-129 slammed into the hard rock beneath the sand, and the impact caused hairline fractures in the grappling arms.

Disappointed but not defeated, the crew tried again, this time grabbing the sub with all five claws. Nine hours later, just 3,000 feet off the bottom, a damaged grappling arm snapped and left K-129 dangling by a hope and thread. Moments later, the fragile sub crumbled like a cookie. The rear two-thirds of the submarine hull broke away and tumbled back to the seafloor. Lost were the conning tower, three missiles, and the room where the codebooks were stored.

The recovery phase of the Azorian project concluded on August 9, 1974. With the entire mission considered a failure by the CIA, the crew of the *Glomar Explorer* looked like a defeated football team as they returned home. Making matters worse, while still at sea, they heard about Richard Nixon's resignation on August 8. In all, the *Glomar* recovered only two nuclear-tipped torpedoes and the bodies of eight Soviet crewmen from K-129. They were buried at sea on September 4, 1974. At least, that's the story told by the CIA. After columnist Jack Anderson divulged classified details about Project Azorian in a *Los Angeles Times* radio program on March 18, 1975, rumors flew that the agency intended to return to K-129's site to bring up the pieces lost in the first attempt. The truth about what really happened remained clouded until the CIA released portions of an in-house journal written by a participant in the operation whose identity remains classified. Although the fifty-page article contains gaps of redacted text, several sections point to an altogether different outcome to Project Azorian than previously reported. On page forty-seven of the journal, an anonymous participant delivers the following account:

> The Soviet tug [SB-10] left. We were going to be able to do the telltale pumpdown operation without surveillance. Our cover story had held: the Soviets had been fooled. Now we could anticipate seeing our prize without being concerned about sharing it with the owner. Everyone wanted to see the first glimpse of the target. [Redacted text.] Those of us waiting anxiously on deck received a reward of a different type. Bobbing up to the surface (luckily in the well) was a brimming full Jerry-can of torpedo juice. It had travelled over three miles to the bottom and back and been subjected to pressures of over 7,000 pounds per square inch without spilling a drop. [Redacted text.]
>
> The Mission Director and his team viewed the scene from a balcony-like portion of the ladder, which led down to the well gates. Radiation monitors had reported readings five times background even at this distance. We knew that we were in for a nasty time. Some of the earlier excitement of the mission was returning to the exploitation party. [Redacted text.] It was going to be difficult—the jumbled hulk was not going to reveal its secrets easily.

A concluding paragraph in the article is also quite revealing, again alluding to full success of the mission in stating: "Thus, the long saga of Azorian came to a conclusion as the HGE [Hughes Glomar Explorer] rested at anchor in the Hawaiian Islands, more than six years since the Soviet G-II-class submarine 722 sank in the Northwest Pacific Ocean. The efforts to locate the site of the sinking and to conceive, develop, build, and deploy the HE system [redacted text] stretched almost as long in time, beginning in mid-1968. And the success that was achieved depended, in the end, on the combined skills of a multitude of people in government and industry who together forged the capability that made it possible to proceed with such an incredible project."

The above statement stands in stark contrast to the dismal admission of near-failure from the CIA in 1975 after the story leaked. A declassified Memorandum of Conversation released by the CIA records a conversation between Secretary of Defense, James R. Schlesinger, and President Gerald Ford during a meeting in the Cabinet Room on March 19, 1975. They were discussing how to deal with Jack Anderson's *Los Angeles Times* radio story and a subsequent article about Project Azorian printed in the *New York Times*.

Schlesinger is quoted as saying "This episode has been a major American accomplishment. This operation is a marvel—technically, and with maintaining secrecy."

President Ford replied, "I agree. Now where do we go?"

After Carl Duckett retired in 1974, he contacted Joe Houston, who had recently returned to Itek as the head of special programs for the Applied Technology Division in Sunnyvale, California. Duckett wondered if ATD might need his services in some capacity. Houston immediately called his boss and convinced him to bring Duckett on board as a consultant. Houston and Duckett worked together for another decade and over those years became close friends.

Duckett never confirmed or denied reports that the CIA story about Project Azorian's failure was a smokescreen to hide the truth. Houston heard through others involved in the project that the salvage operation was hampered by nuclear contamination on some of the recovered objects, which included sixty bodies versus only eight, and a few of the nuclear missiles. In fact, Houston heard that several people, while handling those missiles, were burned by radiation and had to "shower off for an hour" in the decontamination area.

Over a few beers at Houston's house, he asked Duckett if the rumors might be true, that the *Glomar* brought back much more of K-129's remains than previously reported. Duckett's eyes twinkled, and he said, "You know, I was really upset when the CIA said they wanted to return to the K-129 site for a second attempt. That made it sound like we didn't do our jobs and get what we came for the first time around."

CHAPTER FOURTEEN

Never was anything great achieved without danger.
—NICCOLÒ MACHIAVELLI

MANY OF THE DETAILS SURROUNDING PROJECT Azorian became public knowledge after the press leaked the story in 1974. A full year before the *Glomar Explorer* set off on her mission to bring up K-129, Captain James Bradley had another brilliant idea. He figured if the *Halibut*'s *Fish* could locate K-129 on the bottom of the ocean 18,000 feet deep, certainly she could locate a Soviet communications cable in the Sea of Okhotsk at a depth of only 400 feet.

The Soviets likely had no qualms about sending unencoded top-secret messages through their Ministry of Communications cable, given the impossibility—in their estimation—that anyone could tap the signals. But the cable, running between Vladivostok and Petropavlovsk naval bases, could be tapped with a sophisticated recording device to gain access to the Soviet Union's private military conversations. Bradley envisioned a recording apparatus designed to capture weeks of traffic, extracted by the on-station *Halibut*. Russian-speaking analysts could then transcribe the unencrypted tapes, delivered to the U.S. Navy's intelligence headquarters at Suitland, Maryland.

The anticipated traffic might include Soviet military plans, details of maneuvers, unguarded conversations between key military and political figures, and other intelligence delicacies. Perhaps loose Soviet tongues would reveal the location of warheads that often landed near the Kamchatka Peninsula or future plans for ballistic missile tests. U.S. submarines, provided with foreknowledge of pending exercises, could be sent in on special operations to photograph and monitor the tests. The NSA would also be alerted to pending weapons tests by increases in military communications traffic through the cable. Before any of these benefits could become reality, however, a major hurdle needed jumping.

Mammals on the surface breathe roughly an eighty percent to twenty percent mixture of nitrogen to oxygen. Sport divers learn that air compressed into a metal tank can kill or cause serious injury if certain depths are exceeded for too long. Even navy divers are taught to stay above 130 feet for most dives, and although many of us exceeded 200 feet for various missions, we stayed down no longer than the four minutes allotted by U.S. Navy dive tables. Diving to 400 feet compresses air to the point where one lungful draws substantially more nitrogen and oxygen than "landlubbers" breathe. Such concentrations can poison divers and turn them into the equivalent of a town drunk, complete with slurred speech, crazy thoughts, and blurred vision. Although SeaLab underwater habitats experimented with a mix that contained helium versus nitrogen to overcome nitrogen narcosis, divers had to acclimate to the pressure by staying inside a chamber for long days or weeks, and they hated the cold, cramped environment. Still, with helium-induced Donald Duck voices, they often joked that "to air is human, but to HeO_2 is divine."

A serious accident, resulting in the death of one diver, put an end to the SeaLab experiments, but not the divination of HeO_2 saturation diving. Nascent and dangerous, the science moved forward. The art of deploying saturation divers covertly from a submarine in frigid Soviet waters, however, was an altogether different animal that required an innovative new approach. Although navy scientists had envisioned such deep-diving projects, their concepts were never proven in the field. Creating this capability would impel American ingenuity to new heights, and send submariners and divers on the most difficult, dangerous, and decorated missions of the Cold War.

Captain James Bradley pushed through a program to have the *Halibut* outfitted with a pressurized hyperbaric diving chamber mounted on the stern to support deep-sea bottom walking. The chamber was designed to resemble the *Mystic* Deep Submergence Rescue Vehicle (DSRV-1), which became the *Halibut*'s deployment cover story. The DSRV-like chamber, some fifty feet in length and eight feet in diameter, looked like a giant torpedo mounted to the back of a submarine by way of four metal stanchions. Divers entered the chamber from underneath through the boat's escape trunk.

All submarines have one or more small decompression chambers built into the hull. These are called lock-out chambers or escape trunks, and they allow divers to exit or enter the boat while submerged. These escape trunks are also designed to let submariners exit the sub in the event of an emergency—provided that the boat is not more than a few hundred feet deep. While these trunks are adequate for standard compressed air dives, they are not suitable for saturation diving using helium. A hyperbaric diving chamber is needed for this, which is essentially a sealable pressure vessel with hatches large enough for divers to enter and exit, combined with a compressor to raise the internal pressure.

There are two types of hyperbaric chambers, one for decompression and one for recompression. The former is used for deep-sea diving operations. The latter is used to treat or prevent decompression sickness. For both, occupants are "pressed down" or "pressed up" by increasing or decreasing the internal pressure, respectively. Divers are required to remain inside for many long hours, days, or even weeks to prevent serious injury.

In October 1971, the upgraded *Halibut,* carrying a stern-mounted decompression hyperbaric chamber and loaded with deep-sea saturation divers, cruised toward the Soviet Union with Commander "Smiling Jack" McNish in charge. The Soviet communications cable they sought ran thousands of miles under the ocean deep in the Sea of Okhotsk, a large body of water, surrounded by land, that from the air looked like a massive Russian lake. The undersea cable ran from the Petropavlovsk submarine base and missile-testing facilities on the Kamchatka Peninsula, down to Vladivostok, headquarters of the Soviet "Far East" Pacific Fleet.

Commander McNish successfully located the cable's approximate position via *Halibut*'s periscope by finding a small sign posted on a beach warning passersby to avoid running into a buried cable. From there, finding the actual cable in the sand required delicate manipulations of the camera-equipped mini-sub *Fish* using miles of cable lowered from the bowels of the boat. The *Fish* swam on its own power, guided via a tether running into the large bat cave on the bow of the *Halibut*. The tether often fell prey to snags and ocean currents that caused serious jarring of cameras and equipment. *Halibut's Fish* took rolls of film that revealed nothing. Finally, after several days of searching, the image of a three-inch-diameter Soviet communications cable appeared in one color photo.

The sat divers then locked out of the submarine's fake DSRV, tapped the cable, and siphoned off a flurry of signals, but the SIGINT spooks on board were not trained in this type of signal capture. Mc-Nish returned with poor-quality recordings. The *Halibut* made a second run on August 4, 1972, but came home with no better results. NSA and navy officials were disappointed but convinced that intelligence pearls awaited in the cable if skilled experts could tap into them properly. Each mission day cost the navy more than $1 million, given the expense of SR-71 spy-plane overflights to monitor Soviet traffic in the area, as well as decoy submarines poised to create distractions should the Soviets come snooping. In light of the expense and importance of the operation, the NSA escalated the signals capture issue to a priority position.

In the summer of 1974, they contacted one of their own, John Arnold, and tasked him with solving the problem. A navy Mustang officer, Arnold had scratched his way up from seaman to lieutenant commander while gaining expertise in underwater eavesdropping. As a navy spook, he'd previously conducted espionage missions off the Soviet coast aboard the USS *Nautilus,* including the one that used cameras mounted to toilet paper rolls to photograph nuclear test blasts near Novaya Zemlya in 1961. The NSA instructed Arnold to assemble a crack team of spooks, which he found at the "Fred Ten" Wullenweber Bulls Eyes facility in Sabana Seca, Puerto Rico. Arnold's handpicked team consisted of four navy communications technician chiefs, including Master Chief Malcolm "Mac" Empey and Chief Mark Rutherford. A Russian-speaking I-Brancher accompanied the four chiefs.

The five chiefs strode across the gangplank from the pier, seabags slung over uniformed shoulders. The crew did not know these strangers, but most had little doubt as to their function and exchanged excited whispers acknowledging that the "spooks had arrived." Although selected by Arnold, the team reported to their OIC, Tom Crowley, who reported to Captain Augustine "Gus" Hubel. They were officially attached to the clandestine Naval Security Group, but none of the spooks thought of themselves as spies. In fact, they never claimed titles higher than sailors or submariners.

Mac Empey was colorblind, but that didn't hinder his electronic genius. Since the quirky M-Brancher couldn't discern the color coding for electrical wiring or blueprints, that forced him to learn the entire function of each circuit by heart. Those who witnessed Empey troubleshooting a system were entertained by the constant bobbing and weaving of his head as he scanned schematics up, down, and sideways. The other spooks often joked that he looked like a parrot planning an escape.

Mark Rutherford relished his clandestine role as a spook. He loved the "shaken, not stirred" image of James Bond and often imitated his favorite movie character. He had a way with words and women and could charm almost anyone. He also excelled at his job, often demonstrating his technical prowess on every piece of gear used by the spooks.

While the *Halibut* crew trained for more than a year to prepare for their next mission, Empey and Rutherford performed amazing technological feats, including the creation and integration of much of the advanced systems used for the mission. The *Halibut* didn't have CTs of their caliber on the first two runs who understood the nature of the tapped signals. The original R-Branchers performed a "capture all" of every conversation on every wire or "channel" and dumped everything onto a recorder. That was like trying to listen to a conversation in a small room when twelve people were talking at once. Making sense of the recorded cacophony created an NSA nightmare, which is why the first two missions failed. To solve this problem, Empey and Rutherford knew they had to separate each channel. They also needed to make the recorded conversations easier to hear by using better filtering and signal-processing techniques.

To accomplish this, they specified thirty pieces of equipment and lugged them down to a cordoned-off spook area in the torpedo room called the "chart room." I-Branchers came up with the name in reference to the small shack used to store confidential charts aboard Soviet submarines, such as the one on *Foxtrot*-class boats. Over seventy percent of the gear used by the spooks was not military grade, or "mil spec," but came from off-the-shelf manufacturers. The navy relegated maintenance responsibility to the team for this non-mil-spec equipment—most of which was analog versus digital—as well as the task of integrating the systems with each other and with the external cable tapping gear that delivered the signals. Empey and Rutherford did most of the wiring, soldering, tweaking, troubleshooting, calibrating, and worrying.

Much of the equipment needed to interface to a Hewlett-Packard 1730 digital computer with 32 kilobytes of random access memory (RAM), which represented state-of-the-art back then. Today, a small notebook computer can hold upwards of four to eight million kilobytes of RAM. The chief CTs spent four weeks at HP facilities in Loveland, Colorado, learning how to operate and repair the 1730.

By 1975, with five of the navy's top spooks aboard, the *Halibut* was ready to embark on a third mission to the Sea of Okhotsk. All hopes of success were pinned on the improvements made by the spooks, and failure could spell the end of the most promising espionage program deployed during the Cold War.

WHEN FIRST CLASS NAVY DIVER DAVID LeJeune walked across the brow of the USS *Halibut* in 1975, he glanced at the fake DSRV chamber on her deck and wondered if he should have heeded Zipperhead's admonition. His former executive officer warned him that volunteering for Special Projects would mean living in a tiny chamber with three other divers for almost two months at a time. Such an assignment required round-the-clock dives for weeks in enemy waters, where one wrong move could prove fatal not just to the divers, but to the entire submarine crew. LeJeune ignored Zipperhead's warnings and signed up anyway. He'd been saturation diving for years, had earned his instructor's pin, and was a member of the team that set a world record on a mixed gas dive to more than 1,000 feet. He figured if he had the chops to do that, he could muster the balls to ride the *Halibut*.

Born in Fairbanks, Alaska, LeJeune and his family migrated to Santa Fe Springs, California, where he asked his junior high school sweetheart, Cheryl, to marry him. She turned him down, of course, but with a wink and a smile. LeJeune didn't stop trying. He walked Cheryl to the bus stop every day and then, just to impress her, ran the five miles to the school—beating the bus every time. Finally, Cheryl accepted the proposal but insisted they delay the wedding until after their high school graduation. LeJeune waited patiently for the next four years, but just before he could win his lady's hand, Santa Fe Springs High School kicked him out for being a rowdy nonconformist.

LeJeune managed to claw his way to a diploma during summer school, and his reward followed soon after: a draft notice from Uncle Sam. He ignored the notice, bought a motorcycle, and married his first love. LeJeune and Cheryl's wedding bells were accompanied by an unexpected gift: a second draft notice. Having no desire to take a two-year vacation in Vietnam as a ground pounder, LeJeune rode his motorcycle down to the navy recruiter's office. They signed him up on a 120-day delay program, so he passed the time by completing welding school. At the time, he planned only to use that skill on dry land, but he decided to get his feet wet anyway.

LeJeune bought a cheap mask and a pair of fins and enrolled in Los Angeles County's scuba diving program. He paid his $27 for the training and rental equipment and jumped into the water. After cracking the books and completing his pool training, he took a boat ride to Catalina Island for his certification dive. "I puked my guts the whole way," he said, "but once I got into the water, I felt great. I looked around at all the bright colored fish and said, 'This is for me.'"

Convinced that the ocean world held the key to his future, he volunteered for Underwater Demolition Team (UDT) training during boot camp in San Diego. In those days, U.S. Navy SEALs and UDTs were separate commands, and LeJeune wanted to stay in the water, not fight in a jungle. Whether he could pass the prerequisite physical training tests for UDT school was never a concern. He'd run a five-minute-mile since his bus-beating days in junior high, swum like a dolphin in L.A. County's scuba school, and marched off ten pounds in boot camp. He'd also lettered in high school basketball and dunked opponents more than once in water polo. But the line between confident and cocky is sometimes a thin one, which can often throttle one's passion. "I just didn't try hard enough," LeJeune said. "So I came in ten seconds late, and they failed me. I pleaded my case, said I'd do it again, but they turned me down. I damn near cried. I didn't want to be just an ordinary sailor."

Defeated and depressed, he reported to his first duty station at the submarine base in New London, Connecticut. That's when he first heard the term *fleet diver*. He had no idea what that meant, but he knew it had something to do with the ocean, so he filled out a request chit to attend the training. The navy turned him down. Devastated but determined, he put in five more chits. The sixth one got accepted, and he and Cheryl packed up and drove to Norfolk, Virginia—but not before he made the mistake of getting a U.S. Navy diver's tattoo on his arm. When the divers at the navy diving school noticed the unearned tattoo, they punished him by delaying his program start by more than six months. In the mean time, his previous welding experience got him assigned to a detail cutting and torching on the USS *Seawolf*. "Had I known I'd have to dive off that old geezer years later," LeJeune said, "I would have cut a hole in her too big to patch up."

The USS *Scorpion* had gone down with all hands a month earlier, in May 1968, and Cheryl made LeJeune promise that he'd turn down any diving jobs that involved submarines. He agreed, but fate had other plans. In early 1969, Master Diver Billy Kitchens finally let LeJeune suit up for his indoctrination dive—a requirement prior to attending second class divers school to weed out the claustrophobic twitchers. He strapped on his Mark 5 diving rig, complete with a vintage round metal helmet and baggy canvas suit, all the while spewing questions and funny wisecracks. Kitchens never laughed. LeJeune sat in the itchy diving suit for thirty minutes as perspiration stung his eyes and drenched his lips. More bored than irritated, he said, "Aren't you going to give me a test or something?"

Kitchens rolled his eyes and held out his hand. A dime sat in his gloved palm. "See if you can get this dime out of my hand with your glove."

"That's easy," LeJeune said. He smacked the bottom of Kitchens's hand, and the dime went flying into the air. LeJeune caught the dime, opened his glove, and said, "There."

Kitchens manufactured a frown and said, "LeJeune, I really hope you become a master diver someday, and a little wise-ass student like you makes your life a living hell. That would at least give my days in the navy some meaning."

LeJeune spent the next thirteen weeks earning his second-class diver's pin. During that time, Cheryl's name changed to George. The impetus came from being dirt poor without even a spare quarter to attend a movie on the base. That, and a television commercial that depicted two bored vultures with one asking the other, "Well, what do you want to do?"

"I don't know," LeJeune always replied when Cheryl mimicked the vultures. "What do you want to do, George?"

The name stuck.

LeJeune attended first class diver's school in 1970 and graduated from the Washington, D.C., Navy Yard in January 1971. He spent the next several months in boredom, doing little more than inspection dives on the presidential yacht USS *Sequoia* (AG-23). After a transfer to New London and another spate of less than exciting dives, LeJeune volunteered for saturation diver training under the navy's Man in the Sea program. That's when his life changed in ways he could never have imagined.

The navy used an ominous-looking craft called the IX-501 *Elk River* to train divers on the most dangerous and advanced form of diving extent. The IX designator is given to ships that no longer serve in their original capacity. Commissioned in 1945, the *Elk River* once operated as a rocket-launching medium landing ship but now deployed strange space-suited creatures down to depths never before attained by humans. Divers wore one-eighth-inch-thick neoprene wet suits against their skin, but that's where any resemblance to ordinary diving ceased. Over the rubber wet suit they draped a baggy pair of coveralls that resembled a Hazmat suit, but completely enclosed. A hose connected to the back of the suit and pumped in hot water to keep the divers warm in below-freezing conditions. The heated water ran through smaller tubes that branched away from the diver's back and formed a spider's web covering the arms, legs, and chest. Warm water from the tubes coated the neoprene underneath like a sprinkler system on a summer's day, while the skintight wet suit prevented scalding from the heated water.

Divers wore a face mask manufactured by Kirby Morgan that resembled the clear Plexiglas masks worn by firefighters. The masks contained microphones and speakers for communications, as well as readouts for temperature and mixed gas levels. A hose ran from the Kirby Morgan that contained the most important element to all divers: air.

LeJeune learned in saturation diving school that dive durations were limited primarily by two factors: gas supply and diver fatigue. "Saturation diving uses one hell of a lot of gas," said LeJeune. "We only pump in three percent oxygen compared to almost ninety-seven percent helium, but that three percent feels like thirty to the diver."

Extended dives not only used up large supplies of helium, but also placed divers at great risk. The longer a diver stays in the water, the more tired he becomes. That's when mistakes are made. Still, macho saturation divers often held contests to see who'd wimp out first and want to be reeled back up to the surface. "Ten or even twelve hours underwater was not uncommon," said LeJeune. "In fact, if the job only took six, we'd stay down another two just to save face."

LeJeune graduated from sat school and, soon after, requested instructor training. During that program, a master diver asked him to build a system for one of the diving rigs to make it easier for divers to close various valves in an emergency. LeJeune rigged up a pneumatically controlled system that caught the attention of a sat-school student, Lieutenant Commander Ross Saxon, eighteen months later. Saxon—who had a large scar on the back of his head that earned him the nickname "Zipperhead"—graduated and became executive officer of the newly commissioned USS *Ortolan* (ASR-22). He then requested, through official channels, that LeJeune build a gas analysis system to support the ship's thirty divers and two rigs. That system and the emergency valve rig created by LeJeune were later adopted by all navy saturation diving platforms worldwide.

LeJeune stayed on board the *Ortolan* from 1973 to 1975. During that time, he lived the life of hero Dirk Pitt in a Clive Cussler novel. Still starving on navy pay, he and fellow divers Chris Velucci and Don Rodocker came up with a brilliant plan to use saturation diving to hunt for deep-sea treasure. They found some investors, who put up $280,000 to buy the needed equipment and dive boat. Months later, they spent their navy "vacation" leave motoring a less-than-adequate dive boat to a point off Nantucket, Massachusetts, where they dropped anchor. They spent the next forty days diving on the wreck of the *Andrea Doria,* an Italian ocean liner that sank after a collision with another ship in 1956.

The three divers dreamed of finding precious valuables tucked inside the wreck's steel tomb while fending off bandits who tried to rob them at night. During the day, they avoided lurking hungry sharks and militant nosy authorities. The Coast Guard tried to shut them down more than once, claiming that living conditions and life-jacket supplies were inadequate on board the diving boat. "We'd move life jackets from one location to another when those guys weren't looking so they'd think we'd had enough," said LeJeune. "They never did catch on. We also carried a bunch of rifles and fired into the night air to keep the pirates away."

The three treasure hunters cut a large hole in the *Andrea Doria*'s side but found nothing more than a chafing dish, a bottle of perfume, some silverware, and the ship's bell. Their investors were less than pleased. "At least we got a lot smarter about saturation diving on that trip," said LeJeune. "That experience probably saved my ass when I did those Ivy Bells missions off submarines years later."

Which is where LeJeune went next. Tired of working on a sat diving ship while enduring "puke storms," he put in a transfer chit to Special Projects.

"That means living for weeks or months inside a tiny chamber on a submarine," Zipperhead said to LeJeune.

"I know," LeJeune said. "But at least it's calm down there."

That decision brought LeJeune face-to-face with the submarine that had found the K-129 near Hawaii years earlier and hurled him toward a life that gained him more medals than most combat soldiers earn after years on the battlefield. LeJeune didn't know it at the time, but cable-tapping missions, code-named Ivy Bells by the NSA, were considered by the military the most important missions of the Cold War. They were also the most dangerous, requiring direct approval from the president of the United States prior to each deployment.

LeJeune's first clue that these missions were far from ordinary came within hours of reporting aboard the USS *Halibut*. He learned that out of the twenty-two divers who accompanied the *Halibut* on her first two runs, due to the arduous conditions and dangerous nature of the dives, only two volunteered to go again. One of the original divers developed kidney stones on the second run, and Lieutenant Commander "Doc" Halworth had to press down, enter the dive chamber, and shoot him full of Demerol. Given the *Halibut*'s slow speed, the diver suffered for weeks until the submarine returned to port. LeJeune suspected that the incident took a mental toll on many of the other divers, which probably accounted for the high attrition rate.

When LeJeune met Doc Halworth on the *Halibut*, he instantly liked the eccentric dive-trained doctor. "Halworth had a hobby of buying abandoned missile silos," said LeJeune. "He liked to renovate the things and sell them for a profit. He also bought caseloads of overstock shirts to save money. He'd wear an identical shirt damn near every day buttoned all the way to the top. The guy looked like a geek."

Geek or not, Halworth's job aboard the *Halibut* entailed keeping the divers alive. The doctor analyzed the diving math to determine how much gas each diver needed while down. He ensured that mixed gas ratios were accurate and that emergency procedures were adequate to prevent serious injury or death. Still, the divers loved to play tricks on Doc Halworth, including cutting the fingers off the ends of all his rubber examination gloves.

LeJeune's fellow divers included an Asian-Indian with huge "Popeye" forearms, Eugene Diem, whom they nicknamed Gunga Din, a smooth-talking consummate professional named John Hunt, and a short black-haired Italian called Bob "Ginny" Vindetto. Ginny also had a talent for finding and trading spare parts and often joked that if he ever died on a mission, God would let him into heaven because everybody needs a "cumshaw guy."

Twenty-two Special Projects divers competed for four bunks inside the small DSRV-sized hyperbaric chamber, as only four divers made the cut to conduct the actual mission dives in the Sea of Okhotsk. "The master diver selected the four best divers for each mission," said LeJeune. "Divers were disappointed if they didn't get picked, but we were all professional about it and worked as a team. We never forgot that we depended on each other for our survival."

LeJeune spent the next several months with his fellow divers on the *Halibut* conducting at-sea training missions off the coast of San Diego. The *Halibut* set down on the ocean floor using snowmobile-like skis mounted to the underhull of the boat. Teams of three divers deployed from the DSRV on practice runs until every diver had been in the water at least once. The divers wore their neoprene wet suits underneath newer semi-closed Mark 11 saturation diving suits that cost the navy $87,000 each. The MK-11s employed heated water sprinkler systems to keep the divers warm, but also had a small pocket in the front for a thermoluminescent device. The TLDs were similar to the dosimeters worn by nuclear submariners to monitor radiation dosages but were waterproofed down to 1,000 feet.

Inside the diving chamber, divers wore dull brown fire-retardant T-shirts, boxer shorts, pajamas, and socks. Even the towels were brown and fireproof. "The clothing they gave us was hot and itchy and had zero absorbency," said LeJeune. "Using those towels was like drying off with a squeegee."

Helium's light weight increases its conductivity, and the gas will rob body heat from a diver at a rapid rate. In fact, helium is the second-lightest element known. Only hydrogen weighs less. Earth's atmosphere contains just five parts per million of the gas, making helium rare and expensive. Chemically inert, helium offers no odor, color, or taste, which makes the gas ideal for diving as long as strict safety precautions are followed. Just as in scuba diving, where deep divers need to decompress to expel nitrogen from the bloodstream, helium divers must do the same to prevent decompression sickness, commonly referred to as "the bends."

In addition to the bends, saturation divers are at risk for a phenomenon known as high-pressure nervous syndrome (HPNS). The syndrome causes dizziness, nausea, vomiting, tremors, fatigue, body jerks, stomach cramps, decreased mental and physical performance, and insomnia coupled with nightmares. The condition worsens with faster rates of compression or decompression and at greater attained depths. Symptoms were noted in the early 1960s when sat diving came on the scene, and doctors initially referred to them as "helium tremors." The only way to prevent HPNS is to go down and back up again slowly in a hyperbaric chamber while pausing at intervals to allow the body to dissipate the helium from the bloodstream in a similar fashion to decompressing to expel nitrogen.

There were three sections to the cylinder-shaped diving chamber. Like the DSRV it resembled, the fifty-foot-long by eight-foot-diameter chamber had two bulkheads inside that divided the space into three compartments. The smaller aft section housed the watchstanders—a couple of divers who monitored the four mission divers and catered to their every need. The middle section of the chamber had a hatch that opened into a round connector that attached to the escape trunk on the submarine. Divers entered the chamber through this hatch. The area was also used to store extra equipment and served as a pressurizable entry point should someone need rapid access to the divers, like Doc Halworth.

The larger bow section kept the four mission divers alive and served as the exit door out to the sea. Similar to a spacecraft, this area had its own life-support system, and not too unlike the environment found in space, the occupants could not easily come back to the world of non-pressurized air should something go wrong.

Watchstanders funneled food, drinks, and other supplies from their nonpressurized aft section to the divers through an eighteen-inch-diameter "medical chamber" cylinder that ran into the forward pressurized section. "Everything we ate and drank was cold," said LeJeune. "Helium doesn't just rob body heat, it also steals warmth from food." To make up for the cold meals and limpid coffee, watchstanders ran movie projectors that shot flickering light through a Plexiglas window that came to life on the back metal wall. "Sometimes the watchstanders played jokes on us by screwing the lid down too tight on a catsup bottle," said LeJeune. "After it's pressurized, you can't get the damn thing off. We got them back by doing the same thing in reverse. Only when they opened the bottle, catsup exploded in their face."

Helium's superconductivity required that the diver's forward section be heated to ninety-six degrees Fahrenheit, which to the divers felt like seventy-two degrees. Humidity ran high—almost one hundred percent—which exacerbated health concerns. "A diver could go to sleep with a minor earache and wake up with pus dripping from his ear," said LeJeune. "Bacterial infections were a big problem." The divers also shared one very public toilet and a small sink but enjoyed no showers. Fresh water for sponge baths, shaving, and brushing teeth was delivered through the medical chamber cylinder.

Past thirty-five feet, as the helium level rose, divers sounded like cartoon characters with high-pitched voices. To compensate, they spoke into a "helium speech unscrambler." This small box electronically synthesized the diver's helium-induced Donald Duck squeaks, then lowered the pitch and tone and pumped out normal-sounding speech.

Pressing down in the chamber, to equalize the pressure with the ocean outside, took as long as ten to fifteen hours, depending on the depth of the dive. Ten hours was normal for a 400-foot-dive, and getting there any quicker risked exposing the divers to HPNS. "If we went down too fast," said LeJeune, "every joint ached as the pressure pushed out the lubricating joint fluid. Your body needs time to adjust to a high-pressure environment. Most of us got HPNS more than once, which made us act like a drunk with a case of the DTs."

In 1975, LeJeune made his first saturation training dive off the coast of San Diego aboard the USS *Halibut*. He and three other divers locked into the claustrophobic hyperbaric chamber and pulled on their neoprene "undergarment" wet suits. Body odor from the other divers found LeJeune's nose as he sat on a small cot in the divers' section and stared at various indicators, valves, and piping running along the inside of the enclosure. Master Diver Al Frontz, seated outside the diver's section, pressurized the chamber down to thirty-five feet. At this point, the divers still breathed normal air. The "press down" speed used by Frontz to descend to that initial depth determined the partial pressure of oxygen employed for the dive—usually 0.28 to 0.30. Hours passed as Frontz slowly increased pressure, now using only helium. At 160 feet, when the percentage of oxygen in the chamber could no longer support combustion, O_2 sensor alarms were turned off.

A half-day later, with the chamber pressurized to the operating depth of 400 feet, LeJeune stood from his cot and followed the other divers to the back hatch that led to the diving platform. This back area was "blown down" using a small pipe with threaded ends that ran from the divers' section, through the bulkhead, and into the diving platform area. The divers blew helium through this pipe to pressurize the area to the point where the ocean water came up only to the bottom of the exit hole that led to the ocean. They removed the pipe during transit, as it could pose a safety problem for the submarine should the boat experience a flooding incident. If a master diver were to misplace the pipe or forget to bring it along for the ride, the entire mission could be scrubbed.

LeJeune stepped through the thirty-six-inch-diameter opening at the front of the chamber and found his MK-11 diving rig hanging on the bulkhead. He also donned his suit and face mask. The mask was clamped to a fiberglass frame that formed a hydroseal made out of thin rubber. Around the rubber a spongelike material helped hold back the ocean. Divers soaked the material in water to draw out the air, smeared Vaseline on the rubber, then clamped the mask down tight to prevent leaks. "One of the limiting factors on dives was that mask," LeJeune said. "After eight hours in the water, your jaw felt like you'd been punched in the face by Muhammad Ali."

Mask ready, LeJeune pulled on his rubber boots, which contained lead insets and a steel toecap held on with rubber straps. Popeye-armed Gunga Din checked him over to ensure there were no issues. Once suited and checked, LeJeune sat on the edge of the diving platform with his feet dangling in the water. An adrenaline rush ran through his veins as the bottom portion of the platform lowered into the ocean like a hydraulically operated wheelchair ramp in a minivan. The ocean covered his face as he descended into a world of black.

Pitch black.

The blackest of all blacks. Like black hole black.

LeJeune spread his arms and pushed off the curved edge of the submarine. Just like a free-falling skydiver, he floated through the silent ink and fell toward the ocean floor. "This was always my favorite part," said LeJeune. "I never knew if I was falling into a shark's mouth or what, which made the free fall that much more exciting. I used to live for that moment."

Timing his fall, LeJeune braced himself for the impact as his leaded boots hit the murky silt. He heard nothing but the hissing sound of his own breathing. He switched on his light and moved in slow motion toward the other divers. Given the strong currents in the area, he wore two weight belts, which slowed him down even more. Tiny fish darted past his face mask. His light reflected from their shiny sides and made them look like fireflies in an alien world. LeJeune glanced at the two lights inside his face mask near his left eye. The green one was lit, and the red was not. Green was good. That meant his HeO_2 mix was fine. Red was bad. That indicated that his breathing mix had dropped to zero. In that event, he carried a coffee can–sized bottle of mix with enough air for about three breaths. LeJeune often wondered if that can of air would keep him alive long enough to get back inside the boat.

Tubes snaked throughout LeJeune's MK-11 suit, and warm water surged and sprinkled his goose-bumped skin through tiny holes. A two-inch-diameter tethered cord tied him to the boat and delivered the mixed gas, warm water, two-way communications, and power for his lights. One heavy wire in the cable also functioned as a lifeline to reel him back in should an emergency occur.

LeJeune moved toward a line locker located on the underside of the *Halibut*. Gunga Din opened the locker and removed a cord that resembled an oversized automobile jumper cable. The cord could be reeled out to a maximum of 350 feet. Ginny Vindetto grabbed the cord and motioned for the others to help. Only three divers deployed, with the fourth still inside in the event of an emergency. The divers pulled the cord over to the simulated three-inch-diameter communications cable running along the ocean floor. As if they were actually in the Sea of Okhotsk on a live tapping mission, they clamped a "grabber," fashioned to the end of the cord, around the cable. They covered up all evidence of the cable tap with a large sand-colored rubber blanket, then started back toward the *Halibut*.

Near the submarine, Ginny started singing. At first, LeJeune wondered if his ears were playing tricks on him. Then Ginny started cracking jokes, Silly Sally jokes, like the ones he'd told on the playground in grade school.

LeJeune radioed Master Diver Frontz inside the *Halibut* and told him that Ginny was showing signs of HPNS. Frontz acknowledged the report and told them to help Ginny back inside ASAP. LeJeune approached Ginny and reached for his arm. Ginny pulled away. He told LeJeune to "leave me the fuck alone." LeJeune tried again. This time Ginny ripped down the zipper on his MK-11 and let the twenty-six-degree water flood his suit. LeJeune yelled at Ginny to rezip, but Ginny refused. Gunga Din slow-ran over, and the two tried to get the zipper back up, but Ginny fought them off.

Ginny started to sing again, but no more than three words came out before the freezing water temperature finally took its toll. His jaw locked up, and his eyes rolled up into their sockets. He passed out. LeJeune radioed the fourth diver, Cleveland Smith, inside the chamber and told him to start reeling.

LeJeune clung to his friend, while Smith reeled in Ginny's cable line. His arms dangled and flapped about as the currents rushed past, and LeJeune wondered if they were already too late. Inside the chamber, he pulled off Ginny's MK-11 suit, while Smith grabbed the CPR kit. They pumped Ginny's chest and blew air into his lungs. Nothing. No movement. Certain that his friend had just died, LeJeune's heart sank. Finally, Ginny coughed and spewed out saliva. His eyes opened.

He stared at LeJeune for a long moment, then said, "Am I still alive?" LeJeune nodded. "I'm not God."

"I was thinking more like the devil," Ginny said with a smile.

After the incident with Ginny, other divers started showing signs of HPNS. Master Diver Frontz and Doc Halworth shortened the dives and took other precautions, including rotating the divers to see if physical conditioning or other factors might be the cause. Then "all star" diver John Hunt, considered one of the least likely to get the condition, started doing goofy things after only six hours down. Frontz and Halworth decided it was time to start blaming the equipment.

Several divers, including LeJeune, flew to Panama City, Florida. They spent months in simulated environments testing various parts on the MK-11 suits until a culprit emerged. Each suit came with a CO_2 scrubber in the MK-11 backpack that "burped off" one-fifth of the expelled gas from the diver's lungs. The scrubber used baralyme chemicals to scrub the CO_2 and recycle fresh HeO_2 back to the diver. This process also evaporated air bubbles to ensure that none reached the surface, where they could be detected. The scrubber was not heated, and at cold temperatures the baralyme did not scrub efficiently. This caused CO_2 buildup and accelerated the possibility of HPNS. The team redesigned the system and replaced the baralyme with lithium hydroxide—the same chemical used by submariners to scrub CO_2 on diesel boats. One of the drawbacks included the possibility of chemical burns, but as far as the divers were concerned, staying sane took precedence.

WITH HER DIVERS TRAINED AND READY, the *Halibut* made her third voyage in June 1975 to the Sea of Okhotsk, which in Russian means "Mouth of the Bear." Excited about conducting his first live mission in Soviet waters, LeJeune accompanied the boat. Unfortunately, he didn't have the balls to stay on board long enough to make a dive. After only a week out, he developed a condition that caused one of his testicles to swell. Doc Halworth insisted that he depart when *Halibut* pulled in for a minor repair before heading west. Reluctantly, LeJeune complied. The problem turned out to be minor, but by then the *Halibut* had already left on her voyage to the hinterland.

Less LeJeune, *Halibut* arrived on station in the Sea of Okhotsk a month later. The trek to the Western Pacific took almost twice as long as for most other attack boats due to *Halibut*'s geriatric reactor, which could muster no more than thirteen knots. After sneaking past the Soviet fleet near the Kamchatka Peninsula, Commander McNish informed the crew that their mission would be one of the most dangerous in naval history. They'd be well within Soviet territorial waters, and if caught, they might not come home. Explosive charges had been placed throughout the boat to ensure that no equipment or survivors could be captured. McNish stopped short, however, of delivering all the details to the crew about their cable-tapping mission, and only he, his officers, the divers, and the spooks knew anything about the fine print.

Upon relocating the cable, the *Halibut*'s crew used a set of anchors and a winch system to set the submarine on the soft ocean floor atop the two skis. She anchored near the communications cable, and the divers locked out of the torpedo-shaped, aft-mounted diving chamber near the boat's rudder. As they had done during practice runs off California, one diver remained inside, while three divers walked in moonlike fashion to one side of the *Halibut*. Reflective plankton clouds obstructed their vision, while the HeO_2 mix made their tongues dry.

The divers opened the sealed compartment on the *Halibut*'s hull and removed the long electrical cord. They dragged the cord over to the communications cable and located the repeater, a large metallic cylinder that joined sections of the cable every thirty miles and amplified the signals. The spooks knew that the strongest signals could be found there. While attaching the three-foot-long "grabber," which contained a recorder and rolls of recording tape, a large hake fish, attracted by the light, attacked one diver and clamped its razorlike teeth on his arm. Unable to shake the fish loose, the diver removed his knife and jabbed it between the fish's gills. The wounded hake swam away in a flurry of ink.

While returning to the *Halibut*, the divers collected some large crabs for a "mission accomplished" dinner. Inside the boat, in the sectioned-off chart room, three spooks at a time stood watch to accomplish phase one of the Ivy Bells tapping mission. This part, which employed a temporarily placed cable tap, required that the *Halibut* remain on station for several weeks while the spooks recorded signals. The incoming signals needed to be calibrated against data type, time of recording, signal strength, tapped wire (within the bundle), signal anomalies, and to ensure complete continuity and to validate recorded voice and data quality. Unlike the two previous failed missions, two sets of signals were now recorded. One was a "capture all" that contained every signal pulled from the cable, and the other separated out each channel to ensure clear conversations could be heard. The team took this function seriously, as hundreds of men were placing their lives at risk to gather the data, and they had but one shot at getting it right. After the mission, the recording tapes would be delivered to the NSA for full transcribing and analysis—the second phase of the Ivy Bells project. That is, of course, if they managed to capture anything at all.

While the divers set up the tap, M-Brancher Mac Empey and T-Brancher Mark Rutherford waited patiently with Lieutenant Commander Arnold in the chart room, a small converted storeroom forward of the reactor and opposite the radio room. Once the divers finished clamping the cable, the spooks tried in vain to collect the signals from the tap but heard nothing. They flipped switches and rechecked systems, but to no avail. They requested that the divers return to the cable and check the connection. The divers did and found that they'd attached the tap to a "dead" shielded location. They reattached the grabber to an active, unshielded spot, and, using induction, the tap picked up "leaked" signals from the cable and sent them up the wire into the spook's gear.

Empey and Rutherford collected the information and validated parameters, while an I-Brancher strapped on a set of headphones and listened. To his amazement, clear, unencrypted Russian voices filled his ears. He handed a set of phones to one of the spook chiefs, chuckled, and said, "I've got a Soviet admiral talking to his wife on one line and his mistress on the other." Only the I-Brancher could translate the words, so the others took turns listening to the private conversations and imagined other Soviet admirals exchanging top-secret information while sitting at desks in Vladivostok. In essence, the United States had just placed a glass against the Soviet Union's wall to hear their every word.

Reminiscent of a professional baseball team, basking in the glory of winning the World Series, the team of five spooks gathered in the chart room and exchanged excited high fives. The *Halibut* remained on station for another two weeks, filling reels of tape with recordings. Near their departure time, a flooding alarm sounded. A diesel engine pipe had ruptured, sending a gush of ocean water into the boat. The divers, though, were still outside and unaware of the emergency. The *Halibut*'s crew raced to seal up the leak, but freezing cold water numbed fingers and arms as they worked. Commander McNish knew that the extra water made his sub heavy, and that he'd have to do an emergency blow soon to save his boat. He ordered the divers to return to the diving chamber but feared they might not make it back in time.

The salt water rushed in, and McNish now had to make the most difficult decision of his career: blow to the surface to save his sub and leave the divers to die, or wait for the divers and risk the lives of his entire crew. Seconds before he gave the order to blow, the damage control team reported that they'd fixed the leak. McNish breathed a sigh of relief. After the divers were safely inside, he moved the *Halibut* out of the area. The submarine returned home after a short stop in Guam for repairs. Though the ride back to port felt even longer than the one out, the crew sang songs and ate steaks in the crew's mess with the knowledge that their efforts were about to make a significant difference in the duration and outcome of the Cold War.

The *Halibut* returned to the Sea of Okhotsk a month later and came back with another batch of juicy recordings. Emboldened by *Halibut*'s back-to-back successes, Captain James Bradley asked Bell Labs to build the mother of all cable taps. This device also worked through induction but occupied a twenty-foot-long by three-foot-wide container that weighed six tons and could be left on station indefinitely. That would obviate the need for a submarine to remain in the area for weeks. NSA engineers designed sophisticated new recording devices for the pod that could monitor dozens of lines for months at a time. The complex watertight enclosure contained an electronically programmed system for registering the intercepted information, nearly one hundred multichannel magnetic recording units and a miniature plutonium 238 nuclear power source. Divers nicknamed the heavy contraption, which looked like an elongated fifty-five-gallon drum, the "beast."

By 1976, the *Halibut* was ready to be put out to pasture. Two other submarines lined up to take her place, the USS *Seawolf* and USS *Parche*

(SSN-683) but both were in drydock undergoing repairs and modifications. David LeJeune and most of the divers on the *Halibut* packed their seabags and reported to the *Seawolf*. Mac Empey, Mark Rutherford, and the other spooks did likewise. As they had done on the *Halibut,* the team endured months of simulated practice off the coast of California, and the dangers involved in these dress rehearsals were no different than the real thing.

The navy kept the Silicon Valley–built pod-tapping beast on a barge in San Diego. They also kept a nonworking practice beast there made from a large metal drum filled with cement. The 1,700-pound dummy pod, also twenty feet long by three feet wide, came packed in a canvas bag encircled by lifting balloons to allow the divers to move the heavy object around on the ocean floor. Navy diver David LeJeune accompanied an NSA operative named Ronald Pelton on a drive in his brand-new Volkswagen van down the coast to pick up the barge with the fake beast and prep everything for a practice run. LeJeune recalled that one of the divers on the last dry run had overinflated the beast. The massive thing rocketed to the surface and washed up on a beach near Santa Barbara. Fortunately, the civilians who discovered the practice beast had no clue as to its function.

During their pod-prepping trips, LeJeune got to know the charismatic NSA man named Pelton over beers and even started to consider the guy as a potential friend. He had no clue at the time that Ronald Pelton would one day place the lives of submariners and divers conducting Ivy Bells missions in ultimate jeopardy through an act of treason.

LeJeune and Pelton delivered the fake beast to the *Seawolf,* and the crew stored the pod in a large locker on the outside hull of the submarine. As far as LeJeune was concerned, the *Seawolf* looked like the Waldorf Astoria compared to the *Halibut*. Instead of a fake DSRV, she had a shipyard-installed fifty-foot section just aft of the torpedo room designed for diving operations and unmanned aquatic vehicle (UAV) *Fish* deployment. Divers called the section Area D, for "Diving." The space contained a twenty-by-ten-foot hyperbaric diving chamber, similar in function and setup to the DSRV-style chamber welded to the back of the *Halibut*. Inside were four cots, a "piss bucket" toilet, and a single washbasin. The chamber snuggled up against the inside bulkhead of the submarine. Above the divers' heads, angled upward and outward, an inner hatch offered an exit. An outer hatch beyond that, similar to ones used in submarine escape trunks, opened outward to the sea. When the *Seawolf* surfaced, this hatch resided about fifteen feet below the waterline, so it could not be seen by spying eyes.

Near the diving chamber, UAV *Fish* were stored in small racks, similar to the ones used to house torpedoes. A five-foot-diameter tube running through the pressure hull, resembling a torpedo tube, pointed upward toward the ocean. As an added measure of safety, the tube contained three hatches, one inside, one midway in the tube, and one outside to hold back the sea. Operators loaded the *Fish* into these tubes, opened all three hatches, and jettisoned the mini-subs into the deep blue yonder to snoop around for things, such as Soviet communications cables.

The spook's chart room, chockful of electronic eavesdropping equipment, was just forward and dog-legged from the Area D hyperbaric chamber and UAV *Fish*. Although LeJeune met several of the spooks on the *Halibut,* he didn't really get to know Mac Empey until his time aboard the *Seawolf.* LeJeune describes Empey as "having Einstein hair, a scraggly beard, and maybe two good teeth with cigarette stains."

Empey cornered LeJeune near the chart room after a practice run and said, "Howdy! I'm Whacko Maco. Want a tour?" LeJeune's first impression was that Empey "couldn't pour the piss out of a boot if instructions were written on the heel." Born curious, however, LeJeune accepted the invitation.

Empey insisted that LeJeune first needed to watch an indoctrination video before being briefed on any of the equipment in the chart room. LeJeune agreed. Empey pulled back the curtain on the top-secret area and ushered LeJeune inside. The small space contained only two chairs and more than two dozen rack-mounted pieces of equipment that hummed and whirred. A large monitor overshadowed a keyboard.

Empey sat LeJeune down in a chair and said, "This is an important video. Pay close attention to every detail."

The spook pressed a button to start the video. LeJeune's eyes extended with anticipation as he wondered what secrets were about to be revealed. Empey stepped outside and closed the curtain. As the film's title rolled past, LeJeune realized he'd been had. He heard Empey giggling outside the curtain about the same time as Linda Lovelace's name appeared on the credits of the film.

In spite of Empey's warped sense of humor, LeJeune came to respect the rotund man's brilliance. He possessed no degrees, but God had granted the man Albert Einstein's IQ, along with the renowned scientist's tousled hair. He could multitask like a chess master and chain-smoke like a shipyard worker. When dozen's of engineering-degreed contractors couldn't find a problem with a critical piece of equipment, with the entire $150 million Ivy Bells program on the line, Empey saved the day. The engineers rolled their eyes when the chief asked for the schematic and a description of the problem. An hour later, with eyes darting and tongue wagging, the brilliant spook pointed to a spot on the drawing and said, "There. That's where it is." The engineers didn't believe him but checked anyway. Turns out Empey was right, so much so that he received a Legion of Merit from the deputy director of the CIA.

While the *Seawolf*'s crew prepared for their first cable-tapping mission to the Sea of Okhotsk, where they would set up the new beast pod, others honed their clandestine skills, unaware that their destinies would soon collide with those aboard the infamous *Wolf*.

CHAPTER FIFTEEN

All wars are civil wars, because all men are brothers.
—François Fénelon

WHEN THE SOVIET *DELTA*-CLASS SUBMARINE EN-TERED stage right in December 1972, the U.S. Navy shuddered. With a quiver of twelve R-29 ballistic missiles, each capable of delivering megatons of de-struction from almost 5,000 miles away, the *Delta* made U.S. ASW techniques all but obsolete. Prior to the *Delta*'s arrival, Soviet "boom-ers" with shorter-range missiles, in order to get close enough to the United States to launch, needed to transit from Murmansk, curve around Norway, and drop down through the Greenland/Iceland/U.K. gap. A swarm of American ships and planes greeted them there, as well as several fast-attack boats ready to latch on like trained bulldogs. With the advent of the *Delta* class, the Soviets could employ a new strategy: slip under the Arctic ice north of Murmansk and hide for months on end. When and if the order came to fire, they could sneak from under the ice cap, come shallow, and launch. By the time a U.S. submarine found them, it would be too late.

The *Delta* class rivaled America's *Benjamin Franklin*–class boom-ers in every way. In fact, the United States had actually fallen behind in this area, having not launched a new ballistic missile submarine since the USS *Will Rogers* (SSBN-659) hit the water in April 1967. Fortu-nately, U.S. attack submarines were a bit more advanced, but these ca-pable workhorses were ill equipped to find *Delta* submarines hiding under an Arctic ice floe, and no acoustic or signals intelligence yet existed on the Soviet's new boat. Without that intelligence, finding a *Delta* in open waters posed a problem.

The issue of locating these "ice boats" consumed the navy and the NSA, but they were not completely caught by surprise. Anticipating such an advancement from the Soviets, the navy had launched a new ASW program in 1964 that gave the Classic Bulls Eye program some stiff competition for funding. Under the direction of Vice Admiral Charles B. Martell, the Long Range Acoustic Propagation Project (LRAPP) sought to bring heretofore scattered ASW technologies and efforts, including SOSUS, under a single roof. Employing only a few scientists and requiring modest funds in comparison to other programs, LRAPP conducted scientific experiments and exercises that teamed top ASW civilian minds with navy counterparts. Unfortunately, all of the initial experiments failed. The brass threatened to cancel LRAPP until Dr. Marvin Lasky at the Office of Naval Research saved the day.

Lasky helped deploy a miles-long string of acoustic hydrophones towed by a ship or submarine. He called this invention the Interim Towed Array Surveillance System (ITASS), which was, essentially, a movable SOSUS sonar array. The experiments with ITASS that followed proved critical in finding a way to detect Soviet submarines in northern waters near Murmansk. During the height of the program, two opposing scientific camps stood their ground. The LRAPP scientists believed that sound in the ocean traveled in a straight line, or directionally. This meant that sound characteristics would differ depending on where they originated, and submarines could be detected from very far away. Others, including a team of AT&T scientists, argued that noise didn't travel in straight lines, and subs could only be heard if they were not too distant.

Current SOSUS arrays in the Atlantic could not detect *Delta*s in the Arctic. The navy wanted to extend SOSUS into the North Atlantic to fix this problem. AT&T scientists, convinced that their "omnidirectional" sound viewpoint was correct, insisted that this move would be a waste of money. SOSUS arrays wouldn't be close enough to the Arctic to do the job. The LRAPP engineers thought otherwise and challenged AT&T to a duel, with the stakes being the demise of the $180 million SOSUS expansion. If the AT&T scientists were right, the project was doomed. If the LRAPP scientists were correct, SOSUS lived. Moreover, future ASW capabilities could be dramatically improved, tipping the scale back in America's direction.

After a two-week experiment in which ships towed the new ITASS arrays, the LRAPP team took home the gold. They proved that sound propagated directionally. This led to the permanent installation of SOSUS in the North Atlantic and the development of ITASS arrays that could be towed by attack submarines. Soviet vessels could now be detected from far greater distances than ever before. The navy also ordered operational and procedural changes in the world of antisubmarine warfare and increased funding for ICEX Arctic experiments conducted on a tiny ice floe north of Alaska.

Sonar, weapons, and other systems on submarines operate differently in Arctic conditions versus open ocean areas. To better understand these dynamics, every two years, the navy sent two fast-attack submarines to a moving ice floe in the Arctic, a few hundred miles north of Prudhoe Bay, Alaska. Over a two-week span in the warmer month of March, a team of more than sixty naval personnel, along with civilians from the Arctic Submarine Laboratory and Applied Physics Laboratory, teamed with support personnel and a few cooks to build a temporary base camp for Operation ICEX. Two attack boats spent two weeks conducting experiments on new sonar and communications systems designed for Arctic conditions and firing practice torpedoes at each other. Scientists monitored results and used these findings to improve system and weapon designs for Arctic use.

These improved designs found their way into a new class of U.S submarines originally launched on March 3, 1967. These *Sturgeon*-class attack subs boasted vast improvements in quieting technology, coupled with the latest espionage gear. The Soviets had nothing in their attack submarine deck that compared. By 1973, after the success of the LRAPP experiments, the navy expanded SOSUS to twenty-two installations in the Atlantic and Pacific.

American engineers leveraged ITASS and HFDF technologies to design towed sonar arrays and miniaturized versions of Boresight/Bulls Eye detection and DF technologies that could be installed on U.S. ships under the Classic Outboard program. These smaller HFDF systems also found their way onto *Permit* (formerly *Thresher*) and *Sturgeon*-class submarines as integral parts of BRD-6 and -7 ESM systems. Now when Ivan transmitted a burst signal, lit off a radar, turned a screw, or farted into the wind, an American T-Brancher or sonar tech could catch the scent.

IN EARLY 1973, COMMUNICATIONS TECHNICIAN First Class Frank Turban walked across the brow of the *Sturgeon*-class submarine *Flying Fish* (SSN-673) in Norfolk, Virginia. As he stood on the cold deck along with a dozen other spooks, his toes tingled—a sure sign that this mission would be an exciting one. The third submarine to bear the name of the soaring tropical fish with long winglike fins, the *Flying Fish* was fortunate to have Commander J. D. Williams as her commanding officer. That Turban would earn a Navy Commendation Medal on this run while helping Williams earn a Distinguished Service Medal never crossed his mind.

A New York native, Turban dropped out of high school twenty-one days after his seventeenth birthday on August 12, 1963. Escaping the bitterness of a home soured by alcoholic parents, along with the possibility of an army "want you for Vietnam" draft notice, Turban took a cab to the navy recruiters office in Brooklyn. He and twenty other excited kids signed up that day to see the world and experience the navy's promise of an "adventure and not just a job." Even without a high school diploma, Turban scored high enough on his tests for a shot at communications technician school, but with no guarantees.

After boot camp, the navy sent Turban to Pensacola, Florida, for CT and T-Brancher schools. Branded with the unofficial title of "spook," sporting a brown beard, and weighing 205 pounds, he reported to his first duty station at the Pentagon in Washington, D.C. There he worked as a communicator, operating crypto equipment and handling classified messages for the Chief of Naval Operations (CNO) office. Most messages were destined for lower-ranking officers and automatically decrypted, but others were eyes-only for flag officers and came in "off-line." Messages of this nature were usually considered "hot shit" and required manual decryption and personal delivery. A few years after his stint at the Pentagon, Turban got orders to NSGA (Naval Security Group Command) in Kami Seya, Japan, where he did a spy run on the USS *Swordfish* not long after K-129 went down.

On his first day aboard the *Flying Fish,* Turban met Mark Rutherford, an M-Brancher spook who'd just completed a top-secret mission on the USS *Halibut* in the Sea of Okhotsk. Rutherford never said a word about that mission, and Turban didn't probe. As an expert on the BRD-7, Rutherford helped train the spooks on this new gear. Sanders Associates built the BRD-7, and when coupled with two pieces of equipment called Blue Surf and Blue Gill, the integrated system represented the next generation in ESM, along with advanced burst-transmission direction finding and fingerprinting. Turban had been trained on the BRD-6, also built by Sanders, and exposed to the BRD-7, but Rutherford knew the system well enough to preach Sunday sermons.

The BRD-7 captured Soviet burst transmissions and fed them to the Blue Surf/Blue Gill equipment that decoded the tiny data streams that William J. Reed found years earlier in Turkey. The system could not decrypt the information, but built-in transmitter fingerprinting allowed operators to determine the type of platform or, in some cases, the class of vessel transmitting. Unlike stationary Bulls Eye stations, most contacts obtained by the BRD-7 were close enough that sprinting down the bearing line could eventually yield a sonar or visual sighting. To obtain these accurate bearing fixes on remote targets required submarines to capture multiple hits from different bearings by frequently changing course and sprinting to new locations. Given that Soviet subs did not often transmit using the burst signal, such an operation could take days.

With Commander Williams calling the shots in 1973, the *Flying Fish* glided across the Atlantic and crept into the cold waters of the Barents Sea. Once on station, Williams ordered all quiet and popped up the number two periscope for a peak. Turban sat in the spook shack section of the radio room and waited, his thick fingers tapping nervously on his knees. Though he'd conducted a few SpecOps before, this was his first one near the infamous city of Murmansk. Every submariner knew that the location of this port, tucked deep inside the Kola Bay in the far northwest corner of Russia, translated into danger times two. As headquarters for the Soviet Northern Fleet, more submarines and warships entered and exited this port than any other. Remaining undetected while hunting Ivan in these waters was akin to playing hide-and-seek in a small backyard with only a couple of trees to hide behind.

After several days on station, Turban picked up a strange signal on the BRD-7 he'd never heard before. He sat up in his seat and listened. The signature resembled a *Yankee*-class ballistic missile submarine, but not quite. He wondered if this could be a new type of boat altogether. He keyed his microphone and reported the contact to Commander Williams.

The *Flying Fish* sprinted toward the unknown contact. Hours later, Williams slowed to a crawl and brought the boat to periscope depth. He raised the number two scope and peered at the image of an odd-looking submarine resting on the surface. Draping his arms over the scope handles, he let out a long whistle. The contact, designated Master One, looked like a *Yankee,* but with a larger two-step, humped missile compartment behind the sail. Williams reeled off dozens of photos with the 70 mm camera housed in the periscope, then ordered the "under hull" team to report to the control room.

The team consisted of a few handpicked individuals that Williams had "certified" capable of conducting a periscope visual inspection under the hull of a foreign vessel, without destroying the *Flying Fish* as a consequence. Under-hull runs were considered borderline suicidal, given that submarines needed to descend below a Soviet vessel and pass underneath with video cameras reeling while the periscope looked upward. Only a few feet separated the two boats. One wrong twitch, a sneeze, or a glance away from a dial at a critical time could send the *Flying Fish* flying into the underside of the Soviet craft. The impact would be heard in Moscow, and the damage might be bad enough to end a mission, perhaps permanently.

The under-hull team relieved the current watchstanders, while Williams remained glued to the periscope. He glanced away from the scope long enough to call into the radio room, where Turban and the spooks sat behind a curtain in their restricted area. He asked if they had any ideas about the class of this new sub. No one had a clue. He repeated the question to sonar. Since the target remained stationary, they could hear only occasional machinery noises—not enough to determine platform type. Williams had a strong suspicion they were looking at the Soviet's new *Delta*-class submarine. If so, the *Flying Fish* would be the first to collect data on this type of boat. Every submariner dreamed that such a find would come along at least once in his career.

Turban had recently read reports that the Soviets churned out the first Project 667B *Delta I*, designated K-279, in December 1972. As yet, no American submarine had encountered a *Delta* or captured any of her noise or signals information. An upgrade of the *Yankee* class, the new sub resembled an American "boomer" Polaris missile boat but carried twelve versus sixteen ballistic missiles in her tubes. These medium-range R-29 projectiles granted the Soviet Union the ability to launch from almost 5,000 miles away, making the *Delta* a far greater threat than her predecessors.

Williams gave the under-hull order, and the boat crept downward. The dark hull of the *Delta* loomed large through the scope like the underbelly of a sleeping killer whale. Turban and the other spooks could see what the skipper saw via the periscope-mounted video camera that broadcast to several monitors in the boat through a system called Peri-Viz, for Periscope Visual. As the *Flying Fish* passed beneath her prey, barely yards away, rough edges on the *Delta* rushed past, followed by sealed openings and meshed grates. When the scope focused on one of the grates, something shiny flashed. Williams zoomed in on the object, revealing a cylinder about three inches in diameter by five inches long. Some sort of container had wedged itself into one of the grates. One of the I-Branchers in the radio room read aloud the Russian writing painted across the item—a product label belonging to a popular soda pop manufacturer. In essence, they'd just found a Coke can.

Upon completion of the under hull, Williams circled the *Delta* and took more shots with the 70 mm and "snapped-on" 35 mm cameras. Sonar recorded stationary reactor noises, while Turban and company cataloged signals emanating from the masts. Williams moved out to 2,000 yards and ordered the weapons officer to ready tube numbers one and two, which were both loaded with MK-48 torpedoes. He verbally confirmed a few bearings while focused on the *Delta* through the periscope. Sailors sitting on the benches in front of the fire control computers dialed in the bearings.

The weapons officer glanced at Williams and said, "I have a curve and a good solution on the target. Range two thousand yards, target speed zero knots."

"Attention in the conn," Williams said. "Firing point procedures tubes one and two. Horizontal salvo, two degree offset, two-minute interval."

"Weapons ready," the weapons officer said.

"Firing point procedures," Williams said.

A fire-control petty officer sitting on a blue bench in front of a large panel rested a hand on the firing key. With a single order from Williams, the *Delta* would be transformed into a twisted metal coffin. In the spook shack, Frank Turban rubbed his palms together. He was surprised at how dry they were. He thought that coming this close to sinking a Soviet submarine would cause him to sweat like a boxer, that he'd feel something: fear, anticipation, perhaps even sympathy for the submariners on the other boat. Surprisingly, with his mind focused on his job, he felt nothing but a sensation not too unlike the hum of an electric fence in a light rain. He wondered if that hum would turn into a jolt if they actually fired two torpedoes up the ass of the *Delta*.

Williams peeled away from the periscope. He glanced at the fire-control party and said, "Stand down." With the firing exercise concluded, he ordered the diving officer to bring the *Flying Fish* in close, real close. Through the high-power lens in the number two scope, he snapped pictures of Soviet submariners wearing traditional *Ushanka* hats on the bridge, framed against a frigid gray smattering of clouds.

In the spook area, via the PeriViz, Turban watched the Russians light up cigarettes on the bridge of the *Delta* while carrying on an animated conversation. One of the sailors stopped, turned, and stared right at him. Turban held his breath as he stared right back into the sailor's eyes and saw clearly the questioning look of shock on the man's face. The Soviet sailor raised his hand and pointed right at Turban's nose. Turban knew that the Russian wasn't looking at him directly but had spotted the *Flying Fish*'s periscope. Williams slammed down the handles on the scope and lowered the cylinder into its well. He ordered a deep dive, and the *Flying Fish* angled down thirty degrees as she made a rapid descent and sprinted away.

Williams later came shallow again, raised the scope, and afforded the crew the show of a lifetime through PeriViz—complete with popcorn in the crew's mess. The Soviet *Kresta I*–class cruiser *Sevastopol* lit off everything she had, including her Big Net, Head Net-C, and Plinth Net radar. She readied her ten torpedo tubes and started spinning the blades on her Ka-25 Hormone helicopter. Other ships and aircraft came onto the scene with equal ferocity. Given all the action going on, Turban thought for sure that Williams would move the *Flying Fish* outside Soviet territorial waters. Instead, the CO stayed put. Turban knew that *Sturgeon*-class boats were quiet and hard to find and figured that Commander Williams was about to prove that theory correct.

TU-95 helicopters buzzed into the area and dropped sonobuoys. The active sonar pings were loud enough to be heard through the hull. A brand-new *Krivak*-class frigate approached, and the spooks got excited when she lit off her Spin Trough, Eye Bowl, Kite Screech, and Pop Group radars. The more data they could collect on these signals, the better. The linguist I-Branchers were having the most fun, listening to open transmissions between the ships and aircraft as they prosecuted the *Flying Fish*.

One I-Brancher, Phil Cafrey, stayed well past his watch-relief time and refused to turn over his headset. His eyes wide, his tongue wagging, he listened and translated like a madman possessed. Finally, after eight straight hours, he had to relieve himself. He received permission from the XO to use the nearby stateroom head and ran out of the radio room. In his rush to finish up and get back to his headphones, he missed seeing the large sign hanging on the stall door in the head. From across the passageway, Turban heard Cafrey scream in agony.

All submarines keep sanitary waste in large tanks that periodically need to be blown to sea. When this occurs, the tanks are pressurized with air to flush the tainted liquid out to the ocean. Large signs are placed in the heads warning sailors not to pull the flapper valves and flush toilets while sanitaries are being blown. To do so launches the pressurized air, along with a spray of waste, back up through the toilet opening and right into the face of the flusher. Cafrey missed seeing the sign and so pulled the flapper. After the screaming stopped, it took him a half hour to wash the stench from his eyes, nose, and body. From that day on, the crew affectionately called him Flapper Phil.

Through PeriViz monitors, the crew of the *Flying Fish* watched the procession of Soviet ships prowl back and forth. Radio operators received reports from ground stations that they were hearing "stutter nine" transmissions from the Soviets, consisting of all nines in the first group of transmitted characters. This translated to unidentified invader/ contact and almost always meant that the Soviet navy was searching for a U.S. submarine.

Three days later, after all the fun had died down, the *Flying Fish* snuck away and returned to Norfolk so the NSA could study the photos of the new *Delta I*. Six weeks later, Turban got a call from Commander J. D. Williams. The CO wanted to know if Turban would like to ride again on the *Flying Fish* for another northern run. Turban jumped at the opportunity. Three other spooks from the former group of twelve also volunteered, including Mark Rutherford. Turban soon found himself back in cold waters near the North Cape—which is the point where the Norwegian Sea, part of the Atlantic Ocean, meets the Barents Sea, part of the Arctic Ocean. Located on the island of Mag-erøya in northern Norway, the North Cape is not far from Oslo, Norway, where William J. Reed helped set up a Boresight station.

The *Flying Fish* sauntered back to the Barents Sea, only this time the SpecOp bordered on boring for the first few weeks. Turban started to wonder why he'd volunteered but then recalled that after the *Delta*-finding run, he'd made up his mind to follow Commander Williams just about anywhere. Another day passed before Turban got a sniff from a *Kresta II* cruiser, which carried a new type of antisubmarine missile known as the SS-N-14. Accurate and deadly, the N-14 could rip a hole right through the *Flying Fish* within seconds. Turban had no desire to tangle with a *Kresta II,* but the ship continued to transmit to fleet HQ on a frequent basis. The type and frequency of the transmissions seemed odd. Based on a hunch, Turban started scanning the area for known submarine frequencies on the BRD-7. Nothing came up at first, then a faint trace.

Turban almost wrote off the whisper but instead closed his eyes and concentrated. Out of the static came a familiar sound, the scratchy trademark of the burst signal that William J. Reed had heard more than a decade earlier while in Turkey. Too faint to trigger a detection alarm, the trace signal would have gone unnoticed if Turban hadn't become suspicious of the *Kresta*'s repeated transmissions. While analyzing the signal, a neon bulb flashed in Turban's head. He immediately called Williams and explained his theory.

"A hurt *Yankee?*" Williams said as he stood in the tight radio room.

"Sure looks that way," Turban said.

"How so?"

"The *Yankee* sends a short burst signal to the *Kresta*. Right after that, the *Kresta* broadcasts a slightly longer message to HQ. That leads me to believe that the *Yankee*'s not feeling well and needs the *Kresta* to help her out."

"How do we know where to find her?"

Turban explained that while T-Branchers, I-Branchers, and R-Branchers were all considered "spooks," each received entirely different training related to linguistics, radio, radar, and communications signals. T-Branchers, like Turban, were highly trained in Soviet communications, signals, and procedures. Most of the time a T-Brancher could listen to a transmission and didn't have to measure anything to know the type of signal, its location, and its purpose. In this case, Turban knew they were listening to a *Yankee* submarine in the general vicinity of the transmitting *Kresta II*. To help pinpoint that location, Turban needed Williams to sprint to at least two other locations so they could multiangulate a bearing fix.

Williams nodded and hurried back to the control room. The *Sturgeon*-class submarine made a turn and ran for fifteen minutes to a new hole in the ocean. Turban took bearings before Williams cranked up the propeller and drove the *Flying Fish* to another spot, also fifteen minutes away. Turban took more bearings. Satisfied that he had a descent fix, he gave Williams the estimated bearing and range to the new contact.

The *Flying Fish* snuck into the area, whereupon Williams jutted the periscope above the waves for a look. Turban watched on the PeriViz. In the distance he saw smoke billowing from the submarine's sail and open hatches. Sailors darted about the deck in emergency fashion as their *Yankee*-class boat sat motionless upon the ocean. Gray clouds framed the injured submarine as a light rain splattered the deck.

Turban gazed at the bulging shape of the *Yankee* in the monitor. He recalled how he'd felt when the *Flying Fish* cornered the *Delta* and practiced an approach to fire two torpedoes at her. Save for a low-level undercurrent of excitement, not much emotion flowed through him then. Now, something different welled up inside. Something profound. That most Americans considered these endangered sailors the enemy did not diminish the sympathy that consumed Turban as he wondered how many on board had been hurt or killed. But the *Flying Fish* could do nothing to assist.

Fortunately, other Soviet ships arrived and provided assistance as the *Flying Fish* lowered her scope and snuck away into the dark depths of the ocean.

Upon returning to Norfolk, Turban learned that the *Yankee* submarine had endured a serious fire and that some of the crew had succumbed to smoke inhalation. For reasons Turban could not put into words, he felt a deep connection to the fallen. In that moment of grief, those Soviet sailors no longer represented an enemy to be feared or hated. They were fellow submariners and vulnerable in the end, just like him, comrades who shared the unique sacrifices required to work under the sea. Regardless of how cold the war between the superpowers became, Turban would always share a bond with those sailors that only those qualified in submarines could know.

For his achievements in underhulling the *Delta I*, and for locating the distressed *Yankee*, Commander J. D. Williams received a Distinguished Service Medal. Frank Turban earned his first Navy Commendation Medal, and he and Mark Rutherford gained enough notoriety within the NSA to warrant E-ticket rides on two submarines that conducted the most harrowing missions of the Cold War.

CHAPTER SIXTEEN

If we weren't all crazy, we'd just go insane.
—JIMMY BUFFETT

B Y LATE 1975, SOME AMERICANS HATED sailors. They also hated airmen, marines, and soldiers. The last helicopter flew away from Saigon on April 30, and long-haired protesters uncapped their Magic Markers to create new signs. STOP THE WAR changed to BAN THE NUKES and DOWN WITH THE ESTABLISHMENT. Having just enlisted in the navy that summer, I became an officially sanctioned target of the hippies. Almost overnight, friends turned into enemies. Did I not know that the warmongers had fueled a war that cost the lives of 1.3 million Vietnamese and 56,000 Americans? And what about the $584 billion the United States had spent on this meaningless conflict? They asked if I'd lost my mind, had perhaps become as psychotic as Jack Nicholson in *One Flew Over the Cuckoo's Nest*.

I gave little heed to the barbs. I knew things most could not know, perhaps didn't want to know. By then, my father had told me a few hair-raising stories, under an oath to secrecy, about his part in the not-so-cold underwater war against the Soviets. Having enjoyed his years working with the silent service, he suggested that I consider joining the ranks of those committed to finding, and if need be, destroying the new enemy subs entering the fight. What he didn't tell me was that becoming a spy of the deep required learning how to live for months on end without the sun on my face, the love of a woman, or freedom from fear. I think he knew that I'd have to learn these lessons on my own.

My father had retired from the navy years earlier and embarked on his new career as a writer by penning an award-winning biography of the painter Olaf Wieghorst. Olaf's illustrious life included stowing away on a ship from Europe, working for the mounted police in New York, and chasing Pancho Villa along the Mexican border. Olaf's paintings depicted the Old West with such vivid realism that you could almost taste the dust. His sought-after masterpieces commanded high bids and hung on the walls of the wealthy and notable, including those of President Eisenhower and Senator Barry Goldwater, who wrote the foreword to my dad's book. My father took me to visit Olaf's studio in San Diego, where the humble master showed me authentic Indian headdresses and vintage frontier rifles. I told Olaf that I intended to follow in my father's footsteps and join the navy. Olaf said that was admirable but encouraged me not to volunteer for submarines. As an outdoorsman, the very thought of living in a sewer pipe made him cringe.

Ignoring Olaf's admonition, I volunteered for submarines and began my adventure on October 21, 1975, ten days after watching the first episode of *Saturday Night Live,* with George Carlin hosting. That night I stepped off a bus and walked through the gates of the Naval Training Center in San Diego. The navy stripped me of my clothes and ego and over the next eight weeks molded me into a sailor.

After boot camp and six weeks of basic electronics school, I crossed the country to complete submarine school and almost eighteen months of fire-control-technician weapons-systems training at the submarine base in Groton, Connecticut. In the summer of 1977, six months after Jimmy Carter pardoned the Vietnam War draft evaders, the navy promoted me to petty officer second class and flew me to the Philippines to join the crew of the USS *Haddo*. Sonarman Petty Officer Second Class Kenneth "Greenie" Greenawald, also destined for the *Haddo,* flew with me. A medium-built, gregarious guy with an infectious smile and bushy mustache, Greenawald had traveled cross-country from his hometown of Pittsburgh, Pennsylvania, and by now looked like the male version of a mistreated Raggedy Ann doll.

The *Haddo* had just completed a top-secret operation, and the sight of her smooth hull tied up next to the pier in Subic Bay in the Philippines filled me with an equal dose of excitement and dread. As I approached the mostly submerged craft, I noticed that the hull's black steel did not appear at all smooth. The surface contained brick-sized chunks of black rubberized material. I recalled from my sub school training that the material consisted of anechoic tiles, designed to absorb enemy sonar beams. At the time, I didn't give that detail much thought. Weeks later, when we were dodging a bunch of Soviet warships, I praised the genius designers of this technology.

Greenawald and I walked across the brow of our new *Permit*-class submarine and handed a copy of our orders to the topside watch. A warm tropical wind chopped the ocean into wavelets that slapped against the boat and left an oily film. After checking our paperwork, the topside watch pointed toward an open hatch. At the edge of the small hole, I peered downward. The heavy scent of nuclear submarine filled my nose. To this day, I've never forgotten that stench. I learned later that since submarine air systems were "closed loops," the scents generated by everything mechanical, chemical, and personal lingered forever. No amount of ventilating could ever scrub away the acrid, oily, fuel-laden, mechanical, electronic, fart-filled odors that could only be called "the boat smell."

The *Haddo*'s crew consisted of just over one hundred enlisted men and a dozen officers. Commander Norman Mims Jr. served as our CO. He was a seasoned veteran with gray hair and had been on a few diesel boats before progressing to CO of the *Haddo*. Under him, our executive officer projected the haughtiness of an upwardly mobile, by-the-book leader who lived in constant fear of making a mistake.

Like all submarines, our command consisted of several divisions, including engineering (nuclear and machinist disciplines), navigation, sonar, fire control (weapons), and other operational functions. My division chief, FTGC (Chief Fire Control Technician–Gunnery) James T. Lane, took charge of a small team of two fire-control technicians—and now three with my arrival. Lane worked hard and played hard, and had a wide gap between his front teeth, which he flashed often after cracking dirty jokes. One of the other guys on my team included FTG1 (First Class Fire Controlman) Tim Vinson, the leading petty officer (LPO) of our division.

Vinson, a wiry guy from Florida, stood about my height, five feet ten inches, and had slightly thinning hair that he tried to comb forward enough to cover the bald spots. He had a playful, infectious sense of humor that endeared him to the crew. You never knew when to take him seriously. One minute he'd have you engrossed in a serious conversation about intricate technical details, and then suddenly you'd realize the whole thing had been an elaborate ruse.

Vinson gave me my first tour of the boat. In the control room, we moved past the Mark-113 fire-control system panels, around the periscope stand, and over to the ballast control panel. Vinson pointed to a panel labeled MBT VENTS. MBT stood for main ballast tanks. Underneath the panel, several small squares were labeled with various numbers and designators. Underneath these squares, seven smaller squares had lighted round circles in the middle, and underneath these, another seven small squares had nonlighted rectangular lines. My head swam with confusion, and I wondered how anyone could possibly memorize the functions of all those controls.

"What happens if our main ballast tanks fail in an emergency?" I asked.

"Simple," Vinson said. "We die."

That same afternoon, Chief Jimmy Lane handed me a qualification card—a two-sided, eight-and-a-half-by-eleven-inch off-white sheet filled with dozens of check boxes next to every major system, valve, pipe, or piece of equipment essential to submarine operation. Completing my "qual card," Chief Lane told me, could someday save everyone's life in an emergency.

"Suppose you're the only guy in the compartment when there's a big ol' fire," Lane drawled. "You do the wrong thing, and my ass is fish bait. Learn this stuff, nonqual puke, or forever sleep in bad places and forget about movies or leisure-time novels."

Any crewmate who wore the shiny "qualified" dolphin pin on his chest became the tormentor of anyone who did not. Qualifying consumed me for an entire year. I had never wanted to earn anything so badly in my life. Completing my qualification card to earn my dolphins taught me more about discipline and drive than any college or training course that I have ever taken, and never have I felt more accomplished than the day I donned my silver pin. A week after Vinson gave me my first tour through the boat, the *Haddo* slipped out of port.

I'll never forget my first dive. I was sitting on one of the cushioned benches in the control room. We had just cleared Subic Bay and were headed for open sea. Idle chatter ceased as various orders were given and repeated to prepare the boat to dive.

"Dive, dive!" the words echoed through the tiny control room from the diving officer. I expected to hear massive amounts of air blowing as the vent valves of the main ballast tanks popped open. I thought I'd feel the boat shudder as atomized air and water vapor exploded into the tropical air, while the tanks flooded with 800 tons of seawater to make us heavier. I wanted to hear an old-fashioned submarine "auooga" and the sound of Niagara Falls overhead as the submarine slid under the sea. I heard and felt nothing, except a disoriented feeling as the *Haddo* angled down fifteen degrees at the bow and disappeared under the waves.

In the front of the control room, the helmsman and planesman—responsible for steering and diving the boat, respectively—sat comfortably in bucket seats and operated airplane-style half-circle wheels. They concentrated on the analog dials and gauges on their control panels, while the chief of the watch (COW) repeated orders issued by the diving officer. As the boat continued its downward motion, I heard her groan, softly at first, almost inaudibly. As we went deeper, the creaks and moans increased in volume and tempo.

Commander Mims ordered a temporary stop at various depths while the crew checked the boat for leaks. When none were found, we descended deeper. With each one hundred feet of descent, the outside water pressure increased by three atmospheres. At 800 feet, twenty-four atmospheres squeezed in on the boat, and twenty-five tons of pressure per square inch threatened to crush the hull. Finally, at 1,300 feet, test depth for a *Permit*-class boat, our descent into the abyss ceased. Again the crew checked for leaks. I watched the beam of Vinson's flashlight flickering across the piping in Fire Control Alley, a narrow space just behind the control-room equipment where the main digital computer stack was housed. Satisfied that she would again hold back the sea, Mims gave the order to ascend to 800 feet for our transit to Hong Kong.

As the days went by, I was amazed at how fast I adjusted to life under the sea. The focused objective to qualify as fast as possible, so that I would never again have to hear the words *nonqual puke,* helped to keep my mind off the fact that millions of tons of seawater were only a few feet away. Every submariner spends an average of one year "quali-fying" on submarines by studying every switch, valve, system, pipe, and device on the boat until he has all of them memorized: how they work, what they're for, and, most important, what to do with them in case of an emergency.

I spent hours tracing system schematics in the boat's piping-system booklet, and experts tested me on each associated or related item. I lo-cated emergency air-breathing (EAB) manifolds. I learned that in a fire emergency, we needed to don airtight masks, plug our hoses into the EAB manifolds, and commence breathing. Not knowing precisely where to find an EAB manifold in an emergency could cost a submariner his life.

Once I found everything on paper, I performed scavenger hunts to locate them somewhere on the boat. Fortunately, I had help from my boss.

"Did you find the fallopian tubes yet?" Vinson asked me.

"The what?" I said, validating that I had slept through Biology 101.

"They're in the bow, not too far from the diesel. Ask Lieutenant Tomlin; he knows right where they are."

"Okay, I'll ask him."

Tomlin was on watch when I made the mistake of asking, right in front of the entire control-room watch. They all started laughing, and I realized that I'd been had. I felt worse than a high school freshman on his first day of class. More questions followed, all intense. Slacking on qualifications was not tolerated. Lives depended upon everyone know-ing their stuff. I drew each system by memory as an expert watched. He then studied my drawing intently and quickly pointed out even the slightest error.

I had reported on board as an FTG2, the equivalent of a sergeant in the army. Until I earned my dolphins, a *qualified* seaman had every right to denigrate me if I got a qual question wrong. Luxuries were not granted to nonquals. I spent the first six months of my submarine tour sleep-ing on an air mattress on top of a torpedo rack, while qualified sailors two ranks my junior enjoyed quiet bunks in the bow compartment. My mattress had a leak, and every few hours the damn thing deflated. The lights were always on in the torpedo room, and someone always jab-bered away while I tried to sleep. To this day I can fall asleep anywhere and sleep through almost anything. Heaven forbid that anyone should break into my house and steal my bed. I'd probably wake up on the floor sometime later, smiling and unaware that I'd been robbed.

Vinson once told me that if I ever missed sleeping on board a submarine, I should crawl onto the top shelf of my closet and hang a curtain in front. Then I should have a friend come around every two hours, rip back the curtain, shine a flashlight in my face, and say, "Whoops, sorry; wrong guy."

On top of learning what made the boat tick to earn my dolphins, I had to come up to speed on standing watch in the control room and leverage my schooling to keep the fire-control systems operational. Once I did that, I would spend six hours on watch and twelve hours off in a three-man rotation. During my training, I learned about tracking targets, Soviet sub and ship capabilities, and the integration of sonar and periscope information into the fire-control computers. I also learned how to "dance with the scope."

Many of our watches were at night, with the control rigged for dark. The lights were set for red to adjust our eyes to darkness, in case we needed to come to periscope depth. In that event, we might spend several hours with our ESM-infused scope raised so the spooks could monitor radio traffic, or so some of us in the control room could scan the horizon for interesting contacts. During these times, the officer of the deck (OOD) often "conveniently" found alternate duties to perform, requiring the fire-control technician (FT) of the watch—that is, me—to dance with the scope in his stead.

Like the old diesel boats, our submarine had two periscopes, but these were located in the control room in the center of a space no larger than the average two-car garage. The smaller Type 2 "number one" attack scope was controlled manually. Containing simple daylight optical capability, we used this scope for attack maneuvers where a reduced wake diminished the possibility of detection. The larger Type 18 "number two" scope housed all of the neat stuff, like ESM, 70 mm automatic camera equipment, electronically controlled focus and night vision capability, power-assisted rotation, and a special "RAM" (radar-absorbing material) to reduce detection by Soviet radar. Power-assisted rotation made turning the scope easier via a flat thumb-controlled button on the right handle. When the OOD wanted to raise the scope, he turned a large tubular "wheel" encircling the scope. This activated the hydraulics to lift the scope out of the well.

Some of my duties entailed learning about sonar and how these systems interfaced with our fire-control equipment. Sonar was our primary means of gaining contact information related to TMA (target motion analysis). TMA was not too unlike the mental process that a quarterback undertakes to hit a running receiver. Good quarterbacks learn how to do TMA in their heads by estimating the distance to a receiver, along with how fast he's running, to determine where he'll be when the ball arrives. The quarterback then throws the ball to that spot.

In the submarine world, fire-control computers do the calculating for the boat's skipper, who is our quarterback. The receiver is the enemy ship or submarine. Using input from sonar, radar, ESM, and/or the periscope, the computers keep track of the distance, or range, to the receiver, or target. They also track the target's speed, or how fast the receiver is running. Unlike your typical quarterback, however, who must process all this data within seconds to make a decision, back then fire-control computers were painfully slow. Long minutes passed before we could obtain a "good solution" on the target. Only then did our quarterback give the order to "throw the ball" and fire a torpedo.

Our MK-113 fire-control system, which made all these calculations, consisted of two large banks of electronically and mechanically operated position-keeper (PK) analog computers on the starboard side of the control room. Each PK stood as tall as a refrigerator, but twice as wide. The front panel displayed the shape of a ship on two dials, which represented the enemy vessel and our boat, respectively. Dials and number indicators, similar to mechanical dials in a car odometer, kept track of estimated contact bearing, range, and speed. A complicated array of servos, synchros, and gears worked together to calculate a firing solution. The entire not-so-state-of-the-art system could have been replaced by one of the new Hewlett-Packard handheld calculators just coming onto the market at the time, but I suppose that's why toilet seats cost the navy thousands of dollars.

Behind the main MK-113 consoles in Fire Control Alley, the primary digital computer took up a six-foot-long section and fed two smaller consoles on either side of the PKs. These were also as tall as refrigerators but equal in width. They provided essentially the same function as the PKs, but did so using digital versus mechanical technology. Only four contacts could be tracked at a time via the MK-113 system. On either side of the PKs sat the MK-81 consoles, like smaller refrigerators. These digital consoles were used to program and control wire-guided MK-48 torpedoes to their targets after our CO "threw the ball." Just inside these two units sat the main torpedo and SUBROC missile-firing panel. SUBROCs are submarine-launched rockets with a fifty-mile range and nuclear depth charge capable of vaporizing a three-mile section of ocean.

A large handle adorned the midsection of the firing panel, which we used to fire weapons should the need arise. During battle stations, the fire-control party, consisting of all the fire-control techs and several officers and crew assigned to those stations, operated the entire array of MK-113 equipment. We huddled close together on the benches in front of the whirring systems, sound-powered headphones glued to our ears, and cranked information into the gear in an effort to obtain an accurate firing solution. One wrong number on a solution could send the boat headlong into a Soviet submarine or other underwater obstacle, which could in turn send us all to the bottom.

As part of my training, Greenawald and the other sonar techs gave me frequent tours of the sonar shack and briefed me on passive and active sonar systems. Sound travels through seawater at roughly 4,000 feet per second over significant distances but offers only a vague clue as to the whereabouts of a contact. Most submarines employ several types of sonar to sniff out the enemy. The spherical sonar array was located in the bow. This consisted of a large fifteen-foot-diameter sphere that housed both active and passive sonar. It looked like a giant beach ball. The conformal array was a low-frequency sonar that wrapped around the midsection of the bow, like a massive piece of electrical tape with built-in ears.

The towed sonar array, called the STASS, was a by-product of the ITASS system developed by the LRAPP scientists. The "pigtail" array looked like a three-inch-thick, 2,600-foot-long garden hose with thirty-two bumps. Each bump housed a hydrophone that listened for distant targets emitting low-frequency sounds. To deploy the array, divers had to enter the water and attach the pigtail to the boat's stern planes. A small "Mike" motorboat met up with the sub at sea to help "stream the array" for use. The reverse had to be done when entering port.

When sonar located a contact, they designated it Sierra one, two, three, and so on, in a sequential numbering system until the end of the mission. A contact of significance, such as a Soviet sub, usually got an upgrade to a Master number. Visual contacts were designated Victor (not to be confused with a Soviet *Victor*-class sub), and radar contacts were Romeo. Sonar operators spent hours with headphones on ears, turning small wheels on the BQQ2 sonar suite consoles—called the BQS 11, 12, and 13—until they detected a contact aurally.

A waterfall display on an orange monitor for another system, called the BQR-24, visually kept track of the bearing—or the direction—of a sound generated by a contact. The display created a fuzzy narrow line that ran downward as the contact moved. Lines generated by targets shifted left or right as the contact changed direction. More lines equated to more contacts, which created a waterfall effect that resembled raindrops running down a window.

The BQQ2's sensitive passive capability allowed this system to hear surface ships more than 20,000 yards distant (ten nautical miles). Operators used a computer to catalog distinct sounds that were unique to a particular class of vessel, or even the exact vessel itself. A nick in a screw, "singing" propeller shaft, or humming engine noise could become a fingerprint or noise signature that identified a particular sub or surface ship, right down to the hull number. We called this ACINT, for acoustic intelligence.

As part of a Holystone mission, we might spend hours, days, or even weeks trailing closely behind a Soviet nuclear missile or attack submarine, sometimes approaching to within a few dozen yards. We often trailed and listened for hours, days, or even weeks to every noise made by our prey, while high-speed reel-to-reel recorders captured the information for subsequent input into data banks. This "up close and personal" aspect sometimes led to collisions, or near collisions, while shadowing Ivan, especially during a Crazy Ivan maneuver.

I soon learned just what that phrase meant and would come to reason that we called it Crazy Ivan not so much in reference to the shadowed, but perhaps to those of us in the shadow. The baffle area on most submarines spans a sixty-degree arc directly behind the screw. A submarine's sonar is usually mounted in the bow, which creates a "blind spot" in the baffle area. This is where U.S. submarines might be found . . . in Ivan's shadow. We could play there due to our superior sound suppression technology that made it hard for Ivan to hear us.

When U.S. boats cleared baffles, they slowed to one-third, turned ninety degrees to port or starboard, listened for any signs of unwanted visitors, then resumed their previous course. When Ivan cleared baffles, he made a 180-degree turn along his previous track, ran in the opposite direction for a while to ensure nothing suspicious lurked in the shadows, then made another thrilling 180-degree turn to resume his original course. The "thrilling" part related to speed. Unlike Americans, Soviets did not always slow down to clear baffles.

Hearing "Crazy Ivan!" from the sonar shack caused serious nail biting in the control room. The cry meant that better than 5,000 tons of Russian submarine had turned back in your direction, and if she didn't smack you, she might hear you. Either one of those scenarios could end your life. So, I reasoned, perhaps it was not just Ivan who deserved to be called crazy.

AFTER A FEW MONTHS UNDER WAY, I finally learned enough to stand the fire-control watch alone. Excited to be on my own for the first time, I scrambled up the ladder to the main deck of the *Haddo*. The crimson glow of "rig-for-dark" red fluorescent lights bathed the interior of the control room and cast eerie shadows on the faces of the crew. The air, filled with hazy cigarette smoke, smelled like day-old coffee. Gray metal boxes clicked and hummed and blinked like tiny buildings in a metropolis under the sea. As I sipped my bitter coffee, Vinson gave me an update and turned the watch over.

"Who's this?" I asked, pointing to a track on the plot. The plot board, fashioned from a two-by-three-foot metal box, housed a roll of scientific paper. On the checkered paper, dots and lines recorded the bearings and tracks of various contacts reported by sonar. Even though we had mechanical and digital computers to track contacts, we still used low-tech lined paper to plot the movement of all Master contacts. The roll of paper covered the front of the box that hung off the edge of

Few today know that four Soviet *Foxtrot*-class submarines, each carrying nuclear torpedoes, nearly started World War III during the Cuban Missile Crisis in October 1962. *U.S. Navy photograph*

The author with former Cuban Missile Crisis *Foxtrot* submarine commanders in Saint Petersburg, Russia, in March 2009. *Left to right:* RAdm. Victor Frolov (B-130), Lt. Anatoliy Andreev (B-36), Adm. Vladlen Naumov (B-36), W. Craig Reed (the author), Capt. Ryurik Ketov, Capt. Igor Kurdin, and translator Svetlana Verholantseva. *Author's collection*

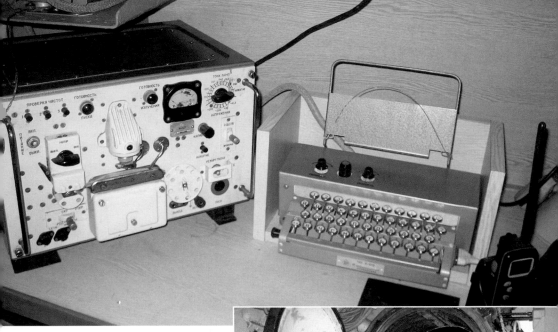

ABOVE: The Soviets installed new SBD "burst" radio transmitters and encryption systems aboard Soviet submarines in 1960. The U.S. Navy was then unable to locate the Soviet subs until William J. Reed, the author's father, discovered the "scratchy" burst signal.

Author's collection, taken on a Whiskey-class submarine in Saint Petersburg, Russia

RIGHT: The torpedo room in a Soviet *Foxtrot*-class submarine. Tube number two, which carried the nuclear torpedo during the Cuban Missile Crisis, is in the center.

Author's collection, Maritime Museum of San Diego, California

The author touring the Soviet D-2 submarine in Saint Petersburg, Russia, in March 2009, with former Soviet submariners from the Saint Petersburg Submariner's Organization. *Author's collection*

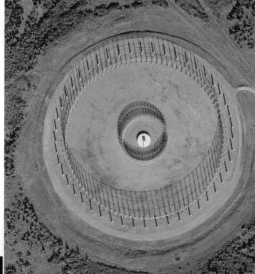

LEFT: During the Cold War, the U.S. deployed more than a dozen "elephant cage" secret listening stations worldwide to locate radio transmissions from Soviet submarines. *U.S. Navy photograph*

BELOW: Author touring the *Los Angeles*–class attack submarine USS *Albuquerque* (SSN-706) in San Diego, California. In the background is the weapons fire-control panel located in the torpedo room. *Author's collection*

Crew of the USS *Haddo* (SSN-604) near Hanauma Bay on Oahu, Hawaii, circa 1979. *Left to right:* Dean Quint, Tim Vinson, Ken Greenawald, unknown, and the author. *Author's collection*

Ken Greenawald and Tim Vinson in the sonar shack of the USS *Haddo* just off the coast of Petropavlovsk, Russia. *Author's collection*

LEFT: Control room of the USS *Florida* (SSGN-728). Helmsman and planesman steer and dive the boat while the diving officer, seated behind them, issues orders. *U.S. Navy photograph*

RIGHT: Control room in a *Sturgeon*-class Cold War submarine. *U.S. Navy photograph by Chief Photographer's Mate Robert Hemmerly, courtesy of www.dodmedia.osd.mil*

BELOW: Fire control technician operating the missile launch console aboard the USS *Seawolf* (SSN-21). *U.S. Navy photograph by Chief Photographer's Mate John E. Gay, courtesy of www.news.navy.mil*

RIGHT: Type 18 "number two" periscope in the control room of a *Sturgeon*-class submarine. *Author's collection, Naval Undersea Museum, Keyport, Washington*

U.S. Navy divers, like the author, deployed from fast-attack submarines during the Cold War to conduct top-secret photographic reconnaissance and surveillance missions. Many of these operations were in concert with Navy SEAL teams.

U.S. Navy photographs by Senior Chief Mass Commuication Specialist Andrew McKaskle and Chief Photographer's Mate Mark Reinhard, courtesy of www.osd.mil

The most dangerous, daring, and
decorated missions of the Cold
War were undertaken by U.S.
Navy deep-sea saturation
divers breathing helium and
oxygen mixtures. These divers,
working from fast-attack
submarines, tapped Soviet
communications cables off the
coast of Russia, at depths
of up to 700 feet.

U.S. Navy photographs by
Mass Commuication Specialist
Fisrt Class Eric Lippmann

U.S. Navy divers, conducting top-secret Ivy Bells cable-tapping missions, spent weeks compressing and decompressing inside small diving chambers mounted on submarines, which resembled DSRVs like the one above. *U.S. Navy photograph*

In April 1981, the author, aboard the USS *Drum* (SSN-677) and deep inside Peter the Great Bay (*below*), was being sent on a mission to take secret photographs of the Soviet *Victor III* submarine K-324 when the *Drum* collided with the other sub. The *Victor* was from the Vladivostok naval base located in the top center of the photograph. The crew of the *Drum*, including the author, barely survived the encounter. *NASA photograph*

one of the large PK computers. One of these dots on the paper, designated Master 21, also had a name.

"Blackjack?" I said.

Vinson shrugged. "Master 21. Greenie gave him that name 'cause of the bet."

"Sonarman Greenawald?" I said. "What bet?"

"Greenie heard a hole in the water on the watch before mine."

"Hole in the water?"

"Yeah," Vinson said. "A dead zone where there's no chirping biologics. As if something might be blocking the sound."

"Like a submarine," I said.

"Exactly. The other sonar jocks couldn't hear it, but Greenie swore something was there. Commander Mims wanted to sprint to a new location, but Greenie insisted he could find this guy."

"So what did Mims do?"

Vinson grinned. "He made a bet that if Greenie could find the contact by the end of the next watch, he could grow his mustache way past regulation standards."

"And what if he lost the bet?"

"Then Greenie would have to shave the thing off."

I pictured my buddy with a clean upper lip and figured he might be better off losing. "So what happened?"

Vinson yawned. "Damned if Greenie didn't stay for two extra watches to find the contact. A submarine, but too faint to tell what kind. Since Greenie won the bet, he nicknamed the contact Blackjack."

"And now Greenawald gets to grow his mustache long."

"Right," Vinson said, stifling another yawn. "I need to crash. The watch is all yours."

"Just don't take my rack," I said.

Knowing that I still slept on an air mattress in the torpedo room, Vinson said, "I'll let all the air out of your rack if you don't keep up on your submarine quals." He flashed a coy smile and turned to leave.

As Vinson strolled out of the control room, Lieutenant Edwin Ladeau Tomlin sauntered in. He was the crew's favorite OOD and had a great sense of humor combined with a quick mind. A graduate of the Naval Academy at Annapolis, Tomlin stood over six feet tall, which often required him to duck as he moved about the tight spaces of the boat. He had jet-black hair, an infectious smile, and a boyish face that belied his age. I watched him relieve the previous OOD and walk over to my fire-control plot for an update.

Master 21, or "Blackjack," remained on a bearing of three-five-two. We were now just outside Vladivostok harbor in the Peter the Great Bay, well inside Soviet territorial waters.

"How we doing?" Tomlin asked.

"Fine, sir," I said.

Tomlin reached up and clicked a lever on the 27MC communications box. "Sonar, Conn—can you give me an update on Master 21?"

Greenawald's voice replied through the box. "Conn, Sonar—Blackjack is still on bearing three-five-two. Contact is classified as a probable submarine, class unknown at this time."

Tomlin leaned in my direction. "Blackjack?"

I shrugged, said nothing.

Tomlin turned toward the front of the control room and issued an order. "Diving Officer, come left to zero-nine-zero, increase speed to two-thirds."

The diving officer echoed the command, and the boat started to turn. Our 270 feet of black stealth cut through the deep in near silence as the rudder brought us to a new heading. Twenty minutes later, Tomlin gave the order to slow, and sonar went back to work.

After a few minutes, Greenawald blared another report through the 27MC. "Conn, Sonar—contact is on bearing three-five-nine making turns for ten knots on two five-bladed props. Designate Blackjack as a possible *Delta III*–class submarine."

In 1977, there were only six *Delta III*s in operation, and they were considered the most formidable threat to Western security at the time. The third generation of the infamous *Delta* ballistic missile sub, this new class came packed with improved stealth and advanced electronics. She was hard to find and even harder to trail.

Tomlin immediately called Commander Mims, who hustled to the control room in a matter of minutes. Mims's crow's feet and forehead wrinkles were a sharp contrast to the young faces in the control room. The CO struck me as a man who might have been gutsy enough to be a boat driver a quarter of a century earlier, when diesel fumes filled the tiny spaces of the original *Haddo*. In those days, submarines fired unreliable Mark 14 torpedoes at massive Japanese warships, then prayed to God they could run and hide before the depth charges started exploding.

"Cap'n," Tomlin said, as Mims stepped onto the periscope stand, "looks like the sonar jocks picked up a *Delta III* headed out of Vlad. She's running about ten knots on two fivers bearing three-five-nine."

"I'll be damned," Mims said, "a *Delta III* coming out of the barn. I have the Conn, Mr. Tomlin, man battle stations. Helm, all ahead two-thirds. Dive, make your depth 1,000 feet, five degree down bubble. Mr. Tomlin, what's our torpedo tube status?"

The battle stations alarm sounded while Tomlin gave the CO an update. I had read about the *Delta III* in the top-secret submarine classification books in the control room and memorized most of the details. Although the Americans and the Soviets were about to sign the Nuclear Non-Proliferation Treaty on September 21, I knew this piece of paper would have no effect on nuclear submarines. The *Delta III* still deployed sixteen nuclear missiles with a range of almost 5,000 miles. Each missile had multiple warheads that could wipe out almost any city in America within a matter of minutes.

Trailing a *Delta III* escalated the tension in the control room as men rushed to battle stations. The fire-control tracking party, consisting of a half-dozen officers and sailors, crammed into the control room. A swarm of bodies lined the benches in front of the Mark 113 target-tracking consoles. Few other priorities trumped the trailing of a Soviet boomer headed out on patrol, and if that *Delta* started flooding a missile tube, one of our Mark 48 torpedoes would be ready to surge from its cylinder and race toward the Soviet boat's twin-screws within seconds. Everyone knew that the longer we could keep a *Delta III* in our sights, the more points we scored in the "keep the world safe" game.

Mims walked over to my plot. His eyes squinting, he examined the Xs I had placed on the paper chart in relation to Master 21's recently reported bearings. He ran his finger along the *Delta*'s plot line, then stepped near the periscope stand. He keyed the sonar 27MC. "Sonar, Conn—give me an update on Blackjack."

Greenawald provided an update, while the tracking party entered data into the fire-control computers. Knobs and gears spun in muted whirs. Before long, the weapons officer (Weps) reported that we had a firing solution. We were ready to spin a torpedo up Ivan's tail, and he had no idea we were there.

"Cap'n," the Weps said, "I have a curve and a good solution on the target. Range 5,000 yards, target speed ten knots."

"Attention in the conn," the skipper said. "We have a good solution on the target. Firing point procedures tubes three and four. Horizontal salvo, one degree offset, one minute interval."

"Weapons ready," I heard myself say, an unsteady hand resting on the firing key. My mind reeled with questions. Was this just a practice run or the real thing? Did our CO know something we didn't? Some secret message from COMSUBPAC just received with orders to sink the *Delta*? Sweat from my palm dripped onto the firing key. I wiped it away before anyone could see it.

Long seconds passed before Mims issued another order. "Fire control team, stand down."

Unsure of what had just happened, my hand stayed glued in place.

Lieutenant Tomlin leaned over and pointed at the panel. "Petty Officer Reed, practice time is over. You wanna let go of that firing key now?"

We tracked the *Delta III* for another several hours as she cleared the harbor and headed toward her assigned station, sometimes closing to within a quarter mile. She crept along at six knots for some time, then suddenly doubled her turns to twelve knots to sprint to a new location. She then slowed for ninety minutes before changing course again by seventy degrees or more. We stuck to her like silent hounds, while sonar recorded every swoosh, nuance, and hiccup. We cataloged the *Delta*'s unique acoustic signature as the SSBN crept into the depths, unaware of the predator lurking in the shadows behind her.

Just as my heart rate subsided, Greenawald reported three more contacts in the area: two *Echo II*–class boats and one *Foxtrot* submarine headed our way. Two *Kresta I*–class guided missile cruisers and the *Moskva* helicopter cruiser also came onto the scene from the other direction, transiting from the base at Vladivostok. The *Delta* managed to glide below a thermal layer and slip away as the *Echo II*s moved into position on an adjacent bearing. I visualized the scene in the sonar shack, with Greenawald sitting in front of the BQS-11 passive sonar stack struggling to hear the final whispers of the *Delta* as the muted swoosh of her spinning propellers melted into oblivion.

Mims decided to bring us to periscope depth for a look around. "Diving Officer, make your depth six-five feet, ten degree up bubble."

The boat angled up as we moved toward the surface.

"Up scope." Mims slapped the number two scope handles horizontal. He positioned his face against the eyepiece and swung the scope around to the bearing of the surface group. The WLR-9 ESM detection warning beeped as the *Kresta*'s Top Sail and Head Net-C search radar signals swept past and caught a whiff of our periscope.

His eye pressed against the socket, Mims said, "Bearing to the *Kresta* on my mark . . . mark! Bearing to the *Moskva* on my mark . . . mark! Down scope."

I knew we couldn't afford to leave the scope up very long for fear that Soviet radar would detect the protruding mast.

"What the hell are these boys up to?" Mims asked no one in particular.

The XO looked up from the navigation plot through a dank swirl of cigarette smoke. Heavy perspiration painted circles under each arm of his blue coveralls. "Must be a decoy to let the *Delta* sneak off to her station."

Mims nodded.

Greenawald's voice broke through the conversation and echoed over the 27MC. "Conn, Sonar—transients in the water. Sounds like torpedo tube doors opening."

Silence. Then another excited chirp. "Conn, Sonar—we've got high-speed screws in the water! I repeat, high-speed screws in the water!"

Torpedoes.

Mims maintained his composure. "Sonar, Conn—bearing to the high-speed screws?"

"Conn, Sonar—bearing to the torpedoes is two-seven-two. They're coming from the *Echo II*."

I recalled that in 1968, sailors on the *Haddo* shadowed an *Echo II* as she transited out of the Mediterranean Sea. Days later, they handed off that contact to the USS *Scorpion*. A few days after that, the *Scorpion* met her demise—probably from a torpedo fired by that *Echo II*. And now, speeding our way, a torpedo from another *Echo II* threatened our home. I shuddered as I thought of those sailors on the *Scorpion,* screaming in fear as a relentless ocean swallowed them whole.

Mims issued another order. "Helm, right full rudder, increase speed to full, make your depth 1,000 feet."

Evasive maneuvers.

"Sonar, Conn—give me an update on Masters 28 and 29."

The two *Krestas.*

"Conn, Sonar—bearing to Master 28 is now two-eight-five, Master 29 is two-eight-eight. They're closing in our direction."

The torpedoes, along with several Soviet warships, sped closer and the tracking party feverishly dialed updates into the fire-control systems. Mims ordered another hard bank to maneuver out of the way of the speeding projectiles. As the boat turned, gravity threatened to suck in coffee cups and loose objects. The control room angled to one side in response to the sharp turn to starboard. The tiny control room closed in around me as I struggled to focus on the panel in front of me. Each breath came slow, the tightness in my chest now like a ton of bricks threatening to crush the wind from my lungs. I wanted to run, to hide from the high-pitched screaming of the deadly propellers headed our way, but only freezing ocean awaited me outside the steel hull of the *Haddo.*

"Conn, Sonar—the torpedoes are now at 2,000 yards and closing. They have not acquired; I repeat, they have not acquired."

Just over one nautical mile away.

I knew that like trained hunting dogs, the sophisticated sonar in the tip of the Soviet torpedoes would soon lock on to the noise generated by our spinning propeller. Mims remained calm, as though he had the situation completely under control. As for me, I was certain that death was a mile away and closing fast.

"Conn, Sonar—torpedoes have closed to 1,000 yards."

One-half nautical mile.

Mims ordered another hard bank, and I wondered what it felt like to drown.

"Conn, Sonar—high-speed screws are slowing; I repeat, high-speed screws are slowing."

I looked up at the overhead of the control room, as if by some mysterious power I might actually be able to see the shiny cylinders in the dark and could will them to stop.

"Conn, Sonar—high-speed screws are silent."

Tomlin stepped up behind me and voiced his opinion. "Damn near skinned our asses on that one."

Whispered cheers rang out as Greenawald reported that the torpedoes had stopped dead in the water. Something didn't seem right to me. We had just barely avoided being blown out of the water, and no one in the control room seemed at all fazed, like this was just another routine day. Did I miss something?

Tomlin noticed my confusion and said, "Soviet torpedo exercise. We just got caught in the middle."

"Oh," I said.

A few minutes later I had to ask: "Were the warheads armed?"

Tomlin shrugged. "Damned if I know."

I had just survived my first encounter with the Red Bear. Now I understood what every fast-attack submariner comes to know: for this profession, the fainthearted need not apply. In this undersea world, risks are commonplace, and lives are gambled every day in a war yet undeclared. Over the next few years, across a half-dozen espionage missions, I almost lost that gamble more than once.

CHAPTER SEVENTEEN

When sorrows come, they come not single spies, but in battalions.

—WILLIAM SHAKESPEARE

AVING REPLACED THE *HALIBUT* AS THE cable-taping spy boat, the USS *Seawolf* left San Francisco Harbor on June 20, 1976. In the expansive Area D diving section, she carried twenty-two navy divers, including David LeJeune. Unfortunately, the *Seawolf* also left port with the same sea gremlins that had plagued Captain Aleksei Dubivko on B-36 during the Cuban Missile Crisis. Breakdowns, fires, and emergency reactor shutdowns—bad things that always came in threes—kept the crew alert, awake, and filled with anxiety most of the time.

Thankfully, Gardner Brown's cousin, Gene Centre, and others like him at Bettis Atomic Power Laboratory back in 1954 took great pride in their work. They went the extra mile during *Seawolf*'s birth to ensure her reactor could exceed design parameters. This functional overbuild kept *Seawolf*'s crew alive more than once. Still, at the ripe age of twenty-two, the old girl had long passed her prime. If one dog year equates to seven human years, then boat years were probably around half that, making the *Seawolf* seem about seventy-seven—just spry enough to keep trying, but debilitated by arthritis and fatigue. Until another boat could take her place, she'd just have to "suck it up" and keep running marathons into the Sea of Okhotsk, and her crew would need to double the frequency of their prayers.

The *Seawolf* carried four remote operating vehicles (ROVs) on racks in Area D near the hyperbaric chamber used by the saturation divers. Operators nicknamed the ROVs *Deep Throat, Happy Hooker, Hawkeye,* and *Shortime*—the latter gaining its name because every time they put her into the water the electronics "shorted out." Although disguised as "oceanographic research" devices, the ROVs had an uncanny resemblance to the *Fish* and *Underdog* mini-subs used on the *Halibut* and *Swordfish,* respectively. Each were tethered to the submarine, powered by an electric motor and propeller, and contained high-resolution cameras to facilitate finding things on the ocean floor, like communications cables or discarded missile parts.

Five communication technician "spooks" also accompanied the *Seawolf* on that mission: Master Chief Mac Empey and Chief Mark Rutherford—alumni from the third *Halibut* mission—along with Chief Frank Courtney and Second Class Bob Ellenwood, who'd also made the last *Halibut* run. The fifth member, Senior Chief Tommy Cox, had just joined the team as a permanent linguistics expert.

After Mac Empey's eye-opening indoctrination video, LeJeune befriended the ACT V gang—the name given to the group of five spooks—and hung out with them in the chart room when he wasn't pressing down, diving, or pressing back up in the dive chamber. LeJeune had recently completed barber's school and gained some skill in the art of cutting the crew's hair under way. He was snipping at the head of a spook when Cox sauntered into the area. The smiling I-Brancher sat on an empty torpedo rack and picked up someone else's guitar. Without asking, Cox proceeded to tune all twelve strings. LeJeune groaned, certain that the guy was about to play "four bars of the only song he knows and then strut off proud and full of himself."

Cox finished tuning, then pulled a large red handkerchief from his pocket. He honked his nose into the hanky and proceeded to wow the small crowd with a combination of well-played original and popular tunes. LeJeune "hired" Cox as his official barbershop entertainer, and the singing spook also became a regular act in the crew's mess, especially when the selected movie sucked or during "spook night," when the ACT V team pitched in and prepared a meal for the entire crew. LeJeune lent a hand for these events, which included serving the officers in the wardroom.

After Mac Empey demonstrated his brilliance by saving the Ivy

Bells program from extinction on a prior run, he received a promotion to warrant officer and moved into the wardroom with the other officers. Rank held little meaning with the ACT V team, and Empey's beer-drinking days with the spooks at the Horse and Cow Saloon in Vallejo, California, were unchanged by the gold bars.

The *Seawolf* set a record in 1958 for the longest submerged transit of sixty days. She broke that record on the first run to the Sea of Okhotsk in 1976 by staying under for eighty-seven days. Although a rough transit, filled with occasional fires, system failures, and turbulent swells, the *Seawolf* completed her wire-tapping run with no serious issues. Prior to her next mission in 1977, T-Brancher Frank Courtney's wife took ill, requiring him to decline future runs. Mac Empey retired, and so the ACT V team needed to recruit two replacements.

There were no strangers in the ACT V group. Everyone knew and vouched for someone. Empey and Rutherford knew each other from the *Halibut*. Empey knew Cox from the USS *Lapon* (SSN-661), where both served with the famous Commander Chester M. "Whitey" Mack. Courtney knew Rutherford and Cox from the *Greenling* (SSN-614), and so it went. The group meshed and worked well together, which had proved critical to the success of previous missions. As such, the ACT V group often circumvented Bureau of Naval Personnel (BUPERS) recommendations and considered only those they could trust.

Angered by this stance, BUPERS insisted that the team at least interview their candidate, Charlie Miller. At first, the ACT V guys refused, having already decided on an M-Brancher in New London. Tommy Cox and Mark Rutherford finally relented and drove to Sugar Grove, West Virginia. There they questioned Miller for hours and found him to be a more worthy candidate. With one down, they still had one more member to select.

Frank Turban did not at all expect a call from Mark Rutherford to discuss the possibility of joining the elite ACT V Special Projects team, currently assigned to the *Seawolf*. He'd heard rumors about that aging boat and had no desire to become a part-time member of her genuflecting, prayer-muttering crew. Having served with Rutherford on the USS *Flying Fish,* where they encountered the first Soviet *Delta* and a wounded *Yankee*-class submarine, Turban trusted his friend and decided to listen. Rutherford said that two members recently left the group, and they needed Turban's advanced T-Brancher skills for an important top-secret mission. Perhaps, Turban commented years later, he should have run from the offer. Instead, he accepted and jogged headlong into disaster.

The *Seawolf* headed west in 1977 with two new spooks aboard. In the crew's mess, Turban overheard sailors talking about the *Wolf*'s constant spate of problems. They said that their tall, easy-going skipper, Commander Charlie McVain, never ran drills. He didn't need to. The *Wolf* endured more than enough real disasters. Turban raised an eyebrow when some of the guys actually placed bets that the old boat wouldn't have the chops to complete her next SpecOp. One petty officer joked that if he lost because the sub went down, he wouldn't have to make good on his bet. Turban dismissed the banter as just talk, but he still tossed and turned nervously in his rack. Six days out of Mare Island, the talk died to a whisper, and Turban finally relaxed.

Then a fire alarm uttered a warning and jolted him to attention. He'd just sat down in the crew's mess and started chewing on a sticky bun when a woman's voice blared over the 1MC. A sweet chunk of the pastry lodged in his throat and damn near choked him to death. The crew called their alarm system the "bitch in the box," as the system came with a soft voice in place of a clanging alarm. Someone back in the fifties, during the *Seawolf*'s formative years, thought they should use a soothing Bell telephone operator's voice for alarms versus the clanks and clangs found on diesel boats. An autographed portrait of the old Bell operator hung on a wall in the wardroom. The soft sound of "Fire, fire in the engine room," whether spoken with Grandma's voice or not, did nothing to sooth Turban even a little.

He initially wondered if the alarm was a drill. Then he smelled the smoke. With the bitch box blaring in his ear, he jumped from his bench, donned an EAB mask, and joined the throng now passing emergency gear aft. Commander McVain announced over the 1MC that a fire had broken out in the reactor compartment. The blood rushed from Turban's face, and his knees shook like branches in a high wind. He knew they were running 100,000 pounds heavier than normal—this to keep the *Seawolf* anchored to the bottom of the ocean during cable-tapping operations once on station. If the fire caused the reactor to scram, they'd lose propulsion. Given the extra weight, they'd lose depth control. And given the depth of the ocean, they could lose their lives.

Turban watched as a horde of damage-control personnel ran past, EAB masks strapped to their faces. Men shouted and pointed. Others responded by running, pulling gear from lockers, or just getting out of the way. Turban became one of the latter by slipping down into the torpedo room and huddling in the chart room with the other spooks. Nearby were racks of torpedoes, long and green and loaded with explosives. Turban wondered if they'd blow up when heated by a fire. As the boat angled down even farther, he glanced at a nearby depth gauge: 350 feet. He knew that the *Seawolf*'s test depth was 750 feet. The needle on the gauge inched downward even more: 400, 450, 500, 550.

Turban looked over at Rutherford, seated on one of the bunks in the torpedo room. His friend said, "We're too heavy. They're probably trying to pump the extra water out of the ballast tanks, but they can't keep up."

"Why not?" Turban asked.

"The reactor's scrammed," Rutherford said. "We don't have enough power while we're running on batteries."

"What happens if they can't get the reactor back online?" Turban said as his worst nightmare unfolded in real time.

Rutherford shrugged, said nothing.

The depth gauge hit 650 and Turban muttered a quick prayer.

Two decks above him, Commander McVain, who'd earned a Ph.D. in nuclear physics, sprinted aft to assist the damage-control party. He left his red-haired, freckle-faced XO in charge. When the boat hit 750 feet, McVain yelled an order up to the control room through a sound-powered phone to "emergency blow!" The XO's freckles disappeared, and his face turned white. Frozen by shock, he did not repeat the order. The chief of the watch maintained a calm head and followed his skipper's orders without waiting for the XO. The gushing sound of air permeated the boat.

Then the *Seawolf* rocketed to the surface, lifted partially out of the water, and crashed back into the ocean. She dove again to more than 200 feet before leveling off on the surface. The crew finally managed to quell the fire and to ventilate the smoke off the boat. Turban later learned that a large fuse in the nuclear compartment had blown. Filled with wet sand, the fuse blew its moist dust into the lube-oil bay, causing a plume of steam. When the reactor operator saw the plume, he thought it was from the propulsion system, so he scrammed the reactor. Without the brute power of her nuclear-propelled blades, the *Seawolf* could not climb the mountain back to the surface, at least not until McVain ordered an emergency blow to force air into the ballast tanks and push out the extra water.

Thankful to still be alive, Turban breathed a sigh of relief. Then he heard the rumor. Apparently, they brought only one spare sand-filled fuse, and they just used that one up. Despite the risk, Commander McVain refused to throw in the towel and head back to port. He figured that the old girl had survived everything thrown at her, and the odds against blowing another fuse had to be one in a million. Then again, thought Turban, Captain Edward John Smith thought the *Titanic* was indestructible.

THERE ARE FEW ICEBERGS IN THE Sea of Okhotsk, but there were plenty of Soviet warships and hunter/killer submarines during the Cold War. Dozens of them strutted across the "Mouth of the Bear's" frigid waters while transiting between Vladivostok in the Sea of Japan and Petropavlovsk on the Kamchatka Peninsula. After the *Seawolf* arrived on station, deep inside Soviet territorial waters, navy diver David LeJeune prepared for the dive of his life. Three other divers joined him inside the twenty-by-ten-foot diving chamber in Area D.

LeJeune pulled on his MK-11 deep-sea diving suit. He then dipped his Kirby Morgan mask into some water to soak the spongelike material—this to better keep the ocean out of his eyes. After smearing Vaseline over the rubber seal on the mask, he clamped it down tight on his face and sucked in a few breaths. Nearby, the other divers duplicated the routine, including David "Whompee Jaw" Sullivan. The lanky diver earned his moniker by the grace of God, who imbued him with an offset jaw. The engineers at Kirby Morgan actually had to design a special mask to fit the man's unique face. After the divers finished suiting up, they completed their checks and opened the inner hatch, but not before LeJeune placed something in his suit pack. Sullivan climbed up through the opening, wrestled with the outer hatch, and disappeared into the black beyond.

The hyperbaric chamber was pressurized to maintain equilibrium with the ocean outside, not too unlike the small space at the front of the DSRV-shaped diving chamber on the *Halibut*. This prevented the seawater from taking up residence in the submarine and allowed the divers to exit to the netherworld. Once outside the sub, darkness engulfed LeJeune as he descended to the ocean floor. Tiny fish fluttered past his face mask, shiny and ethereal in the glow of his underwater light.

Like they had done on countless training dives, LeJeune, Sullivan, and another diver slow-walked the 300 yards over to the twenty-foot-long pod-tapping "beast." The large drum-shaped pod had been placed there on an earlier mission, and thick "jumper cords" now snaked through the sand on their way to the communications cable. The recording tapes in the beast had collected hundreds of Soviet conversations over the past several months, and the divers' job now entailed retrieving those tapes for subsequent perusal by the NSA. After completing this task, LeJeune removed a white oval object from his suit pack. He knew that the USS *Parche* was scheduled to undertake the next mission back to the Sea of Okhotsk and figured he might not be selected as one of the divers to go on that run. As a joke, he wedged a cow's skull under one side of the beast to offer the next set of divers "a good laugh" when they arrived the following year.

While slogging back toward the *Seawolf,* a green glow inside LeJeune's mask verified that his HeO_2 mix was still good. His stomach growled, and he was thinking about devouring a hamburger in the crew's mess. His hunger evaporated when his mix light turned red. He swallowed a lump and called the master diver, who ordered an immediate "reel in." Something tugged hard at LeJeune's back. His feet shot up from the silt as he sailed through the water, pulled along by his lifeline. The fourth diver, inside the boat, reeled him in at warp speed. Tons of ocean between LeJeune and the boat made the seconds tick past like hours. LeJeune reached back and felt for the small can of mix strapped to his back, painfully aware that it held only three breaths. He was more than a football field away.

One breath.

So this is how it ends.

Two breaths.

Would they remember him as a good man?

Three breaths.

His vision blurred as his air ran out. He felt dizzy but at peace. He closed his eyes. He did not see God, but he also didn't see the devil. In fact, he saw nothing but darkness. Then a bright light beckoned, and he wanted to follow.

LeJeune blinked. Sullivan's offset jaw formed a crooked smile inches above his face. "Thought we lost you there, partner."

After decompressing, LeJeune completed a physical examination and sauntered up to the crew's mess. There he ordered a big, fat, juicy cheeseburger smothered with mayonnaise. He figured if he was going to die of a heart attack, it probably wouldn't be today.

The *Seawolf* logged another record that year for the longest-known mission underwater, 106 days all told. She hadn't set out to achieve that milestone, but her aging propulsion system refused to deliver more than ten knots speed, and given the amount of noise she produced, Commander McVain needed to take an indirect course to avoid detection. Those in a position to know full details about upcoming Ivy Bells missions understood that the *Seawolf*'s time was near an end, and the USS *Parche* would be making the next run to the Sea of Okhotsk in 1978.

David LeJeune joined the crew of the *Parche* and made several more dives, some in the Barents Sea at depths reaching 700 feet. He earned his submarine dolphins in 1977 and qualified as master diver in 1978. He and his wife, Cheryl, whom he still calls George, have been married for more than forty years. Submariners on the *Halibut, Seawolf,* and *Parche* referred to the navy saturation divers who undertook these harrowing missions by a special name. They called them heroes.

WHEN DENNIS SMITH GRADUATED FIRST IN his class from navy electronics school in late 1975, he did not envision that Submarine Development Group (DevGru) One would recruit him to help conduct submarine espionage missions. He didn't think his job would entail working on the most top-secret and advanced mini-subs in the world, and he never imagined that fate would place him front and center for the most important underwater missions of the Cold War.

When he arrived in Northern California, DevGru assigned him to the oceanographic ship USNS *De Steiguer* (AGOR-12), skippered by a civilian who reported to Commander Hal Brown, the officer in charge of DevGru One. There he worked with cutting-edge devices called

STOVEs, for Surface Tethered Oceanographic Vehicle Experimental. These camera-filled mini-subs had been resurrected from the USS *Halibut* and painted with "cool colors" such as orange or bright fishlike swirls of red and yellow. On the *De Steiguer,* Smith helped reel out 17,000 feet of cable so the STOVEs could hunt seabeds for interesting tidbits, like spent missile parts. Smith enjoyed his job, but his time aboard the *De Steiguer* abruptly came to an end when a sea mountain damaged the STOVEs beyond repair during a big storm.

ET2 Dennis Smith then received orders to the USS *Parche* in 1976 and reluctantly reported to the Special Projects Rescue Systems (RS) division. He didn't think the *Parche* would have anything as "amazingly cool" as the STOVEs, but as it turned out, he was dead wrong. In the *Parche*'s torpedo room he found the latest in secret underwater snooping technology, known as the System 2090, which incorporated improved optics, sonar, and more. Using tethered mini-subs similar to the STOVE, *Fish,* and *Underdog* UAVs, the *Parche* found stuff on the ocean floor with these advanced devices like nothing had before.

Smith's RS division consisted of new guys who'd not yet gone on a mission, with the single exception of his boss, Senior Chief Al Lusby. Over the next several months, Smith got to know his teammates well. He also met some of the "Rescue Communications" division spooks, including Frank Turban, and a few of the "Rescue Operations" saturation divers, like David LeJeune. Both had just completed a run on the *Seawolf* to the Sea of Okhotsk, and Smith felt a tingle of excitement run down his spine when they talked in hushed tones about *Parche*'s upcoming mission to the same area.

Over the next several months, Smith and the other ETs practiced operating the System 2090 mini-subs. These "*Fish*-like" devices were launched via the torpedo tubes and contained special sonar to help them locate objects underwater. The units were tethered to 400 feet of cable and housed high-resolution television and film cameras and high-beam lights. Despite the fact that the 2090s used torpedo tubes to deploy, the *Parche*'s torpedomen were not allowed to view, let alone work on, any of the equipment. Only the specially trained ETs in the Rescue Systems division were afforded that privilege.

The 2090s resembled stubby torpedoes, right down to the same green paint. They were all custom made by Westinghouse Oceanic Division in Annapolis, Maryland. Another company with offices on Tennessee Street in Vallejo, California, made most of the internal equipment, which included a state-of-the-art plasma display and a videodisc recorder that could store 300 frames of video—about 10 seconds' worth—complete with rewind and slow-speed replay. With 2090s running just twelve feet off the bottom in murky waters, this last feature proved critical in spotting three-inch-diameter communications cables and tiny missile parts. The RS division consisted of ten ETs, with five of them standing six-hour watches at a time. On watch, one ET operated the video gear, and another drove the 2090, while yet another navigated.

As for weapons, the *Parche* carried only four torpedoes—initially a combination of MK-37s and MK-48s, and later just MK-48 ADCAPs (Advanced Capabilities). These torpedoes were specially modified to light off their motors faster after leaving the tube. Standard torpedoes simply would not work on the *Parche* when she was on station. Her cable-tapping missions required that she sit in the mud atop jet ski–like sled runners, with her underbelly only a few feet off the bottom. A normal torpedo, when fired, drops downward by almost fifty feet before the motor lights off and speeds it toward a target. This safety feature, which helps prevent problems with "hot running" torpedoes, is acceptable in the open ocean, but not for scenarios where the sand is less than fifty feet below the sub. Modified torpedoes solved this problem, but also carried the risk of a hot run, which could terminate a mission.

The *Seawolf* remained under repair and tied to the pier while the *Parche* went to the Sea of Okhotsk in 1978, with Commander John H. Maurer in command. The crew took to calling their skipper "Bullet Head" because of his near-bald crewcut. Once the *Parche* came within range of the cable, Smith and the other ETs shot a 2090 mini-sub out one of the torpedo tubes. On a monitor screen in the torpedo room, Smith watched the 2090 speed across the bottom, its bright strobe light illuminating jagged rocks and frightened fish. The shape of a twenty-foot-long barrel came into view. *The beast.* Smith stared in awe at the large cylinder and marveled at the ingenuity. American engineers had actually built a watertight container that housed the latest in sophisticated induction-recording technology powered by a miniature nuclear reactor. If they could do that, what couldn't they accomplish?

Parche spent the next thirty-one days on station while the RS division

divers, including Master Diver David LeJeune, retrieved the recording
tapes and returned to the boat. Given *Parche*'s advanced speed, their
entire mission time was less than the *Seawolf*'s typical transit time. The
Parche started to head home, but Maurer ordered a return to the pod
for overhead mission pictures. The *Parche* hovered over the beast while
a 2090 reeled out and snapped shots, some 300 feet below the submarine.
Maurer maneuvered the *Parche* into a hover, not unlike a helicopter,
where the boat remained completely still. For subs, this is a difficult
maneuver that requires steady hands to pump water in and out of the
trim tanks to keep the boat level and stationary.

Unfortunately, hovering did not sit well with the 2090s. They preferred a moving submarine to ensure an unsnagged tether line. With the
Parche sitting in one spot, the 2090 windmilled out of control and
snapped off from the end of the cable. The ETs had a spare, but it took
ten hours to reel in the old cable, unravel it, splice on a new 2090, and
test the system. The *Parche* then had to use the new 2090 to find the old
one. Once located, the submarine sat down again and redeployed the
divers to grab the spent mini-sub. Having already started to decompress, the divers grumbled but complied.

Dennis Smith's first run into harm's way hooked him into the world
of Special Projects, where he remained for years to come. One of those
missions was *Parche*'s first into the cold northern waters of the Barents
Sea in 1979. Given its proximity to Murmansk, home of the Soviet
Northern Fleet, this area provided a heightened degree of danger and
difficulty for any submarine, let alone a cable-tapping one. Smith and
crew spent days searching for the cable, coming within a few miles of
the Soviet coastline.

Finding the cable took days, with the *Parche* sending out a 2090 to
snoop around the ocean floor some 700 feet down. Once found, they
hovered over the landing spot while dropping 15,000-pound mushroom
anchors. The anchors thudded into the silt fifty feet below the submarine. The *Parche* then inched down slowly until she neared the top of
the anchors. She oscillated back and forth above the two anchors before the winches were engaged to pull the boat toward the bottom. Although the *Parche* had practiced this maneuver dozens of times off the
coast of San Francisco, the actual procedure was fraught with risk. One
wrong move could send the boat slamming into one of the anchors or

rip her free of the tethers and end her ability to stay anchored to the bottom—which would also end the mission.

The *Parche* settled to the bottom and deployed the divers. In the pitch-black sea, they manhandled the massive pod-tapping beast into position near a communications cable. They connected the jumper cord to a repeater box and returned to the *Parche*. T-Branchers then pulled information off the cable, while I-Branchers listened intently to ensure the quality of the tap. Weeks later, the *Parche* returned to California and received a Presidential Unit Citation for operating "in the hostile environment of poorly charted ocean areas." Her crew celebrated the success of their mission at the Horse and Cow Saloon in Vallejo, and all of them signed gag orders saying that they would remain silent about what they had done for at least thirty years.

CHAPTER EIGHTEEN

Success is not final, failure is not fatal: it is the courage to continue that counts.

—WINSTON CHURCHILL

WHEN THE FIRST ULTRA-QUIET SOVIET NUCLEAR-POWERED hunter/killer attack submarine, code-named *Victor I* by NATO, rolled down the ramp at the Admiralty yard in Leningrad in 1968, the U.S. Navy lifted a concerned eyebrow. Sixteen of the *Victor*s splashed into the sea through 1975. By 1978, Seven *Victor II* submarines were added to the Soviet fleet, each carrying sixteen Stallion cruise missiles that could scream through the air at Mach 1.5 and destroy targets seventy-five miles away. Crews quickly manned battle stations whenever a *Victor II* was detected anywhere near a U.S. warship. For NATO forces, however, finding a *Victor II* was not that hard. Soviet engineers were skilled in weaponry but not in the art of stealth.

During the early years of the Cold War, the Soviet Union knew that its boats were noisy as hell. Built to keep up with fast U.S. warships, they had a distinct acoustic disadvantage. U.S. sonar operators often joked that Ivan's subs "whined so loud, they sounded like cats in heat." When the Soviets discovered through their spy network that American subs could easily track *Victor II*s, they halted production to design the *Victor III*. To overcome excessive noise issues, the Soviets reverted to a tried-and-true tactic: they cheated.

During the late seventies and into the early eighties, the Soviet Union secretly contracted with Toshiba and Kongsberg—Japanese and

Norwegian technology companies, respectively—for 140 Kongsberg computers, advanced software, and sophisticated propeller-milling equipment designed to quiet their noisy props. In direct violation of international trade agreements, the two companies, almost overnight, allowed the Soviet Union to produce quiet submarines. In 1979, they did just that with the introduction of the *Victor III*, which swung the Cold War pendulum back in their favor. For U.S. submariners, hunting Ivan just got a lot tougher, and the statement Commander Kinnaird R. McKee made in 1968 when he served as the CO of the USS *Dace* (SSN-607) all but came true.

In short, McKee opined that "eventually, U.S. and Soviet submarine capabilities will converge. . . . It will be blind man's bluff with other submarines . . . because at some point, nobody will be able to find a submarine with anything."

Thanks to Toshiba and Kongsberg, *Victor III*s sported ultra-quiet tandem four-bladed propellers and special sound-reducing shock absorbers to increase stealth. This class of hunter/killer attack submarine also came with something else entirely new. Mounted atop the rudder at the aft of the *Victor III*, a strange tear-shaped pod stretched nearly thirty feet from end to end, with a diameter of eight feet near the front of the pod. From a distance, the odd pod resembled an elongated egg with a sharp taper near the tail. By 1980, the U.S. Navy remained clueless as to what this mysterious pod could be. Obtaining close-up photos of a *Victor III* odd pod, so the NSA could determine its purpose, became a top priority for U.S. fast-attack submarines.

AT THE REQUEST OF SUBMARINE GROUP (SUBGRU) 5 in early 1980, I received a transfer from the USS *Haddo* to the USS *Drum* (SSN-677), a newer *Sturgeon*-class submarine. Like the *Haddo*, the crew of the *Drum* numbered just under a hundred enlisted men and a dozen officers. Our CO, Commander Michael Oliver, was an experienced and proficient leader who often ran headlong into danger like a sword-brandishing swashbuckler after bounty. Our executive officer, Robert E. Fricke, while also professional and intelligent, occasionally pushed the crew a bit harder than needed in an effort to improve efficiency by one more tenth of a percent.

Commander Oliver sat me down in his stateroom a few days after I reported on board and explained why I'd been transferred to the

Drum. He told me that Soviet tactics had changed dramatically since 1975, when the Reds projected power into distant waters and tried to cut off Western sea lanes. Now they focused primarily on protecting their SSBNs by finding and destroying U.S. submarines and ASW ships. The *Victor III* was built with this purpose in mind.

In light of this radical change in Soviet behavior, the NSA had deep concerns about the purpose and capability of the new odd pod mounted on the rudder of the *Victor III.* That fear rolled downhill to the navy and downward still to SUBGRU 5. They examined numerous photographs I'd taken of *Victor IIIs* while on the *Haddo* and liked what they saw. They wanted more—but preferably closer and clearer. Under strict orders not to discuss any of this with the rest of the crew, Oliver told me that the *Drum* would be leaving on a WestPac in the fall to conduct two SpecOp missions to get those photos. We would spend the next several months training and preparing for that run.

I saw a gleam in Oliver's eyes that day and figured that our next deployment represented his best ticket to stardom and perhaps another stripe on his sleeve. Oliver said that he'd been instructed to "exercise every means possible" to obtain better photographs of the odd pod. Since I'd taken and developed more *Victor III* shots than almost anyone else and had also received advanced photographic training, Oliver wanted me practiced and ready to help reel off pictures when needed. As I sat in Oliver's stateroom and listened to his speech, feelings of pride, inadequacy, and fear overwhelmed me. I thought the *Drum* just needed another fire-control technician and navy diver. I hadn't planned on this extracurricular assignment, and I hoped that when the time came, I'd be up to the challenge.

Having completed a few more classes in recon photography and some on-the-job training during local at-sea exercises, I received orders to report to the diving tower at the Navy SEAL training facility on Coronado Island. Commander Oliver requested that I spend some time doing photographic reconnaissance training with a SEAL team there. This type of exercise was not new, and in fact had been commonplace during the Vietnam War. Navy diver photographers like Steve Waterman, author of *Just a Sailor,* undertook such operations all the time. Navy recon divers sometimes swam with SEAL teams to various beaches, near foreign vessels, or within visual range of shore-based facilities to take reconnaissance photos for upcoming operations.

I spent the next few months getting in shape at the obstacle course and exercising with some of the SEALs. The physical conditioning reminded me of navy diver school in Hawaii but paled in comparison to what I knew SEALs endured in BUD/S training. The SEALs also gave me a crash course on the Draeger MK V rebreather. As a diver, I had been using open-circuit SCUBA, usually in the form of heavy twin-nineties that emitted bubbles that could be detected by unfriendly guys. Draegers, which were strapped to your stomach rather than your back, looked like small vacuum cleaners. They emitted no bubbles but could only be used down to a depth of thirty-three feet.

After my land-based training time with the SEALs, the *Drum* conducted an exercise just off the coast of Catalina Island. During this exercise, I teamed with four Navy SEALs—one officer and three enlisted men. We donned wet suits, masks, fins, and standard navy twin-ninety SCUBA tanks. The double-tank rigs were heavy, causing us to bend over as we stood in the bow compartment of the boat just below the forward escape trunk. Someone opened the bottom hatch to the trunk, and I watched one of the SEALs climb up the ladder into the chamber, an oval area about eight feet in diameter and filled with gauges and valves.

This SEAL's job, completed alone, consisted of locking out of the trunk and finding the line locker, near the bow of the boat, where the mooring lines were stowed. There he removed a rubber Zodiac raft to be used for our mission. The SEAL attached the raft to a tether line anchored to the sub, then inflated the Zodiac to propel it to the surface. This delicate operation took about thirty minutes and was done entirely by feel in near-zero visibility. A couple of raps on the trunk signaled that he'd completed the job, and the rest of us could now deploy.

The escape trunk could fit only two at a time with twin-nineties strapped to our backs. Once inside the escape trunk, we turned a few valves to expel the air and fill the space with salt water up to the bubble line—about where our heads jutted above the water. The trunk was now pressurized to match the outside water pressure, which allowed us to open the upper hatch and swim out into the ocean. We hooked onto a lifeline running from the top of the submarine's sail to the foremost line locker. The rest of our equipment, not carried with us into the escape trunk, was stowed in that locker.

The boat slowed to less than one knot, standard operating procedure for diver operations to ensure that none of us would be sucked into the propeller. Our SEAL officer in charge flashed a signal, and we all swam over to the line locker to remove the gear. As I neared the locker, the ocean rushed past my face mask and I shot upward out of control. My lifeline snapped tight with a jolt and prevented me from heading toward the surface. An excruciating pain stabbed at my right ear, and massive pressure squeezed in on my left ear. The sensation felt worse than having a severe head cold while rapidly descending in a plane from 30,000 feet. My stomach fluttered as I tried to clear my ears and kill the pain.

I grabbed hold of the lifeline and pulled myself down toward the SEALs. I could barely see their silhouettes in the dark as a stream of microscopic sea life rushed past. Submarines at periscope depth are supposed to stay at sixty feet as measured to the bottom of the keel. On top of the boat, near the escape trunk, we should have been at around thirty feet. I held my depth gauge up to my mask and nearly let out a yell into my regulator. We were at 200 feet.

I knew that warmer waters tended to make things less buoyant, like 4,000-ton submarines. Sometimes when boats are cruising along in cooler water, and they hit a pocket of warm water, they sink like a rock. An alert planesman in the control room, responsible for keeping the boat level, can often compensate quickly. A not-so-alert planesman can lose depth control and drop a boat by a hundred feet or more in a matter of seconds.

I glanced at my watch. Recalling the navy dive tables, I knew we had only four minutes at 200 feet before we incurred decompression time, not a problem if we regained depth control soon and didn't ascend too fast. I started doing some mental calculations to determine our decomp time if we went past five minutes at 200 feet. While crunching numbers in my head, our OIC held up one hand in the near dark, looking for the okay sign from the rest of us. All of us pointed to our ears first, then gave the okay signal. We all had probably busted our eardrums, but other than that, we were fine.

Five minutes came and went while the boat remained at 200 feet.

We'd now have to decompress. Five more minutes ticked by, and still no movement. I looked over at our OIC and tapped my watch. He shook his head no, and a sick feeling hit my gut. I had no desire to get the bends prior to my twenty-fifth birthday. I wasn't happy about the OIC's decision but figured he had his reasons.

Twelve minutes came and went. The forward escape trunk opened, and another diver appeared, carrying two sets of twin-nineties. He signaled that the boat would ascend to periscope depth and remain there until we decompressed. A minute later, the boat came shallow. Navy dive tables dictated that we needed to spend three minutes at thirty feet, seven minutes at twenty feet, and twenty-seven minutes at ten feet. The extra tanks brought by the other diver would give us just enough air to complete our decomp time. As we clung to the line and sucked in gulps of air, I finally understood why our OIC hadn't ordered us to swim to the surface earlier.

Surfacing while we were still inside the four-minute window, so we didn't have to decompress, would have been too risky. The boat needed to speed up to regain depth control, which increased our chances of getting pulled into the propeller if we unhooked from the line. After the four minutes, we were better off waiting for another diver to deliver extra tanks to ensure we had plenty of air between us to complete our decompression time.

Although I had received extensive training at navy diver school, which in many respects is similar to the first phase of navy SEAL training, I obviously still had a lot to learn about submarine diving. I had spent so much time improving my photographic skills, and keeping up with my regular fire-control duties, I hadn't devoted proper attention to my diving acumen. I made a commitment that day to learn as much as I could from the SEALs and get wet at least once a week to stay proficient. Never once did I consider the possibility that I might actually be called upon to use my navy diver training on an actual mission in Soviet waters.

In between plunges in the ocean, I honed my abilities in the art of periscope espionage photography. Not only did this require learning everything a professional photographer learns, such as adjusting camera settings and lens selection for varying light conditions, setting film speeds, as well as the general physics governing the world of film, but

I also needed to become proficient at what, how, and when to do all this through a periscope—and in a dripping wet suit aboard a bobbing rubber raft while shaking from the cold. Getting up to speed on that took many long hours of studying top-secret Soviet warship manuals, periscope lens functionality, limitations, and lighting, plus time-of-day requirements, then lots of practice during exercises off the coast of California. On one of those runs, I photographed the stern of the aircraft carrier USS *Constellation* (CVA-64). We were so close that we could almost make out the name tags on the uniforms of the deck crews. They never detected us.

Along with taking photos, I was responsible for developing the film and keeping the camera equipment in top condition. We had two types of cameras on our submarine: a standard 35 mm Canon AE-1 camera and a high-speed 70 mm camera built into our number two periscope. Taking photos with the first required connecting the camera's lens directly to the eyepiece on the scope. Taking pictures with the latter was as easy as pushing a small red button on the periscope handle. Theoretically, the 70 mm should have provided twice the quality of the 35 mm, but this wasn't always the case, given that you couldn't adjust settings on the 70 mm the way you could on a Canon.

When it came time to develop the film, things got a lot more complex. Developing 35 mm photos under way was a major challenge. We didn't have digital cameras in those days, so this process required manually dunking the film into several baths of chemicals contained in one-foot-square plastic bins. This operation had to be done in total darkness, usually in the yeoman's shack. The plastic bins were placed wherever one could find a flat, clear space. Heaven forbid we should be at periscope depth. The boat would pitch from side to side in the waves, while the soup in each bath threatened to spill onto the deck. I'd place the film in one tub using a pair of tongs and bathe the sheets until done, then set it in the next tub and so on—a time-consuming process.

In contrast, I used an automated developer for the 70 mm film. This two-by-one-foot metal box did all the work. I simply placed one edge of the film into a side slot and flipped a switch. The machine grabbed the edge of the film and hummed away while feeding the ribbon through a snaked procession across rollers and through baths of developer and fixer solutions. Out the other side, fully developed 70 mm film emerged. This unit simplified the process of film development, as long as the thing wasn't broken. Part of my training included learning how to repair the box when it did break, which was often.

After weeks of periscope and recon photography school in San Diego, coupled with lots of practice at sea, the science of film processing and system repair became second nature. Aboard my previous boat, the USS *Haddo*, I spent several months on a couple of SpecOps taking photos of interesting ships and subs through the periscope using the 35 mm and 70 mm cameras. A good number of these pictures were of the infamous *Victor III*.

I studied the photos I'd taken of the Project 671RTM *Shchuka* submarine and marveled at her sleek race-car design, sloped conning tower, and strange pod mounted atop the rudder. I ran a magnifying glass over the pod and squinted, desperately trying to figure out what this thing might be. The introduction of the *Victor III* in late 1979 caused quite a stir in NATO intelligence circles due to that distinctive oval pod. Some speculated that the egg-shaped housing hid a silent propulsion system, perhaps a magnetohydrodynamic drive unit. Others insisted this had to be an advanced weapons system or a towed sonar array. Nobody knew, and everybody wanted to.

Consumed with finding a *Victor III*, I deployed aboard the USS *Drum* on WestPac in the winter of 1980. We crossed the Pacific and assumed our station near Vladivostok but found nothing during our first multimonth SpecOp. We surfaced outside Diego Garcia, a horseshoe-shaped atoll in the Indian Ocean, and received orders to tie up next to the submarine tender USS *Sperry* (AS-12). Diego Garcia had earned the deserved reputation of being the world's largest toilet bowl. Nevertheless, the tiny island was dry and had beer.

At only one-half mile across, there wasn't much to see on this mass of jungle and sand that served as the only airstrip suitable for long-range bombers anywhere near the troubled Middle East. The Seabees began their largest peacetime construction on Diego Garcia in 1971. The ambitious project took more than a decade and cost $200 million, ending with a massive complex that could accommodate some of the navy's largest ships, military cargo jets, and long-range bombers. The island also housed a contingency of spooks assigned to a Pusher Bulls

Eye HFDF station. The Pusher was a smaller version of the Wullenweber elephant cage antenna array.

We had received reports weeks earlier from the *Sperry* on the VLF radio wire warning us of an intruder lurking near their ship. They claimed to have pictures of the large gray predator shadowing one of the submarine tender's twenty-foot launches. From nose to fin, the massive shark measured as long as the launch. We later learned that some of the cooks on the *Sperry* fed Hector (the nickname they gave the large hammerhead) off the bow of the ship. The shark devoured tons of scraps from the galley, which is probably why it kept hanging around. Having eaten some of the *Sperry*'s meals in San Diego, I wondered how the damned thing was still alive.

When we pulled alongside the *Sperry* during a massive rain squall with high winds, we slammed headfirst into the tender's side and smashed our sonar dome. A team of navy repair divers flew out from Yokosuka a day later to patch up the dent and replace the busted hydrophones. Commander Oliver cordially compelled me to stand shark watch while the divers worked, just in case Hector wandered by. I informed the repair divers that their first sign of imminent danger would be the frantic sound of my fins swooshing past their heads at ninety miles per hour. They all laughed, but not very much.

With repairs completed, we headed back out on our second SpecOp. Our planned liberty in Perth, Australia, was canceled due to the accident. Commander Oliver now had one black mark in his service record, and the brass never overlooked submarine collisions regardless of the circumstances. This collision, however, was rather minor in comparison to the one that almost took my life two months later.

CHAPTER NINETEEN

We're in greater danger today than we were the day after Pearl Harbor. Our military is absolutely incapable of defending this country.

—RONALD REAGAN

THEY SAY YOUR LIFE PASSES BEFORE your eyes just before you die. For submariners, the only thing that passes before our eyes is a wall of water. In early 1981, while aboard the nuclear submarine USS *Drum*, that wall of water visited me more than once in my dreams. One nightmare seemed so real that I jolted straight up in my rack from a deep slumber and pounded my head on the steel-encased fluorescent light above me. I spent the next few days with an aching head and vertical lines on my forehead.

The *Drum* arrived on station near the Vladivostok naval base inside the Sea of Japan, and several uneventful weeks went by despite our proximity to one of the Soviet Union's largest submarine ports. I wondered if we'd come up dry once again on this run and have to head home without any odd pod photos of a *Victor III*, the Soviet's newest fast-attack submarine.

A month went by underwater with no sign of a *Victor III*, and all of us longed for the thrill of the hunt. We stood our watches, cleaned our assigned spaces during "field days," drank green Kool-Aid "bug juice," ate "gedunk" snacks, watched *Saturday Night Fever* until we all hated the Bee Gees, then watched it again. In the after compartments of our

300-foot-long prison, the nuclear-trained "nukes" conducted drills in preparation for an Operational Reactor Safeguard Exam (ORSE) scheduled within weeks of our return to San Diego. They complained about our executive officer, Bob Fricke, who insisted on perfection and so drilled around the clock and then drilled some more.

The nukes got their payback when two of them donned yellow anti-C (antiradiation) suits, grabbed a couple of handheld radiation detection monitors, and paraded down to "officer's of country." They placed a test sample on the bottom of the radiation monitors so they'd tick away as if at ground zero in Hiroshima. Then they waited until the XO drifted off to sleep, crept into his stateroom, and pulled back the curtain on his rack. One of the nukes started poking Fricke in the ribs to wake him up. The XO rubbed his eyes, and his sleepy face wrinkled with perplexity as he pointed at their anti-C suits. The two nukes uncapped their radiation monitors. Both ticked away with ferocity. The XO's eyes widened and resembled the white plates in the wardroom. Playing it to the hilt, one nuke went for the jugular and said, "Damn! I think this one is still alive. We'd better get the doc down here right away."

"Nah," the other nuke said, his voice muffled inside his yellow hood. "Why bother? He won't last long anyway. Way too many zoomies in the ops compartment by now."

Fricke came unglued and started blubbering something unintelligible. One of the nukes said he practically peed in his rack. Laughing hysterically, the two nuclear-trained perpetrators ran from the scene. Since Fricke couldn't tell who'd been inside the thick anti-C suits, he had no one to punish. The following day the XO cut the ORSE drills in half.

Six weeks into our three-month SpecOp, the cooks served butter-drenched lobster and grilled steak for the SpecOp's "midway meal." Savoring the tasty morsels boosted my morale but only temporarily. I wanted to find that *Victor III.* Just over two months into our run, in early April 1981, I finally got that chance. After breakfast I grabbed a cup of coffee from the crew's mess, walked up the ladder into the reddened "rig for dark" control room, and took over as fire-control technician of the watch. My duties now included tracking contacts via the MK-113 fire-control system and, when called for, taking pictures of those contacts through the periscope. Resigned to enduring another boring six-hour watch, I sat on the bench near one of the consoles.

Analog servos and synchros inside the gray metallic enclosure whined and popped as they struggled to keep up with a distant contact.

Downing a gulp of coffee, I glanced around the control room. Cigarette smoke swirled into the stale air and danced with the steam from a half-dozen navy cups. Save for the sound of a jazz band playing Coltrane, the nostalgic scene reminded me of a bar off Market Street in downtown San Diego. In the dim glow, I saw the chief of the watch sitting across and in front of me on the port side of the boat. He faced a gray monolith filled with black panels covered with an array of switches, dials, and gauges. His oversized left arm almost hid the low-pressure blow panel, and his right shoulder all but covered the square snorkel control area. Just above his head, a horizontal row of red indicators validated that we had no hull openings exposed to the sea. I often wondered what might happen if one of those lights ever went from closed to open while we were deep.

The alarm switchboard rested above the COW's left ear, adorned with two rows of ten rectangular red alarm lights underscored by three-way switches. Above the COW's right ear, large silver handles jutted from a gray box with a single indicator and two black signs that read AFT BLOW and FWD BLOW. Of all the panels in the control room, that one sent chills down my spine more than anything else. If the COW ever needed to pull those handles, we'd probably be on our way to the bottom, and our only hope of survival would mean a risky emergency blow to expel the water from our ballast tanks.

To the COW's right, a helmsman and planesman slouched in bucket seats, hands resting at the ten and two o'clock positions on two half-oval steering wheels. Marlboros dangled from their lips as they shared bad jokes. Each focused on two large dials at eye level that indicated the boat's depth. These two yahoos were responsible for maintaining depth control and steering the boat on the right course. When trailing a Soviet submarine, which we did often on Holystone missions, we'd often come within a few dozen yards to record various machinery and propeller noises. One wrong move by either of these sailors could cause a serious accident, possibly sending either or both subs to the bottom.

Above and in between these two, dials depicted rudder, fairwater, and stern plane angles, along with gyro course, speed, and dive bubble—the latter equating to the level of the boat in a similar fashion to a carpenter's level. Just behind the planesman and helmsman, a burly diving officer puffed on a pipe. The grandfatherly smell of his sweet cherry tobacco coated the air and reminded me of home when I was a kid. My dad once smoked a pipe until he switched to cigarettes. The chief had smoker's wrinkles and a bald spot on the back of his head. His teeth had long since turned bitter coffee brown. If he hadn't been wearing a dark blue "poopy suit" pair of coveralls like the rest of us, I might have mistaken him for a homeless person in need of a shopping cart.

To the right of the diving control area, just in front of the MK-113 fire-control gear, a large gray navigation and plot table, covered with a chart of the area near Vladivostok, kept two people occupied: the quartermaster of the watch and the junior officer of the deck (JOOD). A panel flanking the left side of the table had recessed buttons to control various functions, and the top held up a navigation ruler. The plot served a dual role: one, to plot the course to our next destination, and two, to manually keep track of nearby contacts in relation to our track. Making sure we knew our location in relation to the other guys could be critical in preventing a collision. I knew that one wrong calculation or assumption could spell disaster and hoped that such would not happen on my watch.

To my left, on the periscope stand, stood the officer of the deck, Lieutenant Nick Flacco. He'd graduated from Annapolis in 1976 with his eye set on a patrol gunboat squadron based in Naples, Italy. As an engineering major, he ranked in the top twenty percent of his class, and that fact painted a target on his back. While still in his senior year at the academy, the pressure mounted to go nuclear. Officers trained in that discipline questioned Flacco's request for gunboats along with his sanity. During his submarine indoctrination, aboard the USS *John Marshall* (SSBN-611), the navigator turned to Flacco and said, "Whatever you do, don't go submarines. You won't like it." Flacco concurred, stating that he'd already decided on gunboats. Over the next few months, officers pushed and prodded Flacco to select either nuclear subs or surface ships, as the navy needed officers for both.

Flacco chose submarines when he heard that junior officers were sleeping in the brig on aircraft carriers because they didn't have enough staterooms. Based on his class standing, the navy let him choose the

USS *Drum* in San Diego, and he climbed down the hatch in the summer of 1978. An easy-going young officer with a pleasant smile and a get-the-job-done attitude, Flacco was a favorite with the crew. I always hoped that if we did have an emergency, he'd be our OOD at the time.

That night on watch, just over two months into our boring SpecOp, a technical "T-Brancher" spook tucked away in the radio room got a distant sniff on our BRD-7 electronics surveillance system. Faint at first, he almost missed the MRK-50 Series Topol radar, code-named Snoop Tray 2 by NATO. As he analyzed the signal captured by the BRD-7 further, his eyes lit up. At that time, only *Victor III*s and some *Delta*-class submarines used that type of radar.

Our CO, Commander Michael Oliver, did a happy dance in the corridor outside the radio room when he heard the news. I watched his jig from the control room, wondering why he seemed so excited. I learned from Flacco that the spooks reported good news and bad news. The good news: the Snoop Tray 2 signal was not moving, indicating that the *Victor* might be resting at night on the surface, something they occasionally did before an exercise. The bad news: they were inside Peter the Great Bay near Vladivostok, which meant a possible traffic jam of lethal Soviet warships.

Commander Oliver decided to chance the risks and pursue the target. If we could get some close-up shots of that *Victor III*'s mysterious odd pod and under-hull pictures of her sleek frame, there'd be big medals and promotions galore. On the other hand, one small miscalculation could result in a catastrophic collision.

Oliver hadn't slept in a while, so he ordered our XO to take the conn and follow the radar signal, then wake him when we drew close enough for periscope photos and an under hull. As the fire-control technician of the watch, I had the responsibility of keeping a plot of all the contacts we detected. The *Victor III* wasn't moving, so that part of the job was easy. Dozens of other contacts in the area, including several submarines and surface ships, were going to and fro at fast clips, so that part proved difficult. Since our MK-113 fire-control system could plot only four targets simultaneously, I dialed Master Two, our *Victor III* submarine, into one of the digital computer displays, and three other contacts, representing the closest warships, into the other consoles.

We dodged the warships by running slow while weaving our way into Peter the Great Bay. As we neared our contact, just off Popov Island outside Vladivostok harbor, and the signal strength on the Snoop Tray 2 radar increased, the XO had someone wake up Commander Oliver. Our CO strode into the control room a few minutes later. He smelled like Old Spice aftershave as he approached and glanced at my plot board.

"Ready the thirty-five," Oliver said.

"Yes, sir," I said. I opened a locker, removed the 35 mm camera, checked the film status, and waited.

Oliver relieved the XO of the conn and called the under-hull photographic-operations party to the control room. He brought the *Drum* to periscope depth and raised the periscope, spun the metal cylinder back and forth, then stopped. "There she is," he said. "Bearing to Master Two on my mark . . . mark! Range, 900 yards."

Our WLR-9 ESM warning indicator started beeping, signaling that enemy radar had gotten a sniff of our extended masts. Through the small PeriViz monitor mounted in the overhead near the periscope stand, I could see what Oliver saw in full color. Streaks of purple-orange clung to a barrage of gray clouds on the horizon as dawn crept toward sunrise. Against the gray, the dark silhouette of the *Victor III*'s sloping conning tower and extended masts seemed surreal, as if only a picture out of the pages of *Jane's Fighting Ships*. Certainly, the real thing could not be less than a half-mile away. Lights blinked on shore behind the Soviet submarine, intimating that Russians prepared for their day just like we did. I wondered who they were, what they were like, if they loved, laughed, and cried like we did. I wondered what they would think if they knew we were hiding in their front yard.

Oliver pushed the small red button on the scope's right handle. I heard a soft whirring as he snapped a 70 mm photo with each push. He unglued his eye and stepped back from the scope.

He looked my way and said, "You ready to reel?"

"Yes, sir," I said, "the 35 mm is loaded and ready."

"You've got two minutes," Oliver said as the WLR-9 beeped away in the background.

I moved over to the scope well and snapped the 35 mm into position, then settled my eye onto the back of the camera and squinted. Morning light crept across the ocean as the sun peeked above the snow-capped

Sikhote Alin mountain range on the horizon. With moist palms, I gripped the scope handles tighter and tried to slow down my breathing. On low power, the Soviet submarine filled my view. By feel, I adjusted the camera's focus and f-stop setting and lined up the crosshairs on the odd pod. The control room settled into silence, save for the manual snapping of the camera shutter. I snapped a dozen photos, then switched to the highest-power setting. The oval pod took up the entire crosshaired circle through which I gazed. The WLR-9 chirped away in my left ear, now delivering an almost steady procession of tones.

"Let's go, Reed," Oliver said.

The CO's deep baritone pushed my pulse across the red line. My fingers twitched as I swung the view over to the masts and snapped a few more pictures. Moving at light speed, I detached the 35 mm camera and flipped up the scope handles to the vertical position.

"Down scope!" The oily mast lowered into the scope well. Around me the world turned crimson again. I squinted as my eyes readjusted to the dim red lighting.

"Well?" Oliver said. Dark circles underscored the CO's brown eyes.

I shook my head from side to side. "I got some good shots, Cap'n, but with this lighting angle at this distance, I don't think they're good enough. I recommend we move to the other side, draw in closer, and get the light behind us."

I couldn't believe my own words. Nine hundred yards off our port bow sat one of the best attack boats the Soviets had. In nearly every respect, she was comparable, if not better than, our *Sturgeon*-class submarine. Yet here I was recommending that we move in close enough to smell each other's armpits.

"I concur," Oliver said as the XO leaned in close to listen. "The 70 mm shots probably aren't going to cut it either. XO, you have the conn. Reed, follow me."

Oliver walked toward his stateroom. For a brief second, fear and confusion froze my legs. Oliver stopped, turned, and gave me a look.

Feet unfrozen, I followed the CO to his stateroom. Oliver opened the door, and we ducked inside. He sat down at his desk and looked at the floor. I closed the door and stood, waiting for him to speak.

"I may need you to egress," Oliver said, lifting his head.

My heart shot into my throat.

"Egress, sir?" Confused, my thoughts moved in slow motion, as if smothered by cold syrup.

"I may need you to take a Draeger, egress, and get us some better shots of that *Victor III*. We need close-up photos of that pod, unfettered by our periscope optics, to determine what that thing is."

I didn't know what to say. *Take a Draeger? Egress?* That meant donning a bubbleless rebreathing device, locking out of the escape trunk in a Soviet harbor, swimming to the surface while tethered to a line, and taking photographs of a Soviet submarine just a few hundred yards away. Even though I had trained with the SEALs for such a mission, I knew he was asking me to volunteer. I also knew that if I did wind up taking those photos, I could never talk about the ordeal with anyone, not even most of the crew.

I had to push my reply past the lump in my throat. "I'll get the underwater enclosure for the camera and suit up, sir."

Oliver nodded. "When you receive my order, and no sooner, you will egress, stay hooked to the line, surface long enough to take a few photos, and then return. Is that understood?"

"Yes, sir. Understood."

Oliver rubbed his palms together, looked back at the floor, and whispered something to himself that I didn't understand at the time. "If they can send divers to tap cables, then I can damn sure send one to take photos."

Oliver stood and dismissed me.

I left his stateroom and headed toward the bow of the boat. There I suited up and readied my gear, which included placing the 35 mm camera in a watertight enclosure. My father's words, spoken years earlier when I feared stepping to the pitcher's mound in Little League, churned in my head.

Face your fears, son. If you don't, they will own you.

I struck out nine batters in that game.

Once inside the bow compartment, I opened the bottom hatch to the eight-foot-diameter escape trunk. With the help of a seaman trained in escape trunk operation, I climbed up the ladder and squeezed inside the oval. I closed the hatch below my knees and fought off the suffocating fingers of dread that curled about my neck. A small, dim light cast strange shadows about the tiny metal dungeon filled with gauges and valves. Alone and stuffed into my hot neoprene wet suit, I sat on the bottom of the cold trunk and shivered. My eyes focused on the small metal communications box mounted on the bulkhead, from which I knew my orders to go would be delivered. Hundreds of thoughts did somersaults inside my head, all of them dismal.

Will I get good enough photos? Will the Soviets spot me? Will I survive the mission?

Meanwhile, in the control room, things went from bad to ugly.

While Commander Oliver had been talking to me, the XO took the boat deeper to maneuver to the other side of the *Victor* so we could get shots with the sun behind us versus glaring on our scope lens. This had been my suggestion to Oliver before we left the conn together. With the *Victor III* sitting still, sonar remained useless, and our ESM's Snoop Tray 2 radar hits were the only means to determine the target's approximate range and bearing. That information allowed for only a rough idea of the sub's location, despite the previous periscope fix.

The XO, Bob Fricke, ordered Nick Flacco to maneuver the *Drum* to a point opposite our previous location, then bring the boat to periscope depth again. Knowing that doing an under-hull photographic operation might be next, Flacco ran through a mental checklist. As he did, a silent alarm went off in his head. "Shit, the wire."

"What did you say, OOD?" Fricke said.

"We need to reel in our floating VLF radio wire," Flacco said. "It's still out there."

"Dammit!" Fricke said. "Get someone from radio up here now."

Flacco called up a radioman, who sprinted into the control room. The petty officer opened a door at the front of the room and stepped inside the tiny area that led up to the bridge. He undogged the lower hatch, climbed up the ladder, and started bringing in the wire. Meanwhile, Oliver returned to the conn and took over. He approached the number two periscope and waited until Flacco confirmed that the *Drum* had almost reached periscope depth. Oliver wrapped his hand around the orange metal hoop encircling the scope well, then pulled the round bar clockwise. "Up scope."

Hands gripping the scope handles, eyes seated into the rubber socket, Oliver waited for his prize to come into focus. For a brief second a smile played on his lips as he savored the moment. The *Drum* neared the surface and Oliver's smile vanished. He frantically lowered the scope and yelled, "Emergency dive!"

Too late. A thunderous boom shook the boat. The radioman who'd been reeling in the wire tumbled down the ladder and slammed onto the deck. Blood oozed from his head. The sound of metal screeching over metal filled everyone's ears in the control room. The boat lurched forward and angled down at the bow by ten degrees. Flacco glanced at the unconscious radioman, then at the door that led to the bridge. *The lower hatch is still open,* he thought, *if we have flooding now . . .*

Down in the bow compartment, shoved into the escape trunk, I heard a deafening clap above my head, followed by an ear-splitting metal shriek. The dim light in the trunk went out, leaving pitch-black darkness in its wake. The force shoved me head-first into a valve handle. My jaw hit the metal wheel. A stinging pain rippled across my face, and the salty taste of blood filled my mouth. I cupped my palm across my bleeding lip and felt for the communications unit in the dark. My fingers found the square box, and I depressed the key. I spat out a clump of blood and blabbered something unintelligible. Nothing but silence. I keyed the box again. Still nothing. I tried opening the bottom hatch to the trunk using every bit of muscle I could muster, but the wheel would not turn.

Alone in the dark, with the world closing in around me, I wondered if we had suffered a major casualty, wondered if we were on a death spiral toward the bottom. For a brief moment, I contemplated flooding the trunk and escaping through the upper hatch. Then I remembered that we were deep in Soviet territorial waters and I knew secrets. My chest started heaving, and I realized that the oxygen flow to the trunk was probably out.

I figured we must have collided with the *Victor* and the force of the impact near the escape trunk had knocked out the bow compartment communications circuit. A shock wave must have hit oxygen bank number one and ruptured the O_2 valve. Tracing the lines in my head, I saw how this could halt the flow of oxygen through valves O-4 and O-27 that led to the escape trunk. The collision must have also caused a pressure imbalance in the trunk, making it impossible for me to open the hatch from the inside. I spat out some more blood and bit on my Draeger's mouthpiece. The throbbing pain around my bottom lip damn near doubled me over as I sucked in some air. I was now living on borrowed time.

Up in the control room, Flacco had someone drag the radioman away from the bridge door and shut the lower hatch. Someone else called for the doc.

"Why aren't we diving?" Commander Oliver yelled.

Flacco glanced at the depth gauge. Still at sixty feet.

"Chief of the Watch," Flacco said. "Flood forward trim tanks."

The boat surged forward a few feet. She angled down even more but still did not go deep. More screeching and grinding rippled through the control room, followed by several loud thuds.

"I think we're impaled in the *Victor*'s ballast tank," Flacco said. "We're just pushing them sideways."

"All back full!" Oliver ordered.

Metal crunched as the *Drum* moved back several feet. The bow dropped by a few degrees.

"All ahead full," Oliver said. The boat shot forward and downward. The depth gauge registered a hundred feet and descending. Then the flooding started.

Rain poured from the overhead and drenched the scope well. Flacco looked up. One of the scope seals had ruptured in the collision. Cold salt water rained onto the deck and splattered shoes. The flooding alarm sounded.

Oliver clicked the 1MC. "Now flooding in the control room."

The XO called for a damage control party. Auxiliarymen came running with tools and patches. Taking on water, the *Drum* sped toward test depth, 1,300 feet down. Freezing ocean water sprayed out of the scope well as if from a pinched hose. Flacco knew if they couldn't get the flooding under control soon, there'd be serious consequences. Vital equipment might short out, and systems could die, all of which could send the *Drum* to the ocean floor.

"Quartermaster," Oliver said as the A-gangers worked on the scope well leak, "plot a course to Chin Hae."

South, thought Flacco, *to Korea*.

Flacco heard the pinging of Soviet 50 kHz active sonar through the hull. He knew that ASW forces were now hell bent on catching the *Drum* red handed.

"I don't have a Chin Hae chart in here," the quartermaster said.

"Then just take us south!" Oliver yelled.

In the escape trunk, I heard Oliver's flooding report over the 1MC. The announcement meant someone was still alive, but the flooding verified that we had problems. Regardless, I had to get out of the trunk. The air in my Draeger would not last forever. I took out my diver's knife and started tapping Morse code on the metal hatch. My dad taught me the entire alphabet when I was a kid, and at one point I could even keep up with a CW transmission. Now, however, all I could remember were a few letters. It didn't matter; the seaman on the other side of the hatch probably knew less than I did.

I tapped SOS.

No response.

I tapped again louder. Still nothing.

Panic threatened to block what little air I had left from reaching my lungs. I remembered my navy diver training in Hawaii, where they'd harassed me in the water every day for hours. They pulled out my regulator, spun me in circles, and damn near tried to drown me. That training taught me a valuable lesson: how to control my fear. Now, thousands of miles from that tropical paradise, I closed my eyes and said a quick prayer. Then I sucked in a few breaths and tapped again.

Finally, the lower hatch opened, and fresh air rushed in.

Back in the control room, the A-gangers managed to stop the flooding and fix the leak, while the corpsman patched up the radioman. He'd sustained a concussion and a deep cut to his forehead. Flacco watched the doc help the petty officer hobble out of the control room.

Then the Indians showed up and surrounded the wagon. Soviet helicopters dropped sonobuoys that bombarded the ocean with active sonar pings. ASW destroyers and fast gunboats came out of Vlad and started chasing the *Drum* southward. Dozens of propellers chopped at the Sea of Japan, and our sonar jockeys couldn't keep up with all the contacts. Commander Oliver ordered a thirty-degree course change—a zig to remain undetected. Flacco figured the Soviets knew what Oliver knew: that the *Drum* could only head south through a narrow passageway to escape. If they threw enough ships and planes out there, the odds of getting away were about nil.

While Flacco contemplated his odds of survival, a depth charge exploded.

By now I had scrambled out of the escape trunk and sprinted to my rack to pull on my coveralls. I climbed the ladder up to the crew's mess, found the doc, and got a patch for my severed lip. I didn't bother to look in a mirror at the damage. I scrambled up to the control room and slid onto the bench next to a half-dozen officers and sailors in front of the fire-control equipment. The weapons officer (Weps) glanced at the bandage on my face and gave me a look that said, "What the hell happened to you?"

I didn't bother to explain.

Another depth charge exploded, and all eyes looked upward. All lips muttered silent prayers. Weps informed me that we'd rammed into the *Victor III* and probably smashed the entire front end of our sail. The ESM antenna was gone, and both periscopes were useless, not that we needed them now anyway. Flooding occurred but had been contained, and now every Red ship in the Far East meant to do us harm. I wondered if I should have stayed in the escape trunk.

Sonar reported that our closest pursuers were two *Kresta I*–class guided missile destroyers. I dialed them into the fire-control gear, knowing that they could hit a top speed of thirty-two knots. They carried two twin-missile launchers and a Ka-25 Hormone ASW helicopter on the after deck, complete with sonobuoys. Weps figured that the depth charges were probably light warning explosions, but we had no way of knowing for sure.

We zigged and zagged as ASW ships and planes pinged. Oliver had us hug the bottom for the next two days while explosions shattered the silence, some far away, others so close they rattled dishes. My dad once told me that fear does not discriminate. It doesn't care about our nationality, wealth, religious beliefs, or lack thereof. Fear is an equal opportunity employer, and when I looked at my crewmates, I could see the evidence of it ooze from every pore. We knew that if caught, Ivan would show us no mercy. We had entered their territorial waters and rammed one of their boats, and now all bets were off. Oliver would take us below crush depth before he surrendered the boat to the Soviets. All of us understood well the consequences should we fail to escape, yet everyone to a man kept his cool and did his job well. At that moment, I understood what it meant to be a submariner.

With vigilance and creative evasions, coupled with lots of luck, we finally made it out of Soviet territorial waters. The Soviets continued their pursuit anyway, and after another day of dodging dozens of warships and planes in the Sea of Japan, sonar reported the sound of a possible collision, faint and distant. We had no way of knowing who or what had caused the metallic crunch. Weps speculated that two Soviet ships, while pursuing us, had collided. Within hours of hearing the smack, the Soviet prosecution activity dropped in half as a large number of ships and planes headed toward the sound of the faraway collision.

Thanks to the sudden decrease in the number of Soviet vessels trying to find us, we managed to sneak out of the Sea of Japan and back to the safety of Apra Harbor in Guam. We surfaced and entered the harbor in the middle of the night to avoid detection from Soviet satellites. As the boat moved silently through the dark, I watched silver moonlight glitter on the ocean near my favorite dive spot and thanked God for another day on earth. Then I glanced up at the sail. The front masts were bent or missing, and the bridge area had been smashed like an aluminum beer can.

We tied up next to the pier and workers shrouded the entire sail with a cover to ensure that Soviet satellite photos would not reveal the obvious. As was the case with many Cold War SpecOps, the boat's logs were altered. No record remained, save memories, to validate that the USS *Drum* was ever in the vicinity of a Soviet *Victor III* submarine in Peter the Great Bay.

I later learned that two Japanese sailors died so that more than a hundred of us could live. While we were running from the Soviet navy, the nuclear submarine USS *George Washington* collided with the Japanese freighter *Nissho Maru* on April 9, 1981, 110 miles southwest of Sasebo, Japan. When the *Washington* surfaced, she ran into the underside of the freighter and damaged the hull. The Japanese ship sank within fifteen minutes. Two of her crew of thirteen went down with the vessel. The *Washington* sustained only minor damage but then became a decoy of sorts for the *Drum*. The collision distracted more than half of the Soviet forces searching for us, as they assumed incorrectly that the two collisions—the one with the *Victor III* and the one with *Nissho Maru*—were caused by the same boat. The *Washington* managed to sneak away, but the accident sparked a political furor in Japan.

President Ronald Reagan now had two submarine accidents on his hands. Prime Minister Zenko Suzuki blasted him for taking more than a day to notify Japanese authorities about the *Nissho Maru*'s demise, and the fact that a U.S. P3 Orion aircraft circling overhead made no attempt to rescue the survivors. On April 11, President Reagan expressed regret over the accident and offered compensation to the victims while assuring the Japanese that radioactive contamination need not be a concern.

The Soviets threw their own spears. They insisted that a U.S. submarine had collided with K-324, a *Victor III*–class nuclear submarine sitting on the surface in Peter the Great Bay. K-324 sustained severe damage, and some of her crew were hurt in the accident. The United States denied the presence of any American submarines in the area at the time.

Part of the Northern Fleet and stationed out of Saint Petersburg, K-324 had come off the ramp at the Admiralty yard in 1979. She became the seventh vessel of the *Komsomolsk* line. Two years after the incident, thanks to the photos taken in 1981 by the USS *Drum* and other submarines, the United States identified the odd pod on the *Victor III*'s rudder as a housing for a towed sonar array. Proof came later that year when, on October 31, 1983, the USS *McCloy* (FF-1038) and K-324 were ensnarled in each other's towed arrays some 280 miles west of Bermuda. K-324 was monitoring U.S. ballistic missile submarine movements, hoping to find one of our "boomer" SSBN submarines to shadow. The incident severely damaged K-324's propeller. A tug towed her to Cienfuegos, Cuba, for repairs. Two years later, in 1985, the Soviets decommissioned K-324, and her crew never knew who hit them that day in April 1981.

CHAPTER TWENTY

*If we survive danger it steels our courage more than any-
thing else.*

—REINHOLD NIEBUHR

AFTER OUR FATEFUL COLLISION WITH THE *Victor III,*
while nursing a scar on my chin, I no longer took life
for granted. Even the smallest things had more meaning now: a glorious sunrise, a fluttering hummingbird,
and the charming girl I had met more than a year earlier. I married that
girl in June 1981.

In September, a month before my enlistment end date, my navy de-
tailer called. He said that DevGru One—the Special Projects guys—
were interested in talking to me about an opening on the USS *Richard B.
Russell* (SSN-687). I didn't know it then, but the *Russell* was the
Parche's eventual successor for Ivy Bells missions. Around that same
time, a couple of civilian recruiters contacted me from Eastman Kodak
Company. They liked the combination of my electronics, computer,
submarine, and photographic training and offered me a different challenge with a new program starting up in Colorado. I struggled with the
decision, as my time in the navy had been some of the best years of my
life. As one of my crewmates on the *Drum* once put it, "I wouldn't have
taken a million dollars to extend my enlistment after we smacked that
Victor. Then again, I wouldn't have taken two million to forget the experience."

I went to work for Eastman Kodak at the end of 1981 and have never regretted the decision. One of the most valuable lessons I learned during my time on the boats can be summed up in one word: *teamwork*. Having to rely on others for my very survival, who also relied on me, taught me that even if I don't like someone, I need always treat them with respect and value their abilities.

While my time in the navy came to an end, others were still trudging through the Cold War mud, some literally. By late 1980, the USS *Parche* had all but replaced the *Seawolf* for cable-tapping missions. But the old girl still had at least one more run in her, and no one knew that this would be the most difficult and dangerous of her storied career. Former submariner Jimmy Carter held the reins as president of the United States at the time and was up to his eyeballs in international issues. Iran had captured fifty-three Americans and held them hostage for almost a year. Carter authorized the military rescue mission Operation Eagle Claw in April to free them, but the mission failed, raising tensions in the Middle East to a pitch almost on a par with the Soviet Union.

Six months later, in October 1980, James Rule reported to the USS *Seawolf*. He walked through a prison-like maze of gates and approached the guard shack overlooking the drydock enclosure. The *Wolf* looked old but quaint, with her flat teakwood deck, big bull nose, and sleek frame resembling a diesel boat. Bright overhead lights glared off the foreheads of a dozen dungareed sailors who chipped, painted, pounded, and tried to fix the *Seawolf*'s bruised and battered body. Don Langaliers, the *Seawolf*'s chief torpedoman, met Rule at the shack and walked him through orientation, which included a series of fifteen escalating briefings, each one peeling back another layer of the boat's supersecret onion.

As Rule began meeting the crew, he noticed something interesting. The *Seawolf* should have been decommissioned years earlier, or at the very least demoted to a blue-haired geezer status and assigned to a minor role supporting the fleet. Instead, she was still one of only two frontline workhorses with perhaps the weight of the Cold War's outcome resting on her slumping shoulders. Her crew did not complain, however, and none asked for transfers. Not a man whispered jealous remarks as they watched the crew of the USS *Parche*—the other Special Projects boat at Mare Island—disappear through the hatch of a far more modern machine.

Seawolf's sailors worked hard and played hard, the latter occurring most often at the Horse and Cow Saloon in Vallejo. That's where Rule mingled with most of the crew on a casual basis, including Fire Control Technician Third Class Tony Mignon. Rule and Mignon soon became friends and talked in excited whispers about what lay ahead when the *Seawolf* came out of drydock. Rule also met some of the 140 submariners assigned to the USS *Parche*. Due to the *Seawolf*'s constant state of disrepair, the *Parche*'s crew often called their neighbor boat the "Pier Puppy." They laughed and asked Rule if he enjoyed his duty at "Building 575." The kidding had started after the *Parche* made the 1979 Ivy Bells run into the Sea of Okhotsk while the *Seawolf*'s imprisoned crew worked on their aging boat.

Rule took the jabbing from *Parche*'s crew in stride and got along with most of them, who also lived in the wood-framed barracks at an old munitions depot near the docks. Over the next several months, he shouldered a torpedoman's workload on the *Seawolf* alongside the team's leading petty officer. He helped load MK-37 torpedoes, scrub decks, fix gear, and stow several Mobile Submarine Simulators (MOSS) on board, which Rule hoped they'd never need to use. Subs shot MOSS from their torpedo tubes as decoys when the Soviets were trying to prosecute them.

At night, Rule migrated back to the Horse and Cow, where Don Langaliers demonstrated his secret weapon. The man had pierced his penis to connect to a long, thin chain. He ran the chain up through an opening in his shirt. When he wanted to impress the ladies in the saloon, which were the sort that thought perfume should smell like the sump tank on a Harley Davidson, he handed them one end of the chain. When they pulled, the circumcised tip of Langaliers's manhood peaked above his belt buckle and caused eruptions of high-pitched laughter throughout the smoke-filled bar. Although most of the crew thought the chief was "one crazy son of a bitch," and his wife, Bobbi, often shook her head in disgust at his wild antics, she never had to worry about his fidelity or family devotion.

Langaliers considered his torpedo gang a part of that family and often invited Rule and the other torpedomen over to his house for barbecue and beers. The guys played hard but also worked hard, and they respected Langaliers as a leader and mentor. They knew that he'd always have their back, and likewise, they'd always have his. As the months came and went, they became a close-knit fraternity that drank beer at night and worked on weapons of mass destruction during the day.

By the end of 1980, with the overhaul finally complete, the *Seawolf* took a short run up to Bremerton, Washington, to conduct sea trials in Puget Sound. Not a day out to sea, Rule found himself surrounded by smoke in the back of the crew's mess. Someone ran past him at a fast clip, with an EAB mask sucked to his face and a hose dangling in his hand. The bitch-in-the-box alarm alerted the crew to a "fire, fire in the engine room." Rule ran to find his own EAB mask. Still unqualified, he sat in the crew's mess, helpless and scared as blue-suited sailors raced around him to dog hatches, flip switches, close valves, or help stifle the fire.

The *Seawolf*'s CO, Commander Michael C. Tiernan, ordered an emergency blow, and the boat shot to the surface. Ten submariners were overcome by severe smoke inhalation before the crew could ventilate and clear the air. Glad to still be alive, Rule felt ashamed because of his fear. Later he molded that trepidation into a stubborn drive to spend a year qualifying as a submariner so he could contribute during the next emergency.

Limping and wheezing, the *Seawolf* headed back to drydock. She was scheduled to do the next Sea of Okhotsk cable tap, while the *Parche* went to a new tap site in the Barents Sea. With the *Seawolf* back on the blocks, the *Parche* took her place again for the summer Okhotsk run to retrieve the previous year's pod recordings. The *Seawolf* was not scheduled to see the dark of ocean again until late 1981, and for Commander Tiernan, this would be his first run into the heart of danger. Until then, the crew viewed Tiernan almost in the same light as an unproven nonqual and took to calling him "Captain Milquetoast."

The *Seawolf*'s executive officer, J. Ashton Dare, was held in equal esteem. Dare came from aristocratic stock, and as the son of an admiral, he sometimes exhibited an arrogant demeanor. In reference to Dare's strict policies, the crew often exchanged comical remarks such as "Dare'll be no liberty, Dare'll be no fun." For the next several months, while the crew slaved to repair the tired *Wolf,* no one had much fun. Jim Rule watched the *Parche* head out to sea and longed for the chance to go on a mission to the Sea of Okhotsk. Little did he know that he'd soon regret that wish.

WHEN FIRE CONTROL TECHNICIAN THIRD CLASS Tom Ballenger reported to DevGru in June 1981, he wanted to be integrated into the crew of the *Seawolf* right away. He had just completed his training on the antiquated MK-101 fire control system—a predecessor to the MK-113—in Groton, Connecticut. The crisp-uniformed DevGru officers approached him while he was in FT school and dangled the *Seawolf* carrot. He bit and moved his pregnant wife to Northern California. She delivered their son a month before Ballenger walked through the gate at Mare Island to see his new boat. The submarine was still at sea, and the Special Projects folks required a long waiting period to complete background checks. They needed to validate that Ballenger wasn't a "card-carrying commie."

Ballenger spent his six-week waiting period in the "woodshed," a dockside woodworking shop. Each day he gazed at the *Seawolf* in drydock, scaffolding covering her sides, arc welders sparking into the air, and dockworkers carrying pipes, valves, and other parts in and out of her worn hull. After weeks of sawing and sanding, Ballenger finally got called to the SCIF, the secured briefing facility used to orient new members of the *Seawolf* and *Parche* crews.

He signed a decades-long gag order not to talk about anything he'd see or hear about on the *Seawolf,* then endured a battery of fifteen security briefings that discussed the boat's special-projects mission but offered few details. Briefings completed, a yeoman whisked him down to the boat. He walked on board the old girl, thinking that everything he'd learned about nuclear submarines held no meaning here. The *Seawolf* was a unique specimen. As a diesel/nuke hybrid, she resembled both and yet neither. Ballenger felt lost and confused as he followed the yeoman to the boat's sail, into the conning tower, and down through a hatch that dropped into the wardroom, past the CO and XO's staterooms, then officer's country, then forward through the control room and navigation gyrospace area into the crew's mess to meet some of the crew.

Three tables lined up athwartships and two along the starboard bulkhead. The galley was located forward of the tables. Chiefs were allowed to sit at the back table near two main battery breakers, the nuclear-trained "nukes" at the back parallel table and the spooks and divers at the forward table. The yeoman sat Ballenger and the other inductees at a center table. A beefy 300-pound chief with round cheeks approached. He grabbed Ballenger's shoulder, squeezed, and said, "Get a haircut, nonqual puke, or find another boat."

Ballenger offered a shaky grin. "Okay, Chief . . . ?"

"Moorman. Master Chief Dave Moorman. I'm the chief of the boat. Don't ever cross me."

"Yes, sir."

Moorman squeezed harder and rolled his eyes in disgust. "Don't call me sir. I'm a chief, not an officer. Understood?"

"Yes, sir, I mean . . . Chief."

Ballenger spent the next sixty days in the deck division, chipping, painting, and fixing until they allowed him to assume the duties commensurate with his training as a fire-control technician. He then met FT Leading Petty Officer Tony Mignon, who was a nice enough guy but seemed to be wound a bit tight. A wiry but strong southern boy from Alabama, Mignon loved to wear his faded gray Confederate soldier's hat under way. Seems like he never took it off, not even to sleep. Most of the crew liked the guy, who told dumb but funny jokes and talked about his life's goal of unplugging from society and living off the land by himself in the mountains.

Mignon reported to Fire Control Chief Powell, who fit the typical chief profile of semi-dumpy and overweight. Powell had light brown hair, a scraggly beard, and thick glasses. Soft-spoken, the chief knew his stuff about the MK-101 fire-control weapons systems—the only electronics piece of gear situated on the *Seawolf*'s "wet side" in the control room. The port side had blow valves, flood, and drain manifolds, various pipes, and other "wet" mechanical items, whereas the "dry side" held most of the electronics, except the FT gear, which was positioned all the way aft on the port side. An old tube-style power supply, just forward of the ladder into the crew's mess, gave life to the MK-101. The gear consisted of an analog target-tracking Position Keeper and a torpedo control and firing panel that fed data to the MK-37s in the torpedo tubes. Operators selected a tube and programmed firing order via an old rotary-style control device, about as antiquated as a telephone with a dial versus buttons.

Months passed, and Ballenger slowly adjusted to life on the boat. Chief Torpedoman Langaliers enlisted his help, along with that of torpedoman Jim Rule, to travel into the hills near Concord, California, and practice detonating blocks of C4 explosive. They'd jam detonation chords and blasting caps into the aft ends of cow carcasses, find cover, and blow the cows into hamburger. Back on the boat, they rigged blocks of the C4 with three bomb blasters in strategic locations, forward, amidships, and aft. No reasons were given for the explosives, but everyone knew that if they were caught on station, they would not be taken alive.

As a gesture to offset the unease consuming his torpedomen, Chief Langaliers brought a ceramic frog down to the torpedo room. In a strange but funny initiation ceremony that somehow involved his chained penis, Langaliers christened Beauregard the frog as their official voyage mascot and wedged him into an overhead spot near a torpedo tube.

Prior to the *Seawolf*'s departure, Ballenger met some of the navy divers who'd be conducting the cable-tapping missions in Soviet waters. They were all seasoned first-class or master divers who the crew called "heroes." When Ballenger walked past the chart room Special Information Center (SIC) "spook" area, curtains usually blocked his view. Occasionally, he caught a glimpse of the equipment inside, and intuition combined with his electronics knowledge gave him a good idea of its use.

Near the torpedo room, Area D housed the diving chamber, and a section called the aquarium held the four camera-outfitted mini-sub *Fish*. An assortment of video equipment, amplifiers, and other electronic gear also decorated the space. The dog-legged area took up ten feet by almost five feet and could hold up to thirty people. All totaled, up to 190 men could be loaded into the *Seawolf*'s belly, consisting of 125 standard boat crew, 10 or so spooks, and 50 other Special Projects personnel.

Almost a month after leaving port, the *Seawolf* arrived on station over the Soviet communications cable. Under way, Tom Ballenger qualified as a planesman—one of the guys responsible for moving the boat up and down in the water. He sat in the control room in a bucket seat and helped guide the *Seawolf* into position near the twenty-foot-long cable-tapping "beast" pod. On the sub's underbelly were two sledlike skids that the crew called skegs. These allowed the boat to sit on the bottom near the pod. In the control room, Commander Tiernan ordered the chief of the watch to balance the water in the trim tanks to level the *Seawolf* over the cable. The COW flooded the tanks too fast with too much water. The boat lurched forward and downward. She was headed in a beeline for the mud.

The COW yelled at the diving officer, who barked an order to Ballenger. His heart racing, Ballenger tried to unscramble the order in his head. As the planesman, he pulled upward on the half-wheel clenched in his hands. Employing the boat's built-in thrusters—which helped maneuver the sub like the tiny rockets on a space capsule—he tried desperately to compensate for the extra weight in the trim tanks, but the boat was too heavy. He managed to keep the *Wolf* from colliding with the ocean floor, but one of the skegs slammed down on top of the Soviet communications cable.

Commander Tiernan fumed and yelled an obscenity. Ballenger was certain that the skeg whacked the cable hard enough to alert the Soviets, who might at any minute send out a posse to investigate. Tiernan kept the boat silent and still for a long while, waiting and listening. Nothing happened. Satisfied that he'd dodged a bullet, he sent out the divers. The heroes donned their MK-11 deep-sea diving suits and Kirby Morgan masks. They locked out of their pressurized hyperbaric diving chamber and pushed heavy metal boots toward the beast nearby. They retrieved the tapes from the pod, which had been recording conversations from the communications cable for several months, and delivered them to the spooks inside the boat. The T-Branchers collected, verified, and calibrated the signals, while the I-Branchers listened to the Soviet jabber to validate the recordings.

Still outside and near the beast pod, one of the divers felt something tug at his arm. He held his hand out and stared at his glove. Currents moved his arm up and down in a steady rhythm. He contacted the *Seawolf* and asked if they felt anything. They did, but it seemed minor, probably just a light storm overhead stirring the currents. The diver shrugged and continued working.

The *Wolf* was too deep to have the VLF wire floating up near the surface to pick up radio signals. As such, Commander Tiernan and crew did not hear about the twin storms converging from hundreds of miles away as they marched toward the Kuril Islands. They didn't know that every ship in the Sea of Okhotsk had received a warning to leave the area immediately, that high winds were pounding seashores, toppling trees, and threatening to sink vessels as they scurried to safety.

Naval command centers issued urgent reports that two major tempests were now dancing together and kicking up fifty-five-knot geysers of water. When the typhoon hit, since she was 400 feet beneath the gale, the *Seawolf* at first only shuddered. Her divers only sensed a few strong currents. As the perfect storm above turned uglier, the *Seawolf* groaned and rocked from side to side. Her three divers no longer felt mild flutters, but were now upended and pulled across the sand like mannequins in a flood.

The ocean grabbed one diver, and with the vengeance of a wrestler, hauled him about the ocean floor until his leg landed under the edge of a skeg attached to the sub. The *Seawolf's* ski leg had lifted into the air during the boat's rocking. With terrified eyes, the diver lay helpless in the silt as the skeg reached its highest rocking point and then started back down, right on top of him. The skeg brushed his protective suit just as the hand of another diver pulled him free. Thanking God for the miracle, the diver followed his two buddies into the *Seawolf's* chamber hatch.

Inside the boat, cans, plates, cups, utensils, and anything not secured went flying. Sailors ducked as the projectiles whizzed past their heads. Beauregard, the torpedomen's porcelain mascot, jumped from his loft and careened against a torpedo tube. Tiernan had seen enough. With the divers locked back inside, he terminated the mission and ordered the *Seawolf* off the bottom. But before the boat moved an inch, an alarm sounded. A reactor specialist standing next to a heat-exchanger gauge saw something that raised the hairs on his neck. The gauge, which measured the temperature of cycled cooling water for the reactor, had spiked into the red zone. The nuke knew in an instant that something was clogging the intake valves. He checked and discovered they were filled with sand. Grabbing a sound-powered phone, he called the control room.

After hanging up the phone in the control room, Tiernan, Dare, and a flurry of nukes clad in blue ran aft. When they arrived, wet sand had blanketed the compartment. The *Seawolf's* skegs were now buried in the mire and the boat's intake vents, normally kept several feet off the bottom by the skegs, had been pushed flat against the bottom. Now the engines sucked in sand and tiny sea creatures, along with salt water. The uninvited intruders jammed their way into the turbines and generators and refused to leave. What's worse, as the *Seawolf* rocked, she sucked in more and more of the stuff and took on additional weight. Worse still, the storm's incessant pushing and pulling had wedged the *Seawolf*'s skegs deep into the sand. She was now stuck, 400 feet deep in Soviet waters, with no way to break free.

Tiernan knew that the sand could shut down the reactor, which would take days to restart, if it ever did. He huddled his officers and senior "salty" enlisted men together to brainstorm a way to break free. Whatever they tried had to be done with care and speed. Soviet warships patrolled the area, and if the *Seawolf* rocketed to the surface, they'd be a sitting duck. The few MK-37 torpedoes they carried on board were no match against a fleet of missiles, torpedoes, and depth charges. The *Seawolf*'s slow speed and dog-barking noise would make her easy to catch. That left few options that would not take them from the pan into the fire.

They revved the engines and tried a controlled emergency blow, all without luck. The *Seawolf* remained captured, a fly trapped in the ocean's web. Forward and backward they pushed and pulled in an effort to wrest one or both skegs free. Nothing worked. Seventy-two hours ticked by, one agonizing minute at a time. A vital reactor system dropped to one-third efficiency as the vents continued to suck in sand and silt. Carbon dioxide levels rose as many of the ship's systems were shut down, and burner and scrubber capacities diminished. Some of the crew were sent to their racks to conserve air, while others started to come unglued.

Every sailor who volunteers for submarines must pass a battery of psychological tests designed to weed out the claustrophobic, unbalanced, antisocial, and potentially psychotic. Dealing with the unstable on a submarine, where a crew lives in close proximity for three months or longer, can quickly become a nightmare. Occurrences of the "crazies" are a rarity, and when they do happen, captains always fear they could trigger a chain reaction. When the fourth day passed, and the *Seawolf* remained welded to the ocean floor, Commander Tiernan hoped that his crew would keep it together but worried that someone might crack under the pressure.

His eyes glazed, Petty Officer Tony Mignon staggered his gangly, tired body up the ladder into the control room. His Confederate hat covered the top of his head but did little to hide the frightened, distressed look on his face. Chief of the Watch Don Langaliers, while sitting next to the ballast control panel, watched Mignon approach from the back of the control room near the interior communications (IC) switchboard equipment. Langaliers noticed that Mignon's eyes were moist and dazed. The young FT muttered something unintelligible, followed by a loud cry about how they were all going to die, and no one would know what happened.

The XO and a couple of officers near the periscope stand, who were going over ideas to gain freedom from the sand, looked up with stunned faces. Mignon let go another cry, along with a string of profanities. Langaliers leaned back from his station and said, "Tony, chill out. We need you to be quiet right now."

The planesman and helmsman, both gripping steering wheels at the front of the control room, flashed each other puzzled, worried looks. Mignon ignored Langaliers' request and started yelling even louder. Then he banged his head hard against the Ships Inertial Navigation System (SINS) gyroscope housing, all the while screaming obscenities and doomsday chants. The skin on his forehead split open. Blood oozed onto the SINS housing. Commander Tiernan, whose stateroom was nearby around a dogleg from IC alley, stuck his head out into the passageway to see what was going on. Tony lunged at Tiernan, screaming that they were all dead men, and it was the skipper's fault.

Langaliers jumped from his station. He and another crewman, J. K. Branham, grabbed Mignon before he could tear out Tiernan's eyeballs. They wrestled the strong FT to the deck, while the XO called for the doc. A minute later, Don "Doc" Post bolted up the ladder with syringe in hand. Langaliers and Branham held Mignon down, while Doc shoved a needle into the man's arm. Mignon stopped blabbering and started drooling. A couple of guys escorted the sedated sailor down to the torpedo room, where Jim Rule was standing watch.

"Seeing Tony like that hit us all pretty hard," said Rule. "The doc asked me to keep an eye on him in the torpedo room. Once Tony's meds wore off, we had a long conversation, and both of us knew that his submarine days were over. We shared experiences that bonded us like brothers on that boat and swore that we'd maintain our friendship for life. Unfortunately, that never happened."

After the incident with Tony Mignon, Commander Tiernan knew that the sand in the hourglass had all but run out. The hours trapped under the waves had exceeded ninety-six, and it wouldn't be long before the dominoes started falling and others followed in Mignon's wake. *Seawolf*'s nuclear reactor continued to weaken, and air-purifying systems were starting to fail. Tiernan thought about the boat's early days, when teams of shipyard workers had collaborated to overengineer her reactor and equip her with the "heart of a Mack truck and the soul of a '57 Chevy." He recalled discussing the reactor's design capability with his navigator, Lieutenant Commander James Christopher Cane, who had a reputation as the "coolest cat on the boat under fire." The *Seawolf* had so many emergencies that when they did happen, Cane took them in stride and wore an almost annoyed versus concerned demeanor, ordering the crew to take actions using a matter-of-fact tone. Cane had suggested having the divers clear as much sand as possible on either side of the skegs, then push the engines well past the redline while moving back and forth in similar fashion to a truck stuck in the mud. They could cut the anchors loose to reduce their weight and do a normal blow on the ballast tanks until they broke out of jail.

The XO, who'd been involved in the discussion, expressed concerns that a normal blow could cause them to broach the surface. The storm seemed to be subsiding, which meant more Soviet warships might be returning to the area. Cane calmly intimated that the Soviets wouldn't be a problem if the *Seawolf* never broke free. Tiernan put the plan on hold pending the exhaustion of other options. Now he decided that Cane just might be right.

He ordered the divers to lock out and use fire hoses to clear the sand away from the skegs. One diver, while blasting away with a hose, dug a large hole behind him as he went. A mountain of sand built up on the edge of the pit. As the diver worked, the weight of the sand toppled from the crest and covered the diver up to his neck. Only the diver's Kirby Morgan mask could be seen above the mound. The other divers rushed over, dug for an arm, and pulled. Nothing happened. They pulled again. Finally, the diver broke free, but the sand filled in again around the skegs, and they had to start the job all over again, only this time employing more caution.

In the control room, the *Seawolf* groaned in agony as the reactor went into overdrive, and the anchors were cut loose. The nukes ground their teeth as the engines whined, and the spaces aft started to heat up. Jim Rule heard loud screeching through the hull and held his breath. The *Seawolf* struggled and clawed but still barely moved. Commander Tiernan blew the ballast tanks and pushed the reactor well past the design parameters, painfully aware that if his hunch proved wrong, they'd all die together at the bottom of the sea or be caught by the Soviets and probably suffer the same fate.

The *Seawolf* finally nudged, a slight movement at first, followed by fitful screeching, a few trembles, and a rolling wobble. The sub at last broke free and headed for the surface. That was the good news. The bad news was they were going up too fast and would definitely breach the surface, which meant they might be detected by the Soviets. They were also making tons of noise, mostly due to a broken skeg that banged against the hull. Tiernan knew that any nearby warships would have to be deaf not to hear the thing.

"When that loud clanking started," said Rule, "Chief Langaliers and our weapons officer met with us in the torpedo room. Weps said that we should prepare for the worst, as we'd probably be detected. Langaliers didn't say a word. He just dusted off a thick manual, and handed the thing to me. I saw this look in his eyes that said, 'This is what we trained for, so just suck it up.' The manual outlined the procedures for setting off the bomb block charges we'd placed in the boat, launching our MOSS, and if need be, shooting our way out with MK-37 torpedoes."

After hitting the surface, Tiernan fought to bring the boat's sail back under the waves. Damaged and sand-filled equipment almost doubled the *Seawolf*'s noise output as she clawed toward the Kuril Islands and escape. Still, they were free, and the crew cheered. High fives were offered, and palms smacked. Then the bad guys came out, and the cheering stopped. Just a few ASW ships took up pursuit to start, followed by a barrage of ships and planes that dropped sonobuoys and pinged away at the sea around them. In the torpedo room, Rule acknowledged an order from Langaliers, and he and the other torpedomen launched the MOSS units from the tubes. Just before the launch, Rule took a bottle of Wite-Out and painted the words CHITTY CHITTY BANG BANG on the side of the long, ten-inch-wide device in reference to the sound emitted, which simulated the *Seawolf*'s old-jalopy chits and bangs. At first, the MOSS decoy didn't work. Rule wondered if God had allowed him to survive the sand-stuck ordeal so he could experience a more gruesome death at the hands of the Soviets. Luck prevailed and some of the Soviet warships took the bait and peeled away.

After another day of ducking and hiding, the *Seawolf* managed to sneak away and stagger back home to Mare Island. She came back to little fanfare, however, as angry voices in Washington were certain that the *Wolf* had exposed the location of the Ivy Bells tapping pod when she smacked the Soviet cable with her skeg.

"It was a bittersweet return," said Rule. "We were back alive, and pride filled all of us who played a part in that ordeal. Still, the powers-that-be considered our mission a failure. Those of us who were on the *Wolf* during that run disagree. Mother Nature may have forced us to change our objectives, but we successfully completed one of the most dangerous missions of the Cold War."

CHAPTER TWENTY-ONE

During times of universal deceit, telling the truth becomes a revolutionary act.

—GEORGE ORWELL

ON THE MORNING OF JANUARY 15, 1980, the forty-four-year-old Ronald Pelton, a bearded man with blond hair, strolled across Sixteenth Street toward a gated building. He whispered something to the guard at the gate and was ushered in. The FBI agent on the corner noticed the man and signaled to his partner one block away. The agent's partner shook his head and whispered the reply "Unsub" into his ear-mounted communicator, for "unknown subject."

Pelton sat in the outer office and waited. Vitaly S. Yurchenko, the embassy's KGB security officer, walked into the room. He inquired as to the reason for the man's visit. Pelton confessed that he had secrets to sell and wanted to be paid in gold bullion. Yurchenko did not understand, and asked why the man wanted to be paid with cold chicken soup.

Pelton reiterated his request, more clearly specifying *gold* bullion. He removed a document from his pocket and showed Yurchenko, who studied the paper. Pelton explained that he had previously worked for the National Security Agency, and that he had secrets and wanted to be paid well. He informed Yurchenko that he had once been in charge of prepping a large underwater recording pod used by divers to tap sensitive Soviet communications cables. He knew details about how this system worked and where the pod was located. Yurchenko listened, then agreed upon a price for the information.

Three hours later, Ronald Pelton, code-named "Mr. Long" by the Soviets, left from a side entrance, hidden among other Soviet workers. Now clean-shaven, he ducked into a van that drove him to the Soviet residence quarters at Mount Alto. There he consumed a hot meal and received a briefing on how to meet again with a Soviet contact in Vienna. The Soviets returned him to his car, where he drove away into the night, unnoticed by the FBI.

In fact, they didn't notice him for another three years. The breach wasn't discovered until 1984, and the FBI did not arrest Pelton until November 25, 1985. He confessed to revealing details about the Ivy Bells cable-tapping operations being conducted by the *Seawolf* and *Parche* in the Sea of Okhotsk. He is now serving three consecutive life sentences but is scheduled for early release in 2015. When Master Diver David LeJeune heard about what Ronald Pelton had done, he was shocked. LeJeune had worked with the man years earlier and never suspected that his former colleague would stoop to espionage. "He deserves what he got," LeJeune said when asked about the matter. He had no other words to offer on the subject.

By late 1981, based on the information received from Ronald Pelton, the Soviets found the "beast" in the Sea of Okhotsk and brought the pod up from the bottom. The large drum is now on display at a Cold War KGB museum in Moscow. The uncanny timing, in light of the *Seawolf*'s accidental contact with the cable during her last mission, sent the navy into a tailspin. Although Richard Haver, a civilian department head at Naval Intelligence, intimated that a spy might have tipped off the Soviets in a report he filed on January 30, 1982, he and others were concerned that the cable smack might have caused static or an interruption that alerted the Soviets. Regardless, cable-tapping missions to the Sea of Okhotsk came to an abrupt halt.

There was no evidence that the Soviets found the tap placed in the Barents Sea by the *Parche,* so that submarine, with Commander Peter J. Graef in charge, deployed again to northern waters in 1982 for a 137-day trip. Upon her return, President Ronald Reagan awarded the *Parche* a fourth Presidential Unit Citation, which he delivered to Graef along with a box of cigars. The *Parche* went into the yards through 1983, while the noisy *Seawolf* was relegated to finding missile parts. In the wings, the nearly finished USS *Richard B. Russell* awaited her chance to pick up the gauntlet.

PRIDE AND FEAR CONSUMED PETTY OFFICER Second Class Laird Cummings as he shuffled across the brow of the *Russell* in the fall of 1985. The last of the *Sturgeon*-class submarines, it had been built at Newport News, Virginia, and commissioned on August 16, 1975. After a decade in service, she still looked almost new, at least from the outside. A strange rectangular hump jutted from the backside of her sail. Cummings had learned via his fifteen DevGru briefings that this chamber housed the Special Projects divers that the *Russell* employed for various deep-sea missions, replacing the DSRV-like chamber used by other Special Projects boats.

Cummings followed his DevGru escort to the hatch, then held his nose. The strong odor of diesel fuel, amine, body odor, cigarette smoke, burnt coffee, and somebody's butt crack combined to create a smell unlike any other on earth, one that would permanently saturate his clothes for years to come. With timid feet, he climbed down the ladder of his new home and descended into the dungeon.

Within days, Cummings met a good number of the *Russell*'s 112 enlisted men and 14 officers, including her fit-looking skipper, Commander Walter H. Petersen. The six-foot-four COB, Master Chief Wolfert, handed Cummings a qual card and initiated him into the world of *Russell*. When the boat finally went to sea, Cummings slept in the torpedo room alongside the spooks, divers, and other Special Projects personnel. Although curtains were slung across the spook electronics area, he more than once glimpsed the two dozen plus pieces of gear used during their clandestine operations in the Barents Sea. Often while trying to sleep, he'd hear one of the spooks chant, "Comex Run One," followed later by "Finex Run One." Such chatter went on up to number 500 or more. Cummings translated the jargon as "Commence Exercise Run One," in reference to the tapping runs that collected and recorded data from the cables and synchronized times and parameters, this to ensure accurate retrieval and prevent an unnecessary and dangerous rerun.

Cummings often ran into some of the navy divers in the reactor tunnel—the space that divides the forward compartment from the aft—under which resides the nuclear reactor. Despite the fact that enough radiation zoomies sped around inside the tunnel to sterilize King Kong, the divers set up shop there with an exercise bike and a Soloflex workout machine. When Cummings first met these high-risk-takers, he noticed heavy facial lines and deep eye circles—signs of aging far beyond their years. He'd heard from others that the harsh world of saturation diving took its toll on these "heroes," requiring them to press down and decompress for weeks in a small enclosure, eat cold food, and survive the bends at least once in their careers. They'd obviously been through hell and back but still found time to keep in shape.

When Cummings stepped into the reactor tunnel to take readings, he noticed a diver working out on the Soloflex. The beefy guy had placed the machine just below the 400-pound air flask valve that controls the release to the emergency coolant system for the reactor. What Cummings hadn't noticed was that every time the diver raised the bar on the Soloflex, he hit the valve hand wheel just enough to turn it a notch. Eventually, the valve popped off its seat and clanked to the deck. The tunnel filled with the high-pitched screeching of air. The diver froze. Cummings ran from the other side of the tunnel and shut off the air valve seconds before the emergency coolant kicked in, which would have been a very bad thing in the world of reactors.

The diver stared at Cummings with wide eyes. Cummings figured that the hiss of high-pressure air escaping was probably the scariest thing on earth to this macho man. He shrugged, pointed at the Soloflex, and said, "I'd move that thing if I were you."

Later that week the *Russell* found herself stuck in the mud, in similar fashion to the *Seawolf* years earlier. Cummings had heard about that incident from others and began to worry that they might be as unlucky as the *Wolf,* or worse, that they'd be even less lucky and not make it back alive. Ingenuity and experience saved them from an eternal patrol in the Barents Sea as Commander Petersen and crew found a way to dislodge from the bottom after less than a day.

The *Russell* completed her mission and surfaced near Yokosuka, Japan. Cummings was anxious to step on dry land again, but the boat had come up in the middle of a sea state-7 hurricane. Cummings and the crew spent hours sharing seasick trashcans while driving the boat through the squall. Finally, Commander Petersen said, "To hell with this, we're going back down." He dove the boat almost to test depth, where the ocean's fury abated, and the currents were as "smooth as a baby's ass." Cummings strolled into the crew's mess. Someone started a movie. Someone else made popcorn. He smiled, reclined on a bench, and forgot all about the world above. Dry land could wait for another day.

EPILOGUE

When written in Chinese, the word crisis *is composed of two characters. One represents danger, and the other represents opportunity.*

—JOHN F. KENNEDY

ALMOST 200 SUBMARINERS, SPOOKS, AND NAVY divers were interviewed for this book. Not one regretted their time in the navy, and nearly all displayed a genuine love for their country when asked about the selfless sacrifices they made to preserve freedom and democracy. Some may call this corny, but those of us who served use a different word. We call it pride.

With the conclusion of the Cold War, transitions in operational tactics rippled through the submarine force as the bipolar world order of the superpowers shifted to a multipolar collection of interests. The possibility of another world war diminished, but opportunities for regional conflicts, like those in Iraq and Afghanistan, abounded. The makeup and operational posture of the U.S. Navy reflected this change, as priorities migrated from a blue-water stance to a focus on the littorals, or coastal waterways. Intelligence operations shifted from strategic to tactical reconnaissance. Submariners who once maintained long periods of communications silence were expected to exchange information much more frequently with the surface fleet. Invisibility used to be the byword of the silent service, but submarines soon became political weapons brandished publicly and continued as spy platforms and deterrents to nuclear war.

By April 2, 1991, thirteen U.S. submarines were conducting surveillance and reconnaissance operations in support of the first Gulf War. Attack boats USS *Louisville* (SSN-724) and USS *Pittsburgh* (SSN-720) were ordered to launch TLAMs against Iraq. In all, U.S. surface ships and submarines fired 288 land-attack variants of the Tomahawk Land Attack Missiles (TLAMs) during the Gulf War—8 of those coming from the *Louisville* and 4 from the *Pittsburgh*. These launches validated that U.S. submarines could operate as part of an integrated strike force while receiving target and strike data from surface partners. The success of this operation set the stage for future submerged missions in which TLAM strikes could take out early-warning, air-defense, and communications systems to offset threats to U.S. aircraft. Given their superior stealth capabilities, TLAM-loaded submarines could now sneak into attack positions unnoticed, then wreak havoc without warning.

The *Parche* and *Russell* continued to conduct Ivy Bells missions well into the 1990s. When the Mare Island Naval Shipyard closed in 1994, the *Parche* transferred to a new home port at Naval Submarine Base in Bangor, Washington. She received navy unit commendations for her Ivy Bells missions in 1995, 1996, and 1997. The USS *Parche* was decommissioned at the Puget Sound Naval Shipyard on October 19, 2004, as the most decorated vessel in the history of the U.S. Navy. All totaled she received nine Presidential Unit Citations, ten Navy Unit Citations, and thirteen Navy Expeditionary Medals during her thirty years of service. *Parche*'s sail was moved in 2006 to a location near Puget Sound Naval Shipyard.

In 1989, the first *Seawolf*-class attack submarine (SSN-21), intended as a replacement for the aging *Los Angeles* class, rolled down the ramp. Plans called for a fleet of twenty-nine of these new submarines built over a decade, but that number was eventually trimmed to twelve. With the end of the Cold War in 1991, and subsequent military budget cuts, the number dwindled to three.

Quieter, larger, faster, and packing twice the armament of a *Los Angeles*–class boat, the *Seawolf* also came with a bigger price tag. The new subs were intended as a response to the Soviet *Typhoon,* the deep-diving titanium-hulled *Alpha,* and the new super-quiet *Akula.* An advanced design and array of special equipment allowed the new *Seawolf* to operate in shallow waters and deploy up to eight SEALs or divers at a time. Divers and T-Branchers could take advantage of *Seawolf*'s Multi-Mission Platform (MMP), which included an underwater splicing chamber used for fiberoptic cable tapping. Seasoned divers, like David LeJeune, who served on the decommissioned original *Seawolf* during the Cold War and performed Ivy Bells cable-tapping missions, would have been elated with such luxuries.

Submariners claim that the *Parche*'s successor, the USS *Jimmy Carter* (SSN-23), became "Washington's premier spy submarine," conducting cable-tapping missions and retrieving missile fragments from seabeds. Although the *Jimmy Carter* is a formidable Special Operations vessel, packed with advanced combat systems, larger spherical sonar arrays, wide aperture arrays, and new towed arrays, as well as near-silent pumpjet propulsion systems, boats of this class were eventually superseded by newer *Virginia*-class submarines.

The USS *Virginia* (SSN-774) was the first in a line of ultra-modern and sleek submarines designed for a vast array of open-water and coastal missions in virtually any ocean. A less costly alternative to the expensive *Seawolf* class, these new submarines came just in time to replace thirteen decommissioned *Los Angeles*–class boats. The *Virginia* boasted a number of technology and shipbuilding innovations. Gone were the periscopes. In their place came photonic masts mounted outside the pressure hull that housed high-resolution cameras with light-intensification and infrared sensors. Also included were an infrared laser range finder and an integrated ESM array that transmitted signals from the mast's sensors through fiberoptic data lines. Similar to the *Seawolf,* gone also was the traditional propeller, replaced by a quieter pumpjet propulsion system.

As for the submariners who manned these boats after the Cold War, in many ways their roles changed dramatically. In other ways they remained the same. No longer did they hunt the Red Bear's massive carriers and cruisers, and cat-and-mouse Holystone games diminished dramatically. But clandestine espionage operations continued, as did sub-deployed SEAL missions.

To support these missions, several submarines were modified to carry swimmers and equipment more effectively. Attack boats and ballistic missile submarines were retrofitted with special chambers called Dry Deck Shelters (DDSs), which housed Swimmer Delivery Vehicles

(SDVs). Special fittings and modifications to air systems and other features enabled these boats to transport and launch SDVs and deploy combat swimmers. SEALs exited from these chambers while the submarine stayed submerged, then swam to the surface along with their equipment.

The navy recently removed the twenty-four Trident missiles from four of their "boomer" SSBNs and replaced them with 154 Tomahawk land-attack cruise missiles. Converted into SSGNs, these massive bastions of destruction now carry up to sixty-six Special Operations Forces (SOF) that might include the Airforce Special Tactics Squadron (STS), Recon Marine divers, and SEALs. Most missions, however, call for one SEAL platoon of fourteen enlisted men and two officers, along with additional SEALs tasked with mission planning and equipment handling. Two former missile tubes on each SSGN now function as dedicated lock-in/lock-out chambers to allow combat swimmers to exit the boat while submerged. These subs also carry two dry deck shelters and two advanced SEAL delivery systems to support clandestine operations.

Without a superpower to fight, the nature of SSGN operations now focuses on a very different kind of enemy. Rear Admiral William H. Hilarides, program executive officer (submarines), spoke on this topic during a "relaunch" ceremony for the USS *Ohio* (SSGN-726) on February 7, 2006: "*Ohio*'s return to service is truly monumental. Now *Ohio* will conduct missions that will have a direct impact on the ongoing Global War on Terror. . . . SSGNs are truly force multipliers."

Although the Soviet Union vanished almost two decades ago and no longer threatens to undermine the United States' dominance of the world's oceans, and although most terrorist-thwarting activities are land-based, the U.S. Navy continues to operate an extensive fleet of submarines and undertake clandestine missions. In 2008, more than 500 submarines roamed the oceans worldwide. Asian countries commanded 135 of these, while 45 came from the Middle East. The proliferation of advanced conventional submarines is a troubling problem for the U.S. Navy, given that over forty nations rely on advanced diesel submarines. In the Pacific Ocean, over 140 submarines are frequently deployed within striking distance of critical trade-zone choke points.

To monitor potential threats posed by foreign submarines, three classes of American SSNs remain in service, including the *Los Angeles* class, still the backbone of the submarine force, with forty-six in operation. Thirty-one of these are equipped with tubes for vertically launched Tomahawk cruise missiles.

Massive Red Bear ballistic missile boats are no longer the daunting threat they once were, but the task of detecting and neutralizing enemy submarines has actually increased in difficulty for the United States. Most subs are smaller and quieter than ever before and can stay submerged for months on end. Many operate in shallow waters, making detection even harder. The need to locate these boats has led to advances in modern antisubmarine warfare that far surpasses the dismantled Boresight/Bulls Eye technology. The modern ASW arsenal now includes new types of sensors, advanced active and passive sonar acoustics, sonobuoys, and fixed hydrophones.

Communications have also come a long way since the Cold War. In March 2009, two *Los Angeles*–class boats, the USS *Annapolis* (SSN-760) and USS *Helena* (SSN-725), played cat and mouse in the frigid Arctic north of Alaska during a weeks-long ICEX exercise. These boats shot dummy torpedoes at each other, which were later recovered by navy divers via drilled holes in the ice. They also tested a revolutionary new form of underwater communications technology and then surfaced through thick layers of ice to visit the nearby ICEX camp, a smattering of modular huts that house dozens of navy personnel and scientists for weeks at a time.

Jeff Gossett, technical director for the Arctic Submarine Laboratory, said that "we installed an exciting new digital ICOMMS system on the subs that allowed us to type messages into a laptop that were encoded and transmitted to the boats through an underwater noisemaker. Imagine being able to send and receive e-mails to a submarine hundreds of feet deep running at top speed!"

Back in my day we had to park two boats right next to each other and talk through an underwater "Gertrude" device that garbled the sound so bad that boat skippers often gave up in frustration. To most of us Cold War submarine veterans, e-mail at depth and speed still sounds like science fiction.

As for Boresight/Bulls Eye's innovative technology, obsolescence occurred in tandem with the advent of worldwide satellite communications. Not only did satellite transmissions obviate the need to send radio bursts—as Soviet subs once did during the Cold War—but these small orbiting craft can also be used effectively for ASW operations. Today, satellites can image ocean surfaces using optical and radar techniques, which can indirectly detect submarines operating at shallow depths. Thermal imaging is another means of detection, as well as low-flying aircraft. And while SOSUS still plays a role in submarine and vessel detection, many of those arrays have been reassigned to civilian duties for marine research.

IN MARCH 2008, MY DAD AND I sat on his veranda in Puerto Vallarta, Mexico, and talked about the Boresight program as we listened to the tropical rain tap at the plastic patio covering. He scratched at his head and then started laughing. When I asked him what was so funny, he said the itch reminded him of Commander Petersen—his boss at the Karamürsel Huff Duff in Turkey.

"When we lost those Russkie subs," Dad said, smiling, "after we pissed our pants, we all started scratching our heads. But Commander Petersen never stopped scratching. He was going bald and had this skin condition made worse by stress. Captain Mason and I were more worried about Petersen scratching himself to death than we were about the NSA breathing down our necks. Now that I think about it, that's pretty funny. Maybe Petersen's dandruff prevented World War III."

That was the last time I spoke to my dad. He passed away on July 30, 2008.

A week later, one of my colleagues, Joel Harrison, pointed at his computer screen and said, "There."

The chief technology officer at a software company in Silicon Valley and one of the cofounders of a multibillion-dollar storage firm, Joel was once a Boresight/Bulls Eye technician during his navy days in the seventies. I followed his finger to the satellite photo displayed on his computer but saw nothing save an empty circle filled with dirt.

"That's where the Bulls Eye antenna was located," Joel said. "It's gone now, but I still look at that circle and reminisce. That was some pretty cool technology."

Indeed, it was.

Today the large pole antennas used for the last Wullenweber CDAA array built at Imperial Beach near San Diego still remain. Inside the poles, the mortar building no longer houses Boresight systems, but instead is used as a planning facility for advanced U.S. Navy SEAL operations.

As I reflect on the role played by U.S. submariners, navy divers, spooks, and antisubmarine operatives during the Cold War, including the professionals who developed, deployed, and operated Boresight/Bulls Eye stations, I often recall the battle at Rorke's Drift. Fought during the Zulu War, on the afternoon of January 22, 1879, this bloody conflict is one of the most famous in British history. Ninety-five soldiers from Company B of the Second Battalion, led by Lieutenant John Chard, bravely held off 4,000 savage Zulu warriors for twelve hours. They lost seventeen men, while the Zulus lost 600. The British held their ground with .45-caliber Martini-Henry breech loading rifles, while the Zulus pitched spears and swung clubs. Historians credit superior training, technology, and leadership as keys to this victory. A hundred years later, the keys to another victory were quite similar.

During the height of the Cold War, the United States remained greatly outnumbered by the Soviets, with only 123 submarines pitted against nearly three times that number. After the Cuban Missile Crisis, spurred by paranoia, the NSA's voracious appetite for intelligence gathering became a primary espionage motivator. Ascertaining the characteristics and capabilities of Ivan's submarines, many armed with long-range nuclear missiles, became crucial to national security. No cost or danger seemed too high in comparison to potential global destruction.

But the paranoia was more perceived than real. Two years after Nikita Khrushchev became Communist Party first secretary, he ordered Admiral Sergei Gorshkov to dismantle the Soviet fleet. He stopped all heavy cruiser projects in midcourse. By 1957, the Soviet navy had dwindled to just 350 ships. Gorshkov's submarine force consisted mostly of short-range boats designed for coastal defense, not cross-ocean aggression. When the United States launched the world's first nuclear-powered fleet ballistic missile submarine, the USS *George Washington,* in 1960, the Soviets had no answer to this threat. The *Hotel*-class *K-19* launched the year before could hold no candle to the *George Washington. K-19* carried only three nuclear missiles compared to sixteen. She had to surface before firing, and the entire sequence took twelve minutes—more than enough time for a U.S. hunter/killer sub to sink her. She was plagued with problems, including a severe nuclear-reactor failure that cost the lives of twenty-seven sailors.

The Red Bear's attack submarines fared no better. Dozens of nuclear accidents occurred on these boats at the cost of more lives. Although Ivan boasted greater numbers than the United States, a majority of Gorshkov's submarines were old and noisy. For most of the Cold War, until Toshiba and Kongsberg illegally helped upgrade their technology, the Soviets were spear-chucking Zulus fighting rifle-carrying Brits.

On U.S. submarines, enlisted men received many months of training and performed complex and technical functions under way. In contrast, only more senior officers and *michmen*—the equivalent of warrant officers in the U.S. Navy—undertook those duties on a Soviet sub. On an American SSBN, the typical crew consisted of 15 officers and 125 enlisted men. *K-219*, a Cold War Soviet SSBN, had thirty-one officers, thirty-eight *michmen*, and only forty-nine seamen. Compared to the modern-looking U.S. submarine facility in Groton, Connecticut, where I received my training, the Paldiski Soviet nuclear submarine training center resembles a Zulu mud hut.

At the time, though, we didn't know that our opponents were so far behind. By the time the Cold War ended, they'd done a pretty good job of catching up to and, in some cases, surpassing American technology. All but gone were the creaky *Zulu*-class diesel boats and noisy *November*-class nuclear subs. In their place, deep-diving titanium-hulled *Alpha*-class submarines ruled the seas with advanced propulsion systems capable of pushing speeds up to forty-five knots. Our *Los Angeles*–class boats could hit only thirty-five knots on a good day. Had the Soviets not run out of money and patience with communism, the Cold War might have ended differently.

Today, given Russia's current economic state, its fleet has dwindled to a small number of ships and submarines. Those vessels still afloat are often unable to deploy due to a lack of trained crews and resources. Maintenance and repair are dismal. In November 2009, the RIA Novosti news agency quoted retired Russian navy admiral Vyacheslav Popov as saying, "If things remain as they are, we will have to mothball most ocean warships by 2015. That will sharply reduce the navy's capability."

As the Russian navy diminishes, the Chinese navy is increasing and improving at a rapid rate. Given recent confrontations with China's submarines, including a collision with the destroyer *John S. McCain*'s towed array near the Philippines, one wonders if another Cold War is brewing. While the hawks seek more funds to expand the U.S. Navy's submarine fleet, doves argue that there is no need for such weapons in a post–Cold War world. Perhaps these misinformed well-wishers have yet to learn from the past. China now owns a significant portion of America's burgeoning debt and has emerged as an economic power-house in recent years.

In the decade from 1995 to 2005, the Chinese navy launched thirty-one nuclear submarines. Many of these are advanced ballistic missile SSBNs that carry long-range nuclear weapons. Professor Toshi Yoshihara of the Naval War College in Newport, Rhode Island, recently said that, "for at least the next two decades, missile defense . . . will have no an-swer to a capable SSBN patrolling the open ocean. . . . This asymmetry in capability suggests that . . . the only effective response to a capable Chi-nese SSBN is the employment of traditional antisubmarine warfare as-sets, particularly hunter/killer nuclear attack submarines."

If Yoshihara's words are true, American submariners will be ready for the calling. The legacy is strong, and the pride still runs deep.

ACKNOWLEDGMENTS

I WOULD LIKE TO OFFER A special thanks to the following individuals, not already mentioned, for their contributions to the book:

Kenneth Greenawald and T. Michael Bircumshaw (editor of the *American Submariner* magazine), who as "chief submarine technical advisers" caught several errors before we went to print. Likely there are several more, and I'm certain to receive an e-mail from a crusty chief blasting me for "getting it wrong."

Others who provided assistance, input, or stories that did not find their way into this book include: Adam Bridge, Al Burger, Boyce Williams, Brian Lawrence, Chuck Brickell, Chuck Young, Dave Arzani, Doug Bailey, Ernest Harkness, Frank Cawley, George Fraser, George Munsch, Guss Lott, Joel Harrison, Ken Sewell, Pavel Korshunov, Peter Lewis, Ralph Baer, Rich Petsch, Richard Moore, Skip Bauman, Stephen Lozier, and Thomas Hayter.

I would like to thank my agent, James Hornfischer, and my editor, David Highfill of William Morrow/HarperCollins, for their outstanding efforts in helping to mold this into a much better book.

NOTES

Some of the information cited herein was derived via research from numerous books, websites, and documents as outlined under Resources for each chapter, but a vast majority of the content came from interviews with individuals who experienced these events firsthand. Most of those interviewed were in their senior years, and a few recollections were less than optimal. Although every precaution was taken to validate facts, time lines, and names, in a few cases this was not possible. Also, some accounts differed from those of others involved, and in such cases care was taken to determine the most accurate accounting of events based on the facts at hand. Not all stories imparted by submariners, navy divers, and government operatives who were interviewed made the final cut for inclusion in the book. Readers are encouraged to visit the author's website at www.wcraigreed.com to read or download these exciting accounts.

INTRODUCTION

REFERENCES
2007 Submarine Encyclopedia: U.S. Navy Submarine Fleet, Sub History, Technology, Ship Information; Submarine Pioneers, Cold War Technology, Department of Defense, Progressive Management, April 27, 2007. This book and CD-ROM combination contains almost 19,000 pages of in-depth information on almost anything one might want to know about submarines and ASW, with an emphasis on the Cold War. I used this handy reference in almost every chapter of the book.

WEBSITES

http://en.wikipedia.org/wiki/Cold_War

CHAPTER 1

PRIMARY INTERVIEWS

Captain Paul Trejo, USN, Retired, former LTJG on the USS *Blenny* (SS-324).

Gardner Brown, MM3, and other former crew members from the USS *Cubera* (SS-347) and USS *Seawolf* (SSN-575).

Note: Gardner Brown does not know what ancient buildings the USS *Cubera* hid behind while dodging Soviets in the Black Sea, but the explorer Robert Ballard discovered the remains of an ancient structure 311 feet deep near Turkey in 2000 that he believes was flooded during the time of Noah. Brown agrees that the "Main Street" wherein the *Cubera* hid could be a similar city buried under the ocean during the Great Flood (http://news.nationalgeographic.com/news/2000/12/122800black sea.html).

RESOURCES

Cold War Submarines, Norman Polmar and K. J. Moore, Potomac Books, 2004, provided a reference for submarine designations and capabilities

Running Critical: The Silent War, Rickover, and General Dynamics, Patrick Tyler, Harper & Row Publishers, 1986

The Submarine: A History, Thomas Parrish, Penguin Books, 2004

Power Shift: The Transition to Nuclear Power in the U.S. Submarine Force as Told by Those Who Did It, Dan Gillcrist, iUniverse, 2006

WEBSITES

http://en.wikipedia.org/wiki/Hyman_G._Rickover
http://en.wikipedia.org/wiki/USS_Nautilus_(SSN-571)
http://en.wikipedia.org/wiki/USS_Cubera_(SS-347)
http://en.wikipedia.org/wiki/Greater_Underwater_Propulsion_Power_Program
http://en.wikipedia.org/wiki/George_W._Grider
http://en.wikipedia.org/wiki/USS_Seawolf_(SSN-575)
http://www.iwojima.com/

CHAPTER 2

PRIMARY INTERVIEWS

Donald Ross, Ph.D., former DEMON sonar project lead in San Diego, provided excellent information regarding SOSUS capabilities and development of submarine sonar systems. More information can be found in Ross's book noted below.

Note: When the torpedo-propeller design group moved to Penn State in late 1945, Dr. Donald Ross, at the age of twenty-four, presented a paper at the navy's first scientific meeting on hydrodynamic topics sponsored by the Office of Naval Research (ONR). Ross proposed that submarines be built like torpedoes and that the current twin-screw designs could be replaced by a single-screw configuration. He forecast that such a design might empower higher propeller speeds without causing cavitation noise. Albert G. Mumma, the navy captain who'd helped capture German XXI sub designer Helmuth Walter during World War II, shot up from the center of the group and shouted, "The navy will never build something that stupid!"

The crowd nodded their agreement. After all, submarines must have two screws in the event that one fails. Twin-screws are also needed for better maneuverability. Despite pleas to the contrary from Ross, navy officials agreed with the group and dismissed the idea. Years later, when the David Taylor Lab in Maryland successfully demonstrated the new *Albacore*-type submarine, damned if the thing didn't have a single screw. The officer in charge, one Albert G. Mumma, received a naval commendation for his role in furthering single-screw submarine designs. That achievement catapulted Admiral Mumma to chief of the Bureau of Ships—and therefore Hyman Rickover's boss—in 1955, and the *Albacore* single-screw became so successful that most modern navies adopted the configuration for almost all submarine designs.

During one of his runs on the USS *Nautilus,* Ross received an invitation to the wardroom. The *Nautilus*'s skipper, Commander Eugene Wilkinson, flashed a coy smile and pulled out a deck of cards. He laid $20 on the table and said, "You in?"

Ross pulled out a twenty and sat down next to three other officers. Several hours later, after having cleaned out the pockets of everyone in the wardroom, Ross scooped up his winnings and headed toward the door.

Stern-faced, Wilkinson stood from the table and said, "Don't think for a minute I'm going to forget this, Dr. Ross."

Years later, in 1969, Richard Nixon halted all Holystone submarine missions due to a spate of serious accidents. Wilkinson employed Ross and his team to solve a major problem with submarine sonar that appeared to be the root cause of the "run-ins" with Soviet submarines when U.S. boats tried to follow them. Ross solved the problem for Wilkinson and received the highest citation awarded to civilians for his contribution.

RESOURCES
Mechanics of Underwater Noise, Donald Ross, Ph.D., Peninsula Publishing, 1987

Transparent Oceans: The Death of the Soviet Submarine Force, Louis P. Solomon, LRAPP Company, 2003

Cold War Submarines, Norman Polmar and K. J. Moore, Potomac Books, 2004

Hide and Seek: The Untold Story of Cold War Naval Espionage, Peter A. Huchthausen and Alexandre Sheldon-Duplaix, John Wiley & Sons, 2009

Blind Man's Bluff, Sherry Sontag and Christopher Drew, Public Affairs, 1998

Hitler, Doenitz, and the Baltic Sea, Howard D. Grier, Naval Institute Press, 2007

WEBSITES
http://www.globalsecurity.org/intell/systems/sosus.htm
http://www.navy.mil/navydata/cno/n87/usw/issue_25/sosus2.htm
http://en.wikipedia.org/wiki/SOSUS

CHAPTER 3

PRIMARY INTERVIEWS
William J. Reed, LT USN, Retired, regarding HFDF technology and deployment. I interviewed my father in depth and then double-checked the details against other research and interviews.

Robert Lynn Wortman, CD, RCN, RCMP, Retired, regarding World War II Huff Duffs and HFDF sights in the United States and Canada.

Pamela Wallinger Reed (author's sister) offered excellent insights into events in Turkey, Maryland, and San Diego when my father was in the navy.

Joyce Louise Reed (author's mother) offered excellent insights into events in Turkey, Maryland, and San Diego when my father was in the navy.

George Munch, former Boresight/Bulls Eye HFDF systems and Wullenweber array engineer who also helped get most of the joint U.S-Canadian sites up and running.

Frank Cawley, deputy director of the HFDF division at the NSA, former HFDF Boresight/Bulls Eye communications technician.

John Gurley, former communications technician chief, USN, worked on Boresight systems and Wullenweber "elephant cage" arrays.

Gus Lott, Ph.D., founder of Yarcom, Inc. (www.yarcom.com), former Boresight/Bulls Eye systems and Wullenweber array contracting engineer. Gus is now the chief scientist for the Signal-to-Noise Enhancement Program (SNEP). In the 1980s, Dr. Steve Jauregui at the Naval Postgraduate School made SNEP a regular scientific program. The navy returned the program back to the NSA in 2003. Today, SNEP sends experts to sensitive receiver facilities to reduce noise from power lines, computers, and many other emitters that can interfere with signals. Gus provided some outstanding information on the theory and math behind the operation of Boresight/Bulls Eye systems and stations and noted that when these systems were built in the sixties, there were few power lines or other sources of electrical interference. By the nineties, the areas nearby had been built up, and errors increased. Engineers like Gus discovered that interference noise now emanated from lines and laptops and helped correct for such.

RESOURCES

Radio Direction Finding, P. J. D. Gething, Institution of Electrical Engineers, December 1987. This book is considered by most in the field the Bible on HFDF and Wullenweber arrays.

The Canadian History of Signal Intelligence and Direction Finding, Robert Lynn Wortman and George Fraser (self-published), 2006.

The Codebreakers, David Kahn, Scribner, 1996. Chapters 11 and 15 detail interesting information about the use of HFDF Huff Duffs during World Wars I and II.

WEBSITES

http://www.researcheratlarge.com/Pacific/RDF/
http://en.wikipedia.org/wiki/USS_George_Washington_(SSBN-598)
http://en.wikipedia.org/wiki/UGM-27_Polaris
http://rusnavy.com/science/electronics/rv6.htm
http://www.globalsecurity.org/intell/systems/sosus.htm
http://russianfun.net/technology/secret-soviet-submarine-base-in-sevastopol/
http://www.angelfire.com/falcon/usspillsbury-der_133/elequip.html
http://www.jproc.ca/sari/counter.html
http://en.wikipedia.org/wiki/High_frequency
http://en.wikipedia.org/wiki/Battle_of_Jutland

CHAPTER 4

PRIMARY INTERVIEWS

William Reed, LT, USN, Retired.

Robert Lynn Wortman, CD, RCN, RCMP, Retired, former Boresight/Bulls Eye HFDF systems and Wullenweber arrary engineer.

George Munch, former Boresight/Bulls Eye HFDF systems and Wullenweber array engineer.

Frank Cawley, deputy director of the HFDF division at the NSA, former HFDF Boresight/Bulls Eye communications technician.

John Gurley, former communications technician chief, USN, worked on Boresight systems and Wullenweber arrays.

Gus Lott, Ph.D., founder of Yarcom, Inc. (www.yarcom.com), former Boresight/Bulls Eye systems and Wullenweber array contracting engineer.

NOTES

Others involved in the development of the magnetic tape recorder in concert with Robert Misner and Howard Lorenzen were Dr. Hector Skifter and James Gall.

Mack Sheets also accepted the NRL "Top 75 Inventions" award, along with Robert Misner.

Drs. Rindfleisch, Pietzner, Schelhorse, and Wächtler led the effort to build the first CDAA Wullenweber array at Joring, Denmark.

Jurgen Wullenweber's fame grew among the common folk, who revered him as a legendary upholder of the Protestant cause. He died in Wolfenbutiel in 1537 while fighting for that cause, which propelled him to martyrdom.

Mack Sheets assisted Robert Misner with the invention of the AN/FRA-44 recorder/analyzer.

The first successful HFDF fix in World War II came on the morning of July 13, 1943. German U-boat 487 transmitted an update to Berlin. The Tenth Fleet ASW group grabbed the transmission on several Huff-Duffs and fixed U-487's location as just northwest of the Cape Verde Islands. The ASW team hurried the fix to the escort carrier USS *Core* (CVE-13). Wildcat and Avenger aircraft scrambled from the deck of the *Core* and sped toward the U-boat. They found her snorkeling on the surface and blasted the submarine with bombs and cannon fire. U-487 sank to the bottom. The entire operation took less than ten hours and laid the foundation for future ASW operations. After the war, the United States and Soviet Union propelled their respective HFDF programs forward, while shifting their focus on finding each other versus locating German U-boats.

RESOURCES

Radio Direction Finding, P. J. D. Gething, Institution of Electrical Engineers, December 1987.

The Canadian History of Signal Intelligence and Direction Finding, Robert Lynn Wortman and George Fraser

The Soviet Union and the Arms Race, David Holloway, Yale University Press, 1984

U.S. Military Operations Since World War II, Kenneth Anderson, Brompton Books Corporation, 1984.

"Operator's Organizational, Direct Support, General Support, and Depot Maintenance Manual for Antenna Group Countermeasures Receiving Set AN/FLR-7(V7/(V8)," U.S. Army Security Agency, Materiel Support Command, Vint Hill Farms, Warrenton, VA 22186.

"A Wide-Aperture HF Direction-Finder with Sleeve Antennas," Raymond F. Gleason, Robert M. Greene, Naval Research Laboratories Memorandum Report 843, August 20, 1958

"Award for Innovation, 75 Years," Naval Research Laboratory, obtained from the PDF document found at http://www.nrl.navy.mil/content .php?P=75thANNIVESARYAWARDS

WEBSITES

http://www.nrl.navy.mil/pao/pressRelease.php?Y=2000&R=32-00r
http://en.wikipedia.org/wiki/Battle_of_Jutland

CHAPTERS 5–10

PRIMARY INTERVIEWS

William Reed, LT., USN, Retired. Details involving the White House meeting in this chapter were derived from interviews with and memoirs written by my father, William J. Reed. Although I requested information on Boresight and this meeting via the Freedom of Information Act, I received only a rejection notice from the NSA on my first attempt and no response on the second. However, I extensively researched time logs for all the major players and key events of the Cuban Missile Crisis, and based on the reported whereabouts of these individuals, as well as actions taken at specific times, and based on my father's recollections, I believe 5 P.M. on October 25 is when the meeting with President Kennedy and his advisers occurred. My father was not positive about the first names or correct spellings of the last names of Petty Officers Denofrio and Odell, who worked with him on the Boresight project. Apologies to those individuals if the names are incorrect.

Robert Lynn Wortman, CD, RCN, RCMP, Retired, former Boresight/Bulls Eye HFDF systems and Wullenweber array engineer.

George Munch, former Boresight/Bulls Eye HFDF systems and Wullenweber array engineer.

Frank Cawley, deputy director of the HFDF division at the NSA, former HFDF Boresight/Bulls Eye communications technician.

John Gurley, former communications technician chief, USN, worked on Boresight systems and Wullenweber arrays.

Peter Lewis, former communications technician chief who worked on Boresight and HFDF systems during the sixties, offered excellent input regarding this technology.

Gus Lott, Ph.D., founder of Yarcom, Inc. (www.yarcom.com), former Boresight/Bulls Eye systems and Wullenweber array contracting engineer.

Anonymous sources who worked on government programs during the Cold War.

Ryurik Ketov, Captain First Rank, Retired, in Saint Petersburg, Russia, former commander of B-4.

A. F. Dubivko, Captain First Rank, Retired, in Saint Petersburg, Russia, former commander of B-36.

Radm. Victor Frolov, served as the executive officer aboard B-130.

Lt. Anatoliy Andreev, served as assistant to the submarine commander lieutenant aboard B-36.

Adm. Vladlen Naumov, served as captain-lieutenant, navigator, and commander of BCh-1 (navigation) aboard B-36.

Note: I met with the above former Soviet submariners in Saint Petersburg, Russia, via invitation and sponsorship provided by the Saint Petersburg Submariners Organization commanded by Captain First Rank (retired) Igor Kurdin. Igor and his staff, including Ksenya Hohlova, were of tremendous help in arranging meetings and providing transportation, as well as translation and tour-guide assistance. Information offered by these five, in several areas, contrasted with that printed in Peter Hutchthausen's book, *October Fury*. Captain Dubivko, via Adm. Naumov, stated that some of the details printed in *October Fury* were "pure fiction." While Hutchthausen's book offers an interesting and well-researched read, numerous details regarding *Foxtrot*-class submarines and historical accounts, based on expert opinions, are incorrect. Captain Ketov provided in-depth information on the operation and history of the Soviet SBD burst signal radio that spurred the Boresight program.

Charles Skillas, former engineering director at Sanders Associates in New Hampshire, imparted information regarding the navy programs under development during the time my father met with his team in 1962.

Special thanks are due R. J. Hansen, who provided a detailed tour of the B-39 *Foxtrot* submarine located at the Maritime Museum of San Diego. Hansen is considered the most "qualified" American submariner on the *Foxtrot* class, having studied under former B-427 commander

Captain Third Rank Igor Kolosov for two years. Hansen also reviewed the *Foxtrot* details in this book for accuracy, so I blame any remaining mistakes on him, of course. The B-427 is also open for tours and is located next to the *Queen Mary* in Long Beach, California.

Jeff Loman, tour and volunteer coordinator for the Maritime Museum of San Diego, assisted in setting up my tour with R. J. Hansen. Information on the B-39 can be found at www.sdmaritime.org.

Stanley Pearlman, vice president of Newco PTY LTD, LLC, which owns the B-427, provided a few of the *Foxtrot*-class submarine pictures herein, and many of the *Foxtrot* operational details were gleaned from the B-427 tour book. Information on the B-427 and the tour can be found at www.russiansublongbeach.com.

RESOURCES

Body of Secrets: Anatomy of the Ultra-Secret National Security Agency, James Bamford, Anchor, April 30, 2002, offered information regarding John Arnold and Cuban SIGINT missions conducted by the USS *Nautilus* and *Oxford*, as well as details about the USS *Pueblo*.

Jeffrey G. Barlow, "Some Aspects of the U.S. Navy's Participation in the Cuban Missile Crisis," in *A New Look at the Cuban Missile Crisis, Colloquium on Contemporary History*, June 18, 1992, No. 7, Naval Historical Center, Department of the Navy

Command in Crisis: Four Case Studies, Joseph F. Bouchard, Columbia University Press, 1991

The Cuban Missile Crisis, ed. Laurence Chang, National Security Archive, 1992

"The Naval Quarantine of Cuba, 1962," Office of the Chief of Naval Operations, 1963, http://www.history.navy.mil/faqs/faq90-5.html

October Fury, Peter Huchthausen, John Wiley & Sons, 2002

Presidential Recordings: John F. Kennedy, The Great Crises, Vol. III, Philip Zelikow and Ernest R. May, ed., W. W. Norton, 2001

Cables from Cuba History File/U.S. Navy Operational Archives; Deck Logs from Record Group 24, U.S. National Archives

"The Cuban Missile Crisis as Seen through a Periscope," Ryurik A. Ketov, Captain First Rank, Russian Navy (retired), *Journal of Strategic Studies*, vol. 28, no. 2, 217–231, April 2005

"In the Depths of the Sargasso Sea," A. F. Dubivko, Captain First Rank, Russian Navy (retired; unpublished memoir)

"Reflections of Vadim Orlov, We Will Sink Them All, But We Will Not Disgrace Our Navy," V. P. Orlov, Captain Second Rank, Russian Navy (retired; unpublished memoir)

CINCLANT SOSUS contact reports during the Cuban Missile Crisis

"Khrushchev, Castro and Kennedy: Motivation, Intention, and the Creation of a Crisis," Robin R. Pickering, thesis presented to the faculty of Humboldt State University, May 2006

Deck logs obtained for various naval platforms during the Cuban Missile Crisis

Eyeball to Eyeball, The Inside Story of the Cuban Missile Crisis, Dino A. Brugioni, Random House, 1991

Soviet Naval Developments, third edition, foreword by Norman Polmar, The Nautical and Aviation Publishing Co. of America, 1984

Thirteen Days: A Memoir of the Cuban Missile Crisis, Robert F. Kennedy, W. W. Norton & Co., 1969

Submarines of the Russian and Soviet Navies, 1718–1990, Norman Polmar and Jurrien Noot, Naval Institute Press, 1991

Combat Fleets of the World 1980/81, Jean Labayle Couhat, Naval Institute Press, 1980

U.S. Military Operations Since World War II, Kenneth Anderson, Brompton Books Corporation, 1984

One Hell of a Gamble, The Secret History of the Cuban Missile Crisis, Aleksandr Fursenko and Timothy Naftali, W. W. Norton & Co., 1997

"Report Heightens Nuclear Sub Mystery/Torpedo Theory Contradicts Findings of USS Scorpion's Wreckage in 1968," Stephen Johnson, *Houston Chronicle,* December 27, 1993

The Reminiscences of Admiral George W. Anderson, Jr., U.S. Navy (Retired), vol. 2, U.S. Naval Institute, Annapolis, MD, 1983

"Role of Polaris Submarines in the Cuban Missile Crisis," VADM Charles Griffiths, USN (Retired), *The Submarine Review,* July 2009, pp. 69–70

The Face of Moscow in the Missile Crisis, William F. Scott, *Studies in Intelligence,* vol. 37, no. 5, 1994

Command History, ref. (a) OPNAVINST 5750.7, encl., Command History Destroyer Squadron 21, [memo] from Commander Destroyer Squadron Twenty-One to Chief of Naval Operations (OP 29)

CUBEX: U.S. Anti-Submarine Operations during the Cuban Missile Crisis, Norman Polmar, Moscow, September 27, 1994

Chronology (Fleet Operations Report, Cuban Missile Crisis, October 1962), Naval Historical Archive, Washington, D.C.

Cuban Quarantine Operations—Historical Account of, ref. CIN-CLANT Staff Notice 5213 of November 7, 1962 [memo], from Cuba Quarantine Center (O3Q) to CINCLANTFLT Command Information Bureau, Historical Office (J09H)

"Destroyer Forced Red Sub Up," *Evening Capital* (QP), June 14, 1963

History of USS Charles P. Cecil (DD-835), Naval Historical Archive, Washington, D.C.

WEBSITES

http://www.navycthistory.com/index.html

http://www.navycthistory.com/homestead_email_roster.html

http://luxexumbra.blogspot.com/2005/06/frd-10-endangered-species.html

http://www.fas.org/irp/program/collect/an-flr-9.htm

http://www.tscm.com/masset.html

http://jproc.ca/rrp/masset.html

http://www.tscm.com/gander.html

http://www.cubainfolinks.org/webpage/Articles/bejucal.htm

http://www.navycthistory.com/NSGStationsHistory.txt

http://straightwhiteguy.mu.nu/archives/cat_military_stuff.php

http://online.wsj.com/article/SB122619710466311417.html

http://www.thc.state.tx.us/museums/musnimitz.shtml

http://www.navycthistory.com/index.html

http://www.themoscowtimes.com/article/1010/42/372216.htm

http://jproc.ca/rrp/

http://jproc.ca/rrp/grd_6.html

http://www.scribd.com/doc/4566441/Matthew-Aid-The-National-Security-Agencyand-the-Cold-War

http://www.jfklibrary.org/jfkl/cmc/cmc_october28.html

http://www.jfklibrary.org/jfkl/cmc/cmc_october27.html

http://www.gwu.edu/~nsarchiv/nsa/cuba_mis_cri/docs.htm

http://en.wikipedia.org/wiki/Curtis_LeMay
http://avalon.law.yale.edu/20th_century/msc_cuba070.asp
http://en.wikipedia.org/wiki/Maxwell_D._Taylor
http://www.gwu.edu/~nsarchiv/nsa/cuba_mis_cri/photos.htm
http://www.gwu.edu/~nsarchiv/NSAEBB/NSAEBB75/#11
http://www.pbs.org/wgbh/nova/subsecrets/life03householder.html
http://www.gwu.edu/~nsarchiv/nsa/cuba_mis_cri/docs.htm
http://en.wikipedia.org/wiki/Sanders_Associates

NOTES

In addition to the Nakat ESM systems, *Foxtrot* submarines carried a crude but effective Quad Loop radio direction finder that could detect HF radio transmissions and determine rough bearings to the sources. As the name implies, the antenna consisted of four oval pipelike metal loops attached to a center pole jutting out from the top of the sail.

CHAPTER 11

PRIMARY INTERVIEWS

Nihil Smith, former sonarman and navy diver, offered a heart-wrenching account of his dives on the wreck of the USS *Thresher* in 1963.

William Reed, LT., USN, Retired.

Robert Lynn Wortman, CD, RCN, RCMP, Retired, former Boresight/Bulls Eye HFDF systems and Wullenweber array engineer.

George Munch, former Boresight/Bulls Eye HFDF systems and Wullenweber array engineer.

Frank Cawley, deputy director of the HFDF division at the NSA, former HFDF Boresight/Bulls Eye communications technician.

John Gurley, former communications technician Chief, USN, worked on Boresight systems and Wullenweber arrays.

Peter Lewis, former communications technician chief who worked on Boresight and HFDF systems during the sixties, offered excellent input regarding this technology.

Gus Lott, Ph.D., founder of Yarcom, Inc. (www.yarcom.com), former Boresight/Bulls Eye systems and Wullenweber array contracting engineer.

RESOURCES

"An Analysis of the Effects of Feedline and Ground Screen Noise Currents on a Conical Monopole Receiving Antenna," Thomas D. Gehrki, Thesis, June, 1994

"A Wide-Aperture HF Direction-Finder with Sleeve Antennas," Raymond F. Gleason, Robert M. Greene, Naval Research Laboratories Memorandum Report 843, August 20, 1958

"Utilization of a Multiprocessor in Command and Control," Bruce Wald, *Proceedings of the IEEE,* vol. 54, no. 12, December 1966

Cold War Submarines, Norman Polmar and K. J. Moore, Potomac Books, 2004, provided a reference for submarine designations and capabilities.

Soviet Naval Developments, third edition, foreword by Norman Polmar, The Nautical and Aviation Publishing Co. of America, 1984

U.S. Military Operations Since World War II, Kenneth Anderson, Brompton Books Corporation, 1984

Submarine vs. Submarine, Richard Compton-Hall, Grub Street, 1988

Strategic Intelligence for American National Security, Bruce D. Berkowitz and Allan E. Goodman, Princeton University Press, 1989

The Soviet Union and the Arms Race, David Holloway, Yale University Press, 1984

WEBSITES

http://www.lostsubs.com/E_Cold.htm
http://en.wikipedia.org/wiki/USS_Thresher_(SSN-593)
http://en.wikipedia.org/wiki/Bathyscaphe_Trieste
http://en.wikipedia.org/wiki/USS_Preserver_(ARS-8)
http://usnavyphotos.com/wp-content/uploads/2009/04/ars-8-usspreserver1res
    cue-85x11.jpg
http://en.wikipedia.org/wiki/Robert_McNamara
http://www.janes.com/articles/Janes-Military-Communications/US-Navy-
    worldwide-HF-DF-system-AN-FRD-10-or-Bullseye-United-States.html
http://www.jproc.ca/rrp/masset.html
http://www.onpedia.com/encyclopedia/Wullenweber
http://en.wikipedia.org/wiki/Wullenweber

http://coldwar-c4i.net/CDAA/history.html
http://www.r-390a.net/faq-systems.htm
http://groups.msn.com/ctoseadogs/wullenwebers1.msnw
http://www.fas.org/irp/program/collect/an-flr-9.htm
http://www.eham.net/forums/Elmers/201472
http://www.espionageinfo.com/Pr-Re/Radio-Direction-Finding-Equipment
.html

CHAPTER 12

PRIMARY INTERVIEWS

Frank Turban, former communications technician "T-Brancher" chief and spook, provided keen insights into missions conducted by the USS *Swordfish* around the time of K-129's demise.

RESOURCES

Body of Secrets: Anatomy of the Ultra-Secret National Security Agency, James Bamford, Anchor, 2002

Scorpion Down, Ed Offley, Basic Books, 2007

Red Star Rogue, Kenneth Sewell with Clint Richmond, Pocket Star Books, 2005

All Hands Down: The True Story of the Soviet Attack on the USS Scorpion, Kenneth Sewell and Jerome Preisler, Simon & Schuster, 2008

A Century of Spies, Intelligence in the Twentieth Century, Jeffrey T. Richelson, Oxford University Press, 1995

A Matter of Risk, Roy Varner and Wayne Collier, Random House, 1978

The Jennifer Project, Clyde W. Burleson, Prentice Hall, 1977, and the Texas A&M University Press, 1997

Blind Man's Bluff, Sherry Sontag and Christopher Drew, Public Affairs, 1998

The Pueblo Incident, Report No. 76–9, R. C. Spaulding, Naval Medical Research and Development Command, reprinted from *Military Medicine,* vol. 141, no. 9, September 1977

Spy vs. Spy, Ronald Kessler, Pocket Books, 1988

Spy Book: The Encyclopedia of Espionage, Norman Polmar and Thomas B. Allen, Random House, 1997

WEBSITES

http://en.wikipedia.org/wiki/USS_Pueblo_(AGER-2)

http://en.wikipedia.org/wiki/Soviet_submarine_K-129_(Golf_II)

http://www.seattlepi.com/awards/scorpion/scorpion3.html

http://en.wikipedia.org/wiki/John_Anthony_Walker

http://jproc.ca/crypto/kw7.html

http://www.knobstick.ca/museum/kw7.htm

http://www.washingtonpost.com/wp-srv/aponline/20000822/aponline135027_
   000.htm

http://www.lostsubs.com/E_Cold.htm

CHAPTER 13

PRIMARY INTERVIEWS

Joseph Houston, former optics engineering lead on Project Azorian and member of the USSVI submariner's organization in San Jose, California, offered a behind-the-scenes view of the engineering effort involved in bringing up the remains of K-129, as well as a friend's recollection of Carl Duckett, CIA deputy director of science and technology, and of John Parangosky, project director, also known as Mr. P.

To test his photographic "catfish solution" for Project Azorian, Joe Houston used a Pentax camera loaded with Plus-X 35 millimeter black-and-white film while clinging precariously to the end of a large limb that arched over the pond in his backyard. Joe inched to the center of the branch while his fifteen-year-old son, Brant, positioned the targets by tilting them to the best reflective angle of almost 45 degrees. Brant then triggered a Hydro Products strobe lamp after Joe determined that the tree limb had stopped swaying enough to take an exposure. Father and son performed this test after dark in secret seclusion from spying neighbors, and Brant swore an oath to secrecy that he has kept to this day.

David LeJeune, master diver chief, USN, Retired, imparted the only firsthand details ever exposed to the public about top-secret navy saturation diving operations during the Cold War.

RESOURCES

Blind Man's Bluff, Sherry Sontag and Christopher Drew, Public Affairs, 1998

The Jennifer Project, Clyde W. Burleson, Prentice Hall, 1977, and the Texas A&M University Press, 1997

Body of Secrets: Anatomy of the Ultra-Secret National Security Agency, James Bamford, Anchor, 2002

Spy Sub, Roger C. Dunham, Naval Institute Press, 1996

Spying Beneath the Waves: Nuclear Submarine Intelligence Operations, prepared by Hans M. Kristensen, Greenpeace International, August 1994

The Universe Below, Discovering the Secrets of the Deep Sea, William J. Broad, Simon & Schuster, 1997

USS Scorpion (SSN-589)—Court of Inquiry Findings, Opinions, Recommendations, 26 October 1993

"Report Heightens Nuclear Sub Mystery/Torpedo Theory Contradicts Findings of USS *Scorpion*'s Wreckage in 1968," *Houston Chronicle,* December 27, 1993

"Project Azorian: The Story of the Hughes Glomar Explorer," *Studies in Intelligence,* Fall 1985, Secret, Excised copy

Memorandum of Conversation, February 7, 1975, 5:22–5:55 P.M., Confidential, Excised copy, Archival source: Gerald R. Ford Presidential Library; National Security Adviser—Memoranda of Conversation, box 9

Memorandum of Conversation, "[Jennifer?] Meeting," March 19, 1975, 11:20 A.M., Secret, Excised copy, Archival source: Ford Library, National Security Adviser—Memoranda of Conversation, box 10

WEBSITES

http://en.wikipedia.org/wiki/Soviet_submarine_K-129_(Golf_II)
http://en.wikipedia.org/wiki/USS_Halibut_(SSGN-587)
http://www.gwu.edu/~nsarchiv/nukevault/ebb305

CHAPTER 14

PRIMARY INTERVIEWS
David LeJeune, master diver chief, USN, Retired.

Frank Turban, former communications technician "T-Brancher" chief and spook who rode the USS *Seawolf* for several SpecOps.

RESOURCES
Blind Man's Bluff, Sherry Sontag and Christopher Drew, Public Affairs, 1998

The Jennifer Project, Clyde W. Burleson, Prentice Hall, 1977, and the Texas A&M University Press, 1997

Tango Charlie, Tommy Cox, Riverdale Books, 2006

Submarines, Rear Admiral John Hervey, Brassey's Sea Power, Naval Vessels, Weapons Systems and Technology Series, 1994

U.S. Navy Diving Manual, NAVSEA 0994-LP-001-9010, Navy Department, June 1978

WEBSITES
http://en.wikipedia.org/wiki/USS_Halibut_(SSGN-587)

CHAPTER 15

PRIMARY INTERVIEWS
Jeff Gossett, technical director for the Arctic Submarine Laboratory.

Frank Turban, former communications technician "T-Brancher" chief and spook who rode the USS *Flying Fish* and USS *Seawolf* for several SpecOps.

RESOURCES
Submarines of the Russian and Soviet Navies, 1718 to 1990, Norman Polmar and Jurrien Noot, Naval Institute Press, 1991

Cold War Submarines, Norman Polmar and K. J. Moore, Potomac Books, 2004

Combat Fleets of the World, 1980/81, ed. Jean Labayle Couhat, Naval Institute Press, 1980

Transparent Oceans: The Death of the Soviet Submarine Force, Louis P. Solomon, LLRAPP Company, 2003

Submarine, A Guided Tour Inside a Nuclear Warship, Tom Clancy, Berkeley Books, November 1993

Spying Beneath the Waves: Nuclear Submarine Intelligence Operations, prepared by Hans M. Kristensen, Greenpeace International, August 1994

CHAPTER 16

PRIMARY INTERVIEWS

Kenneth Greenawald, former sonar technician aboard the USS *Haddo* (SSN-604).

Lt. Edwin Ladeau Tomlin, former officer aboard the USS *Haddo*.

RESOURCES

Cold War Submarines, Norman Polmar and K. J. Moore, Potomac Books, 2004

The Blue Jackets Manual, twentieth edition, revised by Bill Bearden and Bill Wedertz, United States Naval Institute, 1978

Submarines, Rear Admiral John Hervey, Brassey's Sea Power, Naval Vessels, Weapons Systems and Technology Series, 1994

Anti-Submarine Warfare, W. J. R. Gardner, Brassey's Sea Power, Naval Vessels, Weapons Systems and Technology Series, 1996

Submarine: A Guided Tour Inside a Nuclear Warship, Tom Clancy, Berkeley Books, November 1993

Spying Beneath the Waves: Nuclear Submarine Intelligence Operations, prepared by Hans M. Kristensen, Greenpeace International, August 1994

NOTES

The tradition of the U.S. Navy submarine dolphin insignia, awarded to those who complete submarine qualification, dates back to June 13, 1923, when Capt. E. J. King, commander of Submarine Division Three, forwarded a suggestion to the Bureau of Navigation (now designated secretary of the navy). He recommended that a distinguished device be adopted for those submariners who passed the rigorous requirements to become operationally qualified in all aspects of submarines.

King submitted his own pen-and-ink sketch of the new emblem that submariners spent a year earning. The design consisted of a shield mounted on the beam ends of a submarine, with dolphins forward and aft of the conning tower. The commander of Submarine Division Atlantic endorsed the suggestion, and over the course of several months, the Bureau of Navigation solicited additional "dolphin" designs.

The navy asked a firm in Philadelphia that had previously designed class rings for the Naval Academy to create just the right emblem to dress the uniforms of qualified submariners. The firm came up with two designs that were eventually combined into a single emblem. On March 21, 1924, Theodore Roosevelt Jr., acting secretary of the navy, accepted the dolphins as the official insignia of qualified submariners. Fifty-three years later, on board the USS *Haddo*, I yearned for the day that I could proudly wear the approved emblem on my uniform, just above my heart—a dolphin flanking the bow and conning tower of a submarine.

CHAPTER 17

PRIMARY INTERVIEWS
Frank Turban, former communications technician "T-Brancher" chief and spook who rode the USS *Parche* and USS *Seawolf* for several SpecOps.

David LeJeune, master diver chief, USN, Retired.

Dennis Smith, former electronics technician aboard the USS *Parche* (SSN-683).

RESOURCES
Cold War Submarines, Norman Polmar and K. J. Moore, Potomac Books, 2004

Tango Charlie, Tommy Cox, Riverdale Books, 2006

Blind Man's Bluff, Sherry Sontag and Christopher Drew, Public Affairs, 1998

Submarine: A Guided Tour Inside a Nuclear Warship, Tom Clancy, Berkeley Books, November 1993

Spying Beneath the Waves: Nuclear Submarine Intelligence Operations, prepared by Hans M. Kristensen, Greenpeace International, August 1994

WEBSITES

http://www.navysite.de/ssn/ssn687.htm

NOTES

The ACT V group of spooks originally selected Bill Stringfellow to replace Mac Empey but recanted after interviewing Charlie Miller, the candidate recommended by BUPERS.

CHAPTER 18

PRIMARY INTERVIEWS

Kenneth Greenawald, former sonar technician aboard the USS *Haddo* (SSN-604).

Edwin Ladeau Tomlin, former officer aboard the USS *Haddo.*

Former crew members who served aboard the USS *Drum* (SSN-677).

Anonymous sources who worked on government programs during the Cold War.

Captain First Rank Igor Kurdin, Russian Navy, Retired, president of the Saint Petersburg Submariners Organization.

RESOURCES

Cold War Submarines, Norman Polmar and K. J. Moore, Potomac Books, 2004

Blind Man's Bluff, Sherry Sontag and Christopher Drew, Public Affairs, 1998

Seals in Action, Kevin Dockery, Avon Books, 1991

U.S. Navy SEALs in Action, Hans Halberstadt, Motorbooks International Publishers, 1995

The Power Series; U.S. Navy SEALs, Hans Halberstadt, Motorbooks International Publishers, 1993

Submarines, Rear Admiral John Hervey, Brassey's Sea Power, Naval Vessels, Weapons Systems and Technology Series, 1994

U.S. Navy Diving Manual, NAVSEA 0994-LP-001-9010, Navy Department, June 1978

Submarine: A Guided Tour Inside a Nuclear Warship, Tom Clancy, Berkeley Books, November 1993

Spying Beneath the Waves: Nuclear Submarine Intelligence Operations, prepared by Hans M. Kristensen, Greenpeace International, August 1994

NOTES

Advanced computer-processing capabilities, towed arrays, and superior materials science gave the United States huge advantages over the Soviet Union throughout most of the Cold War. Sound suppression involved a variety of practices, including everything from motor mounts to plant design choices, such as minimizing the amount of rotational equipment required and the use of lower steam pressure. Soviet submarine designs sacrificed stealth for speed, usually employing noisier five-bladed screws that allowed for a few extra knots at the top end. They wanted to keep up with the source of their greatest naval fear—American aircraft carriers. U.S. boats employed secret machining technology to crank out seven-bladed screws that traded top-end speed for stealth. The Soviets were unable to duplicate this until Toshiba and Kongsberg sold the technology required.

CHAPTER 19

PRIMARY INTERVIEWS

Former crew members who served aboard the USS *Drum* (SSN-677).

Anonymous sources who worked on government programs during the Cold War.

Captain First Rank Igor Kurdin, Russian Navy, Retired, president of the Saint Petersburg Submariners Organization.

RESOURCES

Cold War Submarines, Norman Polmar and K. J. Moore, Potomac Books, 2004

Blind Man's Bluff, Sherry Sontag and Christopher Drew, Public Affairs, 1998

Piping Systems, SSN 677, NAVSEA 0903-013-4010, Naval Ship System Command, July 1972

Submarine: A Guided Tour Inside a Nuclear Warship, Tom Clancy, Berkeley Books, November 1993

Spying Beneath the Waves: Nuclear Submarine Intelligence Operations, prepared by Hans M. Kristensen, Greenpeace International, August 1994

WEBSITES

http://wapedia.mobi/en/Soviet_submarine_K-324

CHAPTER 20

PRIMARY INTERVIEWS

James Rule, former torpedoman on the USS *Seawolf* during Ivy Bells missions.

Thomas Ballenger, former fire control technician on the USS *Seawolf* during Ivy Bells missions.

Frank Turban, former communications technician "T-Brancher" chief and spook who rode the USS *Seawolf* for several SpecOps.

David LeJeune, master diver chief, USN, Retired.

RESOURCES

Cold War Submarines, Norman Polmar and K. J. Moore, Potomac Books, 2004

Blind Man's Bluff, Sherry Sontag and Christopher Drew, Public Affairs, 1998

NOTES

Some of the other divers assigned to the *Seawolf* that Petty Officer Tom Ballenger mingled with included Bill Fitzpatrick, Tim Bohl, and Eddie

Baker. Ballenger claims that, in addition to "heroes," the crew often referred to the divers as "3200s." He suspects this was in reference to some of the equipment they used, but he never knew why. Master Diver David LeJeune does not recall anyone using the 3200 term for the divers.

"Beau got a little beat up but survived his fall from the perch," said Jim Rule regarding the smack that the torpedoman's porcelain mascot took during the storm that hit the *Seawolf*. "It was one of those little things that gave us hope that maybe we'd survive. When you're consumed by thoughts of death, you look for anything that can lift your spirits. Beau surviving that fall made us think that maybe we could make it, too."

CHAPTER 21

PRIMARY INTERVIEWS

James Rule, former torpedoman on the USS *Seawolf* during Ivy Bells missions.

Thomas Ballenger, former fire control technician on the USS *Seawolf* during Ivy Bells missions.

Frank Turban, former communications technician "T-Brancher" chief and spook who rode the USS *Seawolf* for several SpecOps.

David LeJeune, master diver chief, USN, Retired.

RESOURCES

Cold War Submarines, Norman Polmar and K. J. Moore, Potomac Books, 2004

Blind Man's Bluff, Sherry Sontag and Christopher Drew, Public Affairs, 1998

A Century of Spies: Intelligence in the Twentieth Century, Jeffrey T. Richelson, Oxford University Press, 1995

Spy vs. Spy, Ronald Kessler, Pocket Books, 1988

Spy Book: The Encyclopedia of Espionage, Norman Polmar and Thomas B. Allen, Random House, 1997

When Dennis Smith returned to the USS *Parche* in 1983, things had changed in ways unimagined. Commander Peter J. Graef had taken command, and a myriad of new faces walked the passageways. Master Diver David LeJeune relieved the previous master diver after he forgot to bring the critical section of threaded pipe that allowed the divers to "blow down" the water from the diving platform in the fake DSRV. An entire mission had to be scrubbed as a consequence.

Smith recalled that previous cable taps were conducted at the repeater locations, the points where two long sections join, and the signal is boosted to last through the next leg in the journey. Tapping the cable at this point seemed most efficient in the early days but came with drawbacks. Soviet cable-laying ships usually pulled cables up for spot checks at repeater locations. Should they hoist up a tap along with the cable, that'd be bad.

A land-based U.S. intelligence operative had managed to obtain a section of the communications cable used by the Soviets from a Russian vendor, which allowed the NSA to examine the design and engineer a more efficient cable-tapping method. Divers now were equipped with special tools that allowed them to penetrate the outer wrapping of the steel wire down to the center dielectric inside the shielding.

The clamp used for the tap also had a breakaway feature. If the Soviets pulled up the cable, the induction tap would simply "pull out" of the cable, allowing it to twist back to normal. Only a barely visible scar remained as evidence of foul play. A new System 9500 came with a small amplifier, buried in the mud, that allowed the large pod enclosure to reside almost a mile away from the recording container. Divers attached 4,500 feet of cable to the amplifier, then returned to the *Parche*. The boat moved to a new location and dropped the other end of the 4,500-foot cable along with a Deep Ocean Transponder (DOT), which allowed the divers to find the cable end and hook up the twenty-foot-long "beer keg" beast. This ingenious system improved cable-tapping efficiency while lowering risk.

During Smith's second run to the Barents, the Soviets detected the *Parche* sneaking into the area. Commander Graef came to periscope depth at night and maneuvered into the middle of a large Russian fishing fleet to hide from the lurking Soviet warships and planes. Sonar reported active pinging and planes and helos scattering sonobuoys all around them. Light "warning charge" explosions clapped in the distance.

His chest thumping, Smith sat in the chief-of-the-watch seat on the port side of the "rigged for red" control room. Graef sounded off bearings to contacts that he spied through the scope. A loud thud, followed by an emergency buzzer, echoed through the compartment. For reasons unknown, the controls for the fairwater and stern planes, which control the depth and angle of the boat, shifted from normal to emergency hydraulics. The noise rattled everyone in the control room, including Smith. The planesman and helmsman jumped in their seats and started muttering expletives. The diving officer followed suit.

Graef peeled his eye away from the periscope and said, "We drill for this all the time, people. It's just the hydraulics shifting to emergency. Get a grip and drive the fucking boat! You guys know how to do this."

With that, everyone calmed down and guided the submarine through the briar patch unscathed. Smith let out a slow breath, certain that his CO had everything under control. Despite the dangers that encircled his home, he knew he'd survive yet another mission into the hinterland and return with something no amount of money could buy: a sense of pride that runs deeper than any other.

EPILOGUE

RESOURCES

"DESI: Diesel Electric Submarine Initiative, a Partnership for Global Security," *Undersea Warfare,* Spring 2006, pp. 19–20.

"Joint Spec Ops: Air Force, Navy Test Rescue Scenario," *Undersea Warfare,* Spring 2006, pp. 9–11.

"The How and Why of Open Architecture," *Undersea Warfare,* Spring 2008, pp. 6–9.

"Managing Modernization: A Fleet First Perspective," *Undersea Warfare,* Spring 2008, pp. 10–12

"Patrolling the Deep: Critical Anti-submarine Warfare Skills Must Be Restored," available at http://www.armedforcesjournal.com/2008/09/3654984/

China's Future Nuclear Submarine Force, ed. Andrew S. Erickson, Lyle J. Goldstein, William S. Murray, and Andrew R. Wilson, Naval Institute Press, 2007

WEBSITES

http://www.sublant.navy.mil/ICEX09/ICEX09.htm

NOTES

Quieter submarines demand shorter passive detection ranges, which has triggered research in low-frequency active sonar. The downside of such systems is the adverse effects on whales and dolphins, which has caused an uproar in environmentalist circles. No such uproar emanates from software engineers, who revere recent changes in modern sonar systems that transcend the closed architectures used in the equipment of my day. In the eighties and nineties, BQQ-5 sonar systems served front and center on submarines and relied on Sperry/UNIVAC UYK-7 processors. Unfortunately, this technology became obsolete not long after installation. Proprietary software came part and parcel with hardware, and decoupling the two could not be done. Software upgrades required drastic hardware plus system changes and vice versa, which translated into expensive, typically in the range of $150 million.

The navy reduced that price tag to under $15 million by incorporating the ARCI program, which stands for Acoustic Rapid COTS Insertion. COTS means commercial off-the-shelf technologies. After ARCI, software is now updated every two years in sonar shacks on modern *Los Angeles-* , *Seawolf-* , and *Virginia*-class submarines, where glittering computer screens resemble a *Star Trek* movie set. Similar changes have impacted fire-control systems. I recall spending hours troubleshooting bays of massive analog gear on *Permit-* and *Sturgeon*-class boats. That's all been replaced by a bank of BYG-1 consoles with readouts that look more like video-game monitors than submarine combat systems. After touring some of these modern marvels, I often miss my time on the boats and wish I was just a few decades younger.

For most submariners, especially those involved in top-secret and highly dangerous Holystone and Ivy Bells missions, dry land often waited for more than ninety days at a stretch. Such was also the case for submariners serving on fleet ballistic-missile "boomer" subs. At sea for months on end, few found time for romance or families. Most missed holidays, birthdays, anniversaries, and their children's first breaths. Still, they served with pride and carried pictures of loved ones while operating in hostile waters.

When Bill Heaton graduated from a small Alabama high school in 1977, uncertainty owned him. The yellow brick road that lay ahead was littered with choices that could lead to somewhere dismal or nowhere at all. His friends suggested the Crimson City, so he followed in their footsteps and enrolled in the University of Alabama at Birmingham. As a starving freshman, he struggled through classes while working nights to pay the tuition. A year went by before the money ran out. Heaton then followed in his father's footsteps and found employment at Alabama Power. Each day after work, he forced his eyes to stay open during night classes at the university, but he soon found it impossible to keep up with schoolwork and earn a paycheck. Out of options and money, Heaton fell to his knees and prayed. His strong Christian faith gave him confidence that an answer would come.

The following day, rays of morning light shot through his window and lit up a book on his desk. He dusted off the cover and smiled. *A book about submarines.* As a kid, he'd read just about everything written about World War II submarines and imagined himself in the role of a Barnacle Bill riding the old smoke boats. Every young boy needs heroes, and for Heaton, submariners were role models worthy of respect. Never was he more excited than when his parents bought him a model of the submarine USS *George Washington.* He played with the missile-launching mechanism so much that he broke the tiny spring. Now, as he stared at one of his old submarine storybooks, he knew that God had answered his prayers.

Heaton's father had served in the Air Force with pride, and although antiestablishment anger still rippled through a society shattered by the Vietnam War, Heaton still got choked up when they played the national anthem at baseball games. Some called his red, white, and blue sentiment corny, but he didn't care. That day, in August 1979, he marched down to the recruiting office and volunteered for submarines. He signed up on the delayed entry program and attended boot camp in Orlando, Florida, in March 1980, graduating on the very day that the mission to rescue the hostages in Iran failed. Eight U.S. servicemen died in that attempt when a helicopter crashed into a C-130 transport plane. Heaton gathered with the other new recruits as the boot-camp company commander informed them of what had happened. Unsure if another war lay just around the corner, Heaton muzzled his trepidation as he stood and saluted, ready to serve alongside his fellow sailors.

After boot camp, Heaton attended interior communications school in San Diego, California, followed by Submarine School in Groton, Connecticut. He learned about how submariners talk to one another, stay on course, and breathe air while submerged through the classes he took about closed-loop communications systems, gyro compasses, and central-atmosphere monitoring systems. Heaton completed C School, added a stripe to his sleeve, and went home on vacation. A friend ridiculed him and insisted that Heaton was "stupid to join the navy," and that "serving in Uncle Sam's Canoe Club was a colossal waste of time." Heaton ignored the insults and reported to his first submarine, the USS *Stonewall Jackson* (SSBN-634). A few years later, while at home on leave again in September 1983, he met the love of his life.

Twenty-three-year-old Beth worked with one of Heaton's cousins, who talked her into a blind date. Beth at first refused, stating that she had no desire to go out with a sailor who'd been at sea for months. Beth still lived at home, and an hour before the date she told her mother that she'd changed her mind and didn't want to go through with the meeting. Her mother insisted that to cancel would be impolite, so Beth reluctantly went upstairs to pick out a pair of shoes. She came back down an hour later in blue jeans and a red V-neck sweater. She had a thing for red shoes and wore a different style on each foot. Near the bottom of the stairs she said, "Mom, I can't decide which of these to wear."

A man's voice responded with "I like the one on the right."

Beth glanced up. A handsome five-foot-eleven, dark-haired sailor stood near the door. As she gazed into Heaton's eyes, Beth felt a glow in her chest.

"When I saw Bill for the first time," Beth says, "bells started ringing, and I just knew that God was telling me he was the one."

It was love at first sight for Bill Heaton as well, but given his at-sea schedule on the *Jackson,* he saw Beth for only a few weeks at a time during the next nine months, which made each day all the more precious. Following a brief visit during the July Fourth weekend, Heaton called Beth from the pier at Kings Bay, Georgia. His voice shook as he talked.

"What's wrong?" Beth asked.

"We're going out to sea tomorrow," Heaton said, "and I don't want to spend another day without you by my side."

"I don't understand," Beth said. "I am by your side."

"I mean," Heaton said, pausing for a few seconds, "I want you by my side forever. I want you to be my wife."

Beth had expected a kneeling proposal at a fancy restaurant complete with violins and a diamond ring, but she figured that "ordinary" was not in the cards for a submariner. Tears streaming down her cheeks, she said yes. "I decided that it didn't matter if he proposed in a restaurant, my living room, or from Timbuktu. I loved him and wanted to be his bride. We knew he was due back in port in September, so we made plans for the weekend following the boat's arrival."

Beth spent the next three months planning and inviting and giggling with anticipation. Then she got a call from her church pastor. He couldn't do the ceremony that weekend, so they needed to change the date. Beth panicked. How could she get word to Heaton, who was now 500 feet deep somewhere off the coast of Russia? E-mail did not exist in those days, so she sent a "family gram" with the message "Wedding date changed—stop—details later—stop—new date September 15—end."

Bill Heaton married Beth in September 1984. He nicknamed her Dusty in reference to an episode when she had tried to reuse an old vacuum cleaner bag because they couldn't afford to buy a new one. The bag split open and covered her from head to toe with dust. Heaton later transferred to the USS *Batfish* (SSN-681), a fast-attack boat that took him to the Barents Sea on SpecOps more than once. During one run, they conducted an "under hull" on an infamous *Alpha*-class Soviet boat, at the time one of the most advanced, fastest, and deepest-diving submarines in the world. Several times on that mission, when they were almost caught by the Soviets, Heaton wondered if he'd ever see Beth again. He relied on God and his fellow submariners to bring him home. "I learned to trust others implicitly in submarines," says Heaton. "Not just to throw me the ball or keep my car safe, but with my life. That kind of trust changed me forever."

Heaton became the Protestant lay leader on the *Batfish,* conducting Sunday services and Bible studies with others of faith. "I was humbled and honored to serve my country as a submariner," says Heaton. "To this day my friends still call me 'Navy Bill.' "

INDEX

ABC News, 147–48, 156
Acoustic intelligence (ACINT), 273
Active sonar systems, 18, 272–73
Agafonov, Vitali, 96, 172–73
Akula-class submarines, 338
Alabama Power, 376
Alantika, 115
Aleksandrovsk, 64
Alpha-class submarines, 338, 344
Alvarez, Floyd, 220
Amagin, 137–38
Amelko, Ivan, 209
American Seaside Club, 25–26
America's Cup (1962), 78
Anadyr, 58–61. *See also* Operation Kama
Anderson, George W., 102, 106, 116–18, 125–29, 137, 147, 149
Anderson, Jack, 225–26
Anderson, Rudolph, 149
Andrea Doria, 237–38
Andreev, Anatoliy, 108–10
Anechoic tiles, 266
AN/FLR-9, 47
AN/FRA-44, 43, 47
AN/FRD-10 ("Fred Tens"), 47–49
AN/GYK-3, 191–92
Annapolis, USS, 341
Applied Physics Laboratory, 254
Applied Technology Division (ATD), 227
Apra Harbor, 316
AQA-7 signal processors, 143
ARCI program, 375*n*
Arctic Submarine Laboratory, 254, 341
Arkhipov, Vasily, 73, 151–55, 176–77
Arnold, John, 62–63, 231–32, 247
AT&T, 253–54
Aurora 7, 181
Austin, Mary, 22

B-4, 67–77
    assuming combat readiness, 93–94, 107
    avoiding detection, 171–74
    heading to Cuba, 73–77, 93–94
    mission details, 69–71, 75–76
    special weapon onboard, 68–70
    storm troubles, 94–95
B-36, 65–67, 73–77, 165–71, 174–78
    assuming combat readiness, 167–70
    avoiding detection, 85–86, 110–15, 118–21, 149–50, 160–61, 165–68, 170–71

bad luck onboard, 174–76
    heading to Cuba, 73–77, 82–86, 88–93, 95–100, 107–15, 118–21, 149–50
    mission details, 69–71, 75–76
    Pankov operation onboard, 82–84, 90–91
    returning home, 176–78
    special weapon onboard, 66–67, 69, 71, 88–89
B-37, 188
B-59, 70–77, 107, 120, 149–55, 158–59
B-130, 65, 69–77
    avoiding detection, 164–65, 167–68
    engine troubles, 130–34, 164–65
    heading to Cuba, 73–77, 107, 120
    mission details, 69–71, 75–76
Bache, USS, 150
Baer, Ralph, 146–47
Bailey, Don, 202–3
Baker, Eddie, 371–72*n*
Ballard, Robert, 350*n*
Ballenger, Tom, 322–26, 371–72*n*
Bangor Naval Submarine Base, 338
Banks, Bill, 19–20
Baralyme, 245
Barb, USS, 211
Barbel, USS, 184–85
Barron, James, 203–4
Batfish, USS, 378
Bathyscaphes, 181
Bathythermograph (BT), 6
Baud, 38
Baudot, Jean-Maurice-Émile, 38
Bayern, 44–45
Bay of Pigs, 179
BBC Radio, 120
Beale, USS, 150
Beatty, David, 45
Beauregard (mascot), 324, 326
Bekrenyev, Leonid, 129
Bellini, Ettore, 44
Bell Telephone Laboratories, 16–19, 248
Bends, 240, 335
Benjamin Franklin-class submarines, 252–53
Benton, Hugh, 211
Besugo, USS, 2–3
Bettis Atomic Power Laboratory, 14, 282
Birinci Inonu, 52–56
Blake, Gordon, 136–37
Blandy, USS, 106–7, 164–65
"Bleed air," 1–2
Blenny, USS, 1–9

Blind Man's Bluff (Drew and Sontag), ix
Blue Gill, 256
Blue Surf, 256
Bob (CIA operative), 196–97
Bohl, Tim, 371–72*n*
Bolo lines, 158
Bol'shevik Sukhanov, 116
Bondville Station, 47
Bonesteel, Charles H., 203
Boresight project, 47–51, 188–91
  calibration and signal analysis, 50–51, 53
  installing and testing, 49–50, 81–82, 87
  McNamara and, 127–28, 137, 138, 188
  NSA funding and naming, 44, 47–48
  refinements and expansion, 118, 141–47, 149,
    189–94
  success of, 179
Boston Naval Shipyard, 181
Bowles, Ethel, 32
Bowles, Hoyle, 32
Boykin, Dennis B., III, 14
BQR-24, 273
Bradley, James F., Jr., 215–17, 221, 228–31, 249
Branham, J. K., 328
BRD-6, 209, 256
BRD-7, 256–57
Brill, USS, 51
Brown, Aubrey, 101
Brown, Gardner, 10–15, 282, 350*n*
Brown, Hal, 289–90
"Brute Force" plan, 220–21
Bryant, James S., 2–9
Bucharest, 126, 134
Bucher, Lloyd, 201–4
BUD/S training, 297
Buffett, Jimmy, 264
Buinevich, Doctor, 82–84
Bulls Eye, 44, 45, 47–49, 188–91, 197–98
Bundy, McGeorge, 87, 136–37, 148
Bureau of Naval Personnel (BUPERS), 284
Burke, Arleigh, 126
Burroughs Corporation, 191
Burst signals, 33–38, 41–42, 50–51, 123–24,
    144–45
Byham, Donald, 4–6

Cable-tapping missions, 228–31, 233–34, 238,
    245–51, 287–89, 291–93, 319, 324–27, 332–36
Cafrey, Phil, 260
Caicos Passage, 107
Caltech, 25–26
Canadian-U.S. Atlantic HFDF network, 197–98
Cane, James Christopher, 329
Carbon dioxide (CO2) poisoning, 8, 9, 12–13
Carlin, George, 265
Carlson, Rich, 17–18
Carpenter, Scott, 181
Carter, James Earl "Jimmy," 15, 265, 319
Castro, Fidel, 59, 157, 176
Catch-22 (Heller), 35
Cavala, USS, 19–20
Celik, Captain, 51–56
Central Intelligence Agency (CIA)
  Cuban Missile Crisis and, 62, 63–64, 103–5,
    106, 148
  Foreign Broadcast Information Service, 200–201
  Greek incident, 194–97
  Project Azorian, 217, 219–28
Centre, Gene, 14–15, 282

Cesium clocks, 145
Chaffin, Leo, 7
Chard, John, 343
Charles P. Cecil, USS, 107, 159–64, 168–70
Chebanenko, Admiral, 65–66
Cheprakov, Lieutenant, 133–34
Chesapeake, USS, 203–4
Chicca, Robert, 203
Chinese navy, 345
Churchill, Larry, 17–18
Churchill, Winston, vii, 294
Circularly Disposed Antenna Array (CDAA),
    45–46, 47
Clarinet Bulls Eye, 191
Clarke, Arthur C., 215, 223
Classic Bulls Eye, 191, 197–98, 253
Classic Outboard program, 254–55
"Clementine," 224–26
Clower, Lieutenant, 80–81, 122
CODAR (COastal raDAR), 142
Cold War Medal Act of 2007, ix
Combat Information Center, 159
Compass rose board, 29–30, 124, 145
Constellation, USS, 300
Cony, USS, 107, 129, 134, 150, 157–59
Coral Sea, USS, 180
Core, USS, 355*n*
Courtney, Frank, 283–84
Cousins, Rich, 35–38
Cox, Tommy, 283–84
Crandal, Charles, 203
Crazy Ivan maneuver, 273–74
Cretan Star, 119
Crews, Harry, 156
Crowley, Tom, 231–32
Cuban Missile Crisis, 57–179. *See also* Operation
    Kama
  first U.S. casualty, 149
  Kennedy-Khrushchev deal, 176
  Kennedy's secret memo to Khrushchev, 156–57
  Khrushchev's letters to Kennedy, 156–57
  Soviets halt advance of cargo ships, 148
  Soviets launch Operations Kama and Anadyr,
    57–61
  U.S. ExComm meetings, 87, 102, 118, 129–30,
    178
  U.S. military preparations, 115–18, 147–48
  U.S. quarantine measures, 64–65, 106–7, 115,
    116–17, 120, 125–30, 150
  U.S.-Soviet standoff, 149–50, 156–57
  U.S. surveillance, 78–82, 86–88, 101–7, 115–18,
    122–25, 130, 134–40, 148–49
  U.S.'s initial discovery of Soviet operations,
    61–65
Cubera, USS, 10–13, 15, 350*n*
Cummings, Laird, 334–36
Cumshaw, 35
Cussler, Clive, 237

Dace, USS, 217–18, 296
Dare, J. Ashton, 321–22, 326–27
Dartmouth College, 11
Davis, John, 62, 78
"Dead zone" for Soviet torpedoes, 169
Decompression sickness, 230, 240
Deep Ocean Transponder (DOT), 373*n*
Deep Submergence Rescue Vehicle (DSRV),
    229–30
Defense readiness condition (DEFCON), 117–18

Delta-class submarines, 252–53, 257–58, 276–81
Demodulation, 19, 20
DEMON sonar systems, 20, 21
Denofrio, Tommy, 81–82, 141–45
Depth charges, 152–55, 315–16
Design collapse depth, 6–7, 162
De Steiguer, USS, 289–90
Developing film, 300–301
Diego Garcia, 301–2
Diem, Eugene "Gunga Din," 239–40, 243, 244–45
Diesel-powered submarines, 8–10, 13, 14, 15
Dipping sonar, 166–67
Directional Frequency and Ranging (DIFAR), 142–44
Directional sound viewpoint, 142–43, 253–54
Diver fatigue, 237
Dive tables, 229
Diving depths, 6–7, 162, 167
Dobrynin, Anatoly, 102
Dolphin insignia, 180, 367–68*n*
Donovan, Robert, 157
Drew, Christopher, ix
Drum, USS, vii, 295–317
    collision with *Victor III,* 311–14
    evasive maneuvers, 314–16
    exercises, 297–99
    outside Diego Garcia, 301–2
    recon photography, 296–97, 299–301
    spotting *Victor III,* 304–11
Dry Deck Shelters (DDSs), 339–40
Dubivko, Aleksei, 65–67, 73–77, 165–71, 174–78
    avoiding detection, 85–86, 110–15, 118–21, 149–50, 160–61, 165–68, 170–71
    background of, 65–66, 69
    bad luck onboard, 174–76
    heading to Cuba, 73–77, 82–86, 88–93, 95–100, 107–15, 118–21, 149–50
    mission details, 69–71, 75–76
    Pankov operation onboard, 82–84, 90–91
    readying the weapon, 167–70
    returning home, 176–78
    special weapon onboard, 66–67, 69, 71, 88–89
Duckett, Carl, 216–17, 221, 222–23, 227
Dulles, John Foster, 78
Dygalo, Viktor A., 205

Eastman Kodak Company, 318–19
Eaton, USS, 150
Echo-class submarines, 193, 212, 214
Edzell Bulls Eye, 48–49, 87–88, 189
Eisenhower, Dwight D., 265
Electric Boat Corporation, 9–10, 14–15
Electronic countermeasures (ECM), 41, 42
Electronic intelligence (ELINT), 40
Electronic surveillance measures (ESM), 5, 52–55, 114, 151–52, 200
Elephant cages, 47
Elk River, 236
Ellenwood, Bob, 283–84
Emergency air-breathing (EAB), 269
Empey, Malcolm "Mac," 231–33, 247–51, 283–84
Enterprise, USS, 105, 107
Escape trunks, 230, 297–99, 310–15, 340
Essex, USS, 105, 106–7, 120, 134
ExComm (Executive Committee of the National Security Council), 87, 102, 118, 129–30, 178

Fénelon, François, 252
Fifteenth Submarine Squadron, 205

Film processing, 300–301
Fingerprinting, 256
Fire control, 3
Fish, 215–16, 230–31, 249–50
Fitzpatrick, Bill, 370*n*
Flacco, Nick, 306–7, 311–15
Flasher, USS, 11
Flying Fish, USS, 255–63, 284
Fokin, Vitali, 58–61, 70–71
Follow-on Test (FOT), 116
Fomin, Aleksandr, 147–48, 156
Ford, Gerald, 227
Ford Motor Company, 190
Fort George G. Meade, 4, 22, 50, 102–6, 130
Foxtrot-class submarines, 62, 68, 150, 162, 357–58*n,* 360*n*
Fred (CIA operative), 194–97
Fricke, Robert E., 295–96, 304, 311–12
Frontz, Al, 242, 244–45
FUNNEL, 62
Fursenko, Aleksandr, 157

Gagarin, 149
Gall, James, 354*n*
Gas supply, and divers, 237
Gates, Thomas, 118
Gauss, Karl, 144
Gaussian, 144
Gearing, USS, 134
Geneva Convention, 46
George Washington, USS, 116, 189, 193, 316–17, 343–44, 376
Gibran, Kahlil, 199
Gilpatric, Roswell, 125, 127–29, 136–37
GIUK gap (Greenland, Iceland, United Kingdom), 83
Glomar Explorer, 223–28
Gnomonic projection, 30
Goldwater, Barry, 265
Golf II, 205, 207–8, 210, 212, 216, 224
Goniometers, 48, 197
Gorshkov, Sergei, 31, 57–61, 64–65, 70–71, 82, 188, 343–44
Gorski, Anatoly, 157
Gossett, Jeff, 341
Governor Dummer Academy, 11
Graef, Peter J., 333–34, 372–74*n*
Grant, Ulysses S., 135–36
Graznyy, 134
GRD-6 stations, 43, 49, 87–88, 105–6, 134, 149, 188–89
Greater Underwater Propulsion Power (GUPPY), 10
Grechenov, Major, 148–49
Grechko, Andrei, 177–78
Greenawald, Kenneth "Greenie," 265–66, 272, 275–81
Greenling, USS, 284
Grider, George W., 11–13, 15
Griffiths, Charles, 116
Grigorievich, Valentin, 153–54
Gromyko, Andrei, 102
GT&E Sylvania Electronics Systems, 46, 47
Guardfish, USS, 211–12
Gulf War, 338
Gurley, John, 122–25
GYK-3 computers, 145, 147, 191–92, 206

Haddo, USS, 265–81
    crew of, 266–67

Haddo, USS *(continued)*
  first dive, 267–68
  life under the seas, 268–81
  sonar systems, 271–74
  tracking and photographing submarines, 212,
    276–81, 301
  transfer from, 295–96
Halibut, USS, 229–33, 238–51
  cable-tapping mission, 228–31, 245–51
  searching for K-129, 216–17
  upgrade, 215–16, 230
Halidere (Turkish neighbor), 23–24
Halworth, "Doc," 238–39, 245–46
Hansen, R. J., 357–58n
Hanza Bulls Eye, 87–88, 189
Harrier, USS, 223
Harrison, Joel, 342
Harvard Underwater Sound Laboratory, 17
Harvey, John Wesley, 180, 185–86
Haver, Richard, 333
Hayden, Edgar, 46–47
Heaton, Beth, 377–78
Heaton, Bill, 376–78
Helena, USS, 341
Helium diving, 229, 230, 240–42
Heller, Joseph, 35
Helms, Richard, 217
Hendrix, Jimi, 180
Hensley, Jimmy, 34–37
Heyser, Richard S., 86–87
High frequency (HF), 27
High frequency direction finding (HFDF), 23,
  26–29, 41, 50–51, 355n
High-pressure nervous syndrome (HPNS), 240,
  242, 244–45
Hilarides, William H., 340
Hitchcock, Alfred, 220
Hodges, Duane, 203
Holser, Alex, 220
Holystone, ix, 273, 305, 375
Homestead Station, 78–82, 86, 105–6, 116,
  122–25, 134
Hotel, 193
"Hot running" torpedoes, 291
"Hot shit," 255
Houston, Brant, 364n
Houston, Joseph, 217–24, 227, 364n
Hubel, Augustine "Gus," 232
Hughes, Howard, 219
Hughes Glomar Explorer, 223–28
Hunt, John, 239–40, 245
Hunter, John, 159–64, 168
Hutchthausen, Peter, 357n
Hybla Valley Coast Guard Station, 44
Hydro Products, 222, 364n
Hyperbaric diving chambers, 230, 240, 242, 249,
  287–88

IDKCA ("rise to the surface"), 126
Ilyushin Il-28 bombers, 104, 176
Independence, USS, 172–73
Indigirki, 64
Intelligence operatives (I-Branchers), 29, 262
Interim Towed Array Surveillance System (ITASS),
  253–55, 272–73
Ionospheric hop, 87–88
Iran hostage crisis, 319, 376–77
Iraq, and Gulf War, 338

Itek Corporation, 217–24
ITT Federal Systems, 48, 189–90
Ivanov, P. K., 177
Ivy Bells, ix. *See also* Cable-tapping missions
Iwo Jima, 11

Jamming systems, 40–41, 50
Jauregui, Steve, 353n
Jefferson, Thomas, 135
Jimmy Carter, USS, 339
John Marshall, USS, 306
John S. McCain, USS, 345
Johnson, Lyndon, 203
Juliett-class submarines, 193
Jupiter missiles, 148, 156–57
Just a Sailor (Waterman), 296–97

K-3, 188
K-8, 188
K-19, 188, 343–44
K-129, 205–12
  Halibut search for, 216–17
  Project Azorian salvage plan, 217, 221–28
  rogue theory about, 207–8
  sinking of, 211–12
  Swordfish stalking of, 207–12
K-219, 344
K-324, vii, 304–17
  Drum collision with, 311–14
  Drum spotting of, 304–11
Kalugin, Oleg, 204–5
Kami Seya Naval Security Group Activity, 202–3,
  209–10, 255
Karamürsel Station, 22–23, 26–39, 49–50
Kaye, Jack, 78–79, 87, 102–6, 118, 130, 134–40,
  147, 188, 192
Kelley, Edward, 164
Kelly, Joseph, 16
Kennedy, John F.
  address to nation (October 22, 1962), 107, 115
  assassination of, 193–94
  briefing with Reed about Boresight, 134–39
  on crisis, 337
  deal with Khrushchev, 176
  ExComm meetings, 102, 118, 129–30
  invasion preparations, 147–48
  quarantine measures and, 102, 105, 106, 107,
    118, 125–26, 129
  secret memo to Khrushchev, 156–57
  Soviet offensive weapons and, 62, 63–64, 87,
    101–2, 104, 115, 120, 176
  Soviet submarines and, 102, 106, 129–30,
    134–35, 149, 150
  World War II and PT 109, 137–38
Kennedy, Robert, 117, 148
Ketov, Ryurik, 67–77, 93–95, 107, 171–74, 357n
Key West Naval Base, 115–16
KGB, 157, 204, 332
Khabarovsk Krai, 46
Khrushchev, Nikita, 343
  deal with Kennedy, 176
  halts advance, 148
  letters to Kennedy, 156–57
  nuclear arms in Cuba, 59–60, 64, 148
  submarine mission, 82, 89, 117, 118, 134
King, E. J., 367–68n
Kirby Morgan masks, 236, 287
Kissinger, Henry, 188

Kitchens, Billy, 235
Klimov, Yuri, 175–76
Knox, William, 134
Kobyakov, Lieutenant, 174–75
Kobzar, Vladimir, 205–7
Kodiak Station, 123
Kola Bay, 256–57
Kolosov, Igor, 358n
Komiles, 149
Komsomolsk, 317
Kongsberg, 294–95
Kopeikin, Arkadyi, 67, 74–76, 96–99, 110,
    174–75
Korean War, 2, 5, 42
Kresta II, 261–62
Kurdin, Igor, 357n
KW-7 devices, 200, 204–5, 208

Lacy, Gene, 202–3
Lane, James T., 266–67
Langaliero, Bobbi, 320
Langaliers, Don, 319–21, 324, 328, 330–31
Laning, R. B., 15
La Pérouse Strait, 2–3
Lapon, USS, 284
Lasky, Marvin, 253
LeJeune, Cheryl, 233–36, 289
LeJeune, David, 233–46
    Andrea Doria dive, 237–38
    background of, 233–35
    on Halibut, 233, 238–46
    on Ortolan, 237
    on Seawolf, 235, 249–50, 282, 283–84, 287–89,
        292
    training, 234–35, 237
Leninsky Komsomol, 193
Libbert, John, 40
Limiting depth, 162
Lindsay, Frank, 219
Lithium hydroxide, 245
Lock-out chambers (escape trunks), 230, 297–99,
    310–15, 340
Loman, Jeff, 358n
Long Range Acoustic Propagation Project
    (LRAPP), 253–55, 272
Lorenzen, Howard Otto, 40–44, 50
Los Angeles-class submarines, 339, 340–41, 344,
    375n
Los Angeles Times, 225–26, 227
Lott, Gus, 353n
Louisville, USS, 338
Lovelace, Linda, 251
Low Frequency Array (LOFAR), 18–19, 142
Low-frequency sonar, 16, 272–73, 375n
Lusby, Al, 290
Lyubimov, Engineer, 85–86

McCloy, USS, 317
McCone, John, 62, 63–64, 106, 148
Machiavelli, Niccolò, 229
Mack, Chester M. "Whitey," 284
McKee, Kinnaird R., 295
McMahon, Knight, 45
McNamara, Robert, 62, 102, 118, 125–30,
    136–39, 149, 150, 188–91
McNish, Jack, 230, 246, 248
McVain, Charlie, 285–87, 289
Magerøya, 261

Magnetic anomaly detection (MAD), 163
Maintenance personnel (M-Branchers), 29
Malinovsky, Rodion, 58–61
Manganese nodules, 218–19, 224
Marconi, Guglielmo, 44
Mare Island Naval Shipyard, 338
Marianas Trench, 185
Maria Ulyanova, 59
Marine Technology Society, 222
Mark 11 saturation diving suits, 239–40, 242–45
Martell, Charles B., 253
Maslennikov, Ivan, 155
Mason, Frank, 25, 31–33, 37–39, 49, 50
Masset Station, 197–98
Maurer, John H., 291–92
Mayans, viii
Mignon, Tony, 320, 323, 327–29
Mikoyan, Anastas, 120
Military grade ("mil spec"), 232–33
Miller, Charlie, 284–85
Mims, Norman, Jr., 266, 268, 275–81
Misner, Robert, 42–44, 47
MK-11 diving suits, 239–40, 242–45
MK-113 system, 271–72
Mobile Submarine Simulators (MOSS), 320
Mondrian, Pieter, 220
Moore, C. Edward, 216
Moorman, Dave, 323
Mueser, Roland, 16–19
Multiangulation, 28, 30, 45, 49, 138
Mumma, Albert G., 351n
Murray, USS, 150
Mystic, 229–30

Naftali, Timothy, 157
Napier, Russ, 159
National Photographic Interpretation Center,
    86–87
National Press Club, 157
National Security Council (NSC), 118
National Security Operations Center (NSOC), 62
Naumov, Sergei, 75, 85, 97, 108–9, 113, 166
Nautilus, USS, 8–10, 13–14, 15, 20–21, 62–63,
    180, 193, 231, 351n
Naval Electronics Laboratory, 185
Naval Reactors Branch, 9–10
Naval Research Laboratory (NRL), 40–45, 50,
    145, 147, 191, 253
Naval Reserve Officers Training Corps (NROTC),
    11
Naval Scientific and Technical Intelligence Center
    (NAVSTIC), 20
Naval Security Group (NSG), 22, 146, 232
Navy Expeditionary Medal, 9, 338
Navy SEALs, 234, 296–99, 338–39, 340
Nea Makri Station, 212–13
Nelson, Elroy, 163, 168
Net Control (NC), 29–30
New York Herald Tribune, 157
New York Times, 227
New York Yankees, 100
Nicholson, Jack, 264
Niebuhr, Reinhold, 318
Nissho Maru, 316–17
Nitrogen narcosis, 229
Nitze, Paul, 216
Nixon, Richard, 80, 217, 225, 352n
Noisemaker torpedoes, 163, 164

Norfolk Naval Communications Area Master
  Station, 199–200
North American Aerospace Defense Command
  (NORAD), 191
North Korea, and USS *Pueblo,* 200–205
North Pole, 15, 193
Novaya Zemlya, 62–63, 139, 231
November-class subs, 188, 193
Nuclear Non-Proliferation Treaty, 277
Nuclear-powered submarines, 8–10, 13, 14,
  20–21, 180

Odell, Carl, 81–82, 141–45
Office of Collection and Signals Analysis, 40–44
Office of Undersea Warfare, 215
"Off-line," 255
Ohio, USS, 340
Okinawa Bulls Eye, 48–49
Oliver, Michael, 295–96, 302, 307–16
Omnidirectional sound viewpoint, 142, 253–54
One Flew Over the Cuckoo's Nest (movie), 264
One Hell of a Gamble (Fursenko and Naftali),
  157
Onslow Beach, North Carolina, 176
OP-20-G (Office of Chief of Naval Operations),
  45
Operational Reactor Safeguard Exam (ORSE), 304
Operation Anadyr, 58–61. *See also* Operation
  Kama
Operation Eagle Claw, 319
Operation Falling Leaves, 118
Operation ICEX, 254
Operation Kama, 57–77, 82–86, 88–100, 107–15,
  118–21, 130–34, 148, 149–55, 157–78.
  See also B-4; B-36; B-59; B-130
Operation Paperclip, 46
Operation Sand Dollar, 216
Operations specialists (O-Branchers), 29
Orel, Vice Admiral, 66
Orestes, 200
Orlov, Pavel, 150–55
Ortolan, USS, 237
Orwell, George, 332
OSNAZ, 77, 94, 121, 133
Oxford, USS, 61–62, 63, 86, 101, 106

Packard, David, 221
Palm Beach International Airport, 115
Pancho Villa, 265
Pankov, Lieutenant
  avoiding detection, 108–15, 161, 170–71
  operation onboard, 82–84, 90–91
Parangosky, John, 217, 220–23
Parche, USS, 289–93
  cable-tapping missions, 248–49, 288, 291–93,
    319–22, 333–34, 338, 372–74n
Parshin, Viktor, 131–33
Pasha (cat), 72–73
Passive sonar systems, 16–21, 272–73
Patton, George, 188
PCS-1380, 139–40
Pearlman, Stanley, 357n
Pearson, J. W., 209
Pelton, Ronald, 249, 332–34
Pennsylvania State University's Ordnance Research
  Laboratory, 16–17
Periscopes, 270–71, 299–300
Peri-Viz (Periscope Visual), 258, 259–60

Permit-class submarines, 254–55, 266, 375n
Petersen, Walter H., 30–31, 33, 37, 49, 334–36,
  342
Petropavlovsk Naval Base, 2, 228
Philadelphia Naval Shipyard, 10
Photographic intelligence (PHOTINT), on the USS
  Blenny, 4–6
Piccard, Auguste, 181
Pittsburgh, USS, 338
Plato, 40
Plato Sea Mount, 213
Pliyev, Issa, 59
Polaris missile program, 189, 258
Pollack, USS, 11
Pomilyev, Alexander, 66–67, 88–89, 167–70
Popov, Vyacheslav, 344
Portsmouth Naval Shipyard, 185
Position-keeper (PK), 271–72
Post, Don "Doc," 328
Potapov, Lieutenant, 121, 166, 174–75
Preserver, USS, 180–84, 186–87
Price Waterhouse, 190
Project 627, 20–21
Project 629A, 205. *See also* K-129
Project 641, 66, 68, 71, 83. *See also* B-4; B-36;
  B-59; B-130
Project 667B, 257–58
Project Azorian, 217, 219–28, 364–65n
Project Boresight, 47–51, 188–91
  calibration and signal analysis, 50–51, 53
  installing and testing, 49–50, 81–82, 87
  McNamara and, 127–28, 137, 138, 188
  NSA funding and naming, 44, 47–48
  refinements and expansion, 118, 141–47, 149,
    189–94
  success of, 179
Project Bulls Eye, ix, 44, 45, 47–49, 188–91,
  197–98
Project Colossus, 17
Project Corona, 217–18
Project Jezebel, 16–17
Prokov, Johnny, 157
Pronin, Vladimir, 93–95
Propeller designs, 17, 351–52n
Proust, Marcel, 57
PT 109, 137–38
Pueblo, USS, 200–205
Puget Sound Naval Shipyard, 338
Pump seals, 14–15

Quantity vs. quality of submarines, 187–88
Quittner, Arnold, 139–40

Radar-absorbing material (RAM), 270
Radiation shielding, 14–15
Radio-Countermeasures Sound Recorder-
  Reproducer (IC/VRT-7), 42–43
Radio direction finding (RDF), 44–45
Radio Liberty, 120
Radio Moscow, 148
Radio operators (R-Branchers), 28–29, 50, 200,
  232, 262
RAF Chicksands, 47
Randolph, USS, 107, 120, 150, 152–53, 159
Reagan, Ronald, 303, 317, 333–34
Red November, 20–21
Reed, Joyce Louise, 23–24, 26
Reed, Lon, 32

Reed, Pamela Wallinger, 24, 26, 34
Reed, W. Craig
  on the *Drum,* 295–317
  at Eastman Kodak Company, 318–19
  on the *Haddo,* 265–81
Reed, William J., 22–44
  on the *Birinci Inonu,* 52–56
  Boresight system refinement and expansion,
    118, 141–47, 149, 189–94, 194–97
  at Fort George G. Meade, 102–6, 130
  Greek incident, 194–97
  at Homestead Station, 78–82, 122, 123
  at Karamürsel Station, 22–23, 26–39, 49–50,
    342
  Kennedy assassination and, 193–94
  at Masset Station, 197–98
  at NSA's Office of Collection and Signals
    Analysis, 40–44
  promotion to A22 head of field operations, 192
  retirement of, 264–65
  at Sanders Associates, 141–46, 189–90
  Scratchy and, 23–26, 34
  at Skaggs Island Station, 146–47, 149
  on success of Boresight program, 179
  at Vadsø Station, 192–94
  at White House for Boresight briefing, 134–40,
    188, 356*n*
Richard B. Russell, USS, 318, 334–36, 338
Rickover, Hyman G., 8–10, 14
Rigsbee, John, 208–9
Rindfleisch, Hans, 46
Robert E. Lee, USS, 116
Rodocker, Don, 237–38
Rogers, Warren, 157
Room 40, 44–45
Roosevelt, Franklin, 136
Roosevelt, Theodore, 136, 368*n*
Rorke's Drift, 343
Ross, Donald, 16–20, 351–52*n*
Rossokho, Anatoly, 70–71
Rounds, H. J., 44–45
Rozier, Charles, 159, 161–64
Rule, James, 319–22, 324, 328–31, 372*n*
Rules of engagement, 70–71, 76, 93, 102, 128,
  129, 150, 152
Rusk, Dean, 62, 102, 148
Rutherford, Mark, 231–33, 247–48, 249, 256,
  261, 263, 283–86
Rybachiy Naval Base, 205
Rybalko, Galena, 61
Rybalko, Leonid, 57–61, 68, 70–71, 82
Rybalko, Natasha, 61

Safety of submarines, 187–88
Safford, Laurance F., 45
Sanders, Royden, Jr., 141–42
Sanders Associates, 141–46, 189–90, 209
San Diego Naval Training Center, 265
San Francisco Giants, 100
Santa Fe Springs High School, 234
Saparov, V. G., 75, 84–86, 96, 99, 110–14, 164,
  166–67
Sargasso Sea, 61, 65, 93–100, 108
Satellite communications, 341–42
Saturation diving, 229–30, 237–51
  training, 237, 242–45
Saturday Night Live (TV program), 265
Savitsky, Vitali, 70–77, 107, 125, 149–56, 158–59

Saxon, Ross "Zipperhead," 233, 237, 238
Sayda Bay, 65–68, 88, 115, 176
SBD radios, 61, 93–94, 133–34, 206, 357*n*
SC-35, 201–2
Scali, John, 147–48, 156
Schade, Arnold F., 212
Schlesinger, James R., 227
Scorpion, USS, 62–63, 212–14, 235, 279–80
Scratchy (bear), 23–26, 34
Seadragon, USS, 180
SeaLab, 229
SEALs, 234, 296–99, 338–39, 340
Sea of Okhotsk, cable-tapping missions, 228–31,
  233–34, 238, 245–51, 287–89, 291–93, 319,
  324–27, 332–36
Sea Robin, USS, 180
Sea Scope, 223
Seawolf, USS, 13–15, 282–89, 291, 319–31
  cable-tapping missions, 248–51, 287–89,
    324–27
  sand-stuck ordeal, 325–31
Seawolf-class submarines, 13–14, 193, 338–39,
  375*n*
Sequoia, USS, 236
Sevastopol, 11–12, 52
Sevastopol, 259–60
Shackleton aircraft, 85–86
Shaddock, 193
Shakespeare, William, 282
Shchuka, 301
Sheets, Mack, 354*n*
Ships Inertial Navigation System (SINS), 328
Ship submersible nuclear (SSN), 8–10
Shkval, 86, 115
Shot lines, 158
Shumkov, Nikolai, 60, 65, 69–77, 107, 130–34,
  164–68, 172, 175
Signal intelligence (SIGINT), 5
Signal-to-Noise Enhancement Program (SNEP),
  353*n*
Situation report (SITREP), 117
Sizov, F. Ya., 177
Skaggs Island Station, 48–49, 87–88, 105–6,
  146–47, 149, 189
Skean, 87
Skifter, Hector, 354*n*
Skillas, Charles, 141–47, 357*n*
Skipjack, USS, 11, 21
Skory, 54–56
Skylark, USS, 185–86
Skywave propagation, 27
Slattery, Francis Atwood, 212–13
Slaughter, Gary, 157–59
Smith, Cleveland, 244–45
Smith, Dennis, 289–93, 372–74*n*
Smith, Edward John, 287
Smith, Gaines "Whirly," 12
Smith, Nihil, 180–87
Solomatin, Boris A., 204–5
Sonar arrays, 16–17, 253–55, 272–73
Sonar systems, 16–21, 208, 254, 271–74, 375*n*
Sonar transducers, 167
Sonobuoys, 111, 142–43
Sonographs, 34, 35–37
Sontag, Sherry, ix
SOSUS (Sound Surveillance System), 16–17, 103,
  117, 253–54
Sound suppression, 274, 370*n*

Special Warning 32, 126
Sperry, USS, 301–2
"Spot report," 29
SS-N-14, 261–63
Stalin, Joseph, 58, 59, 122
Stanford Research Institute, 47
Stone, John, 44
Stonewall Jackson, USS, 377
Straub, Herman, 17–18, 20
Stringfellow, Bill, 369n
Sturgeon-class submarines, 254–55, 295, 334, 375n
Subic Bay, 265–66
Submarine Development Group (DevGru) One, 289–90
Submarine Towed Array Surveillance/Sonar System (STASS), 272–73
SUBSAFE, 187
Sullivan, David "Whompee Jaw," 287–89
Surface Tethered Oceanographic Vehicle Experimental (STOVEs), 289–90
Surface-to-air missile (SAM), 63–64
Suzuki, Zenko, 317
Sweeney, Walter, 102
Swimmer Delivery Vehicles (SDVs), 339–40
Swordfish, USS, 207–12, 215, 255, 283
System 2090, 290–91, 292–93

Tapping of cables. See Cable-tapping missions
Target motion analysis (TMA), 271
Technical Extracts of Traffic Analysis (TEXTA), 101
Technical specialists (T-Branchers), 29, 101, 123, 209, 211, 255, 262, 284–85, 307
Terek, 86
Test depth, 6–8, 56, 268, 286
Thermal imaging, 342
Thermal layers, 6, 56, 108, 121, 165
Thermoluminescent devices (TLDs), 240
Thetis Bay, USS, 175
Thresher, USS, 180–87
    demise of, 185–86, 187
    recovery of parts, 183, 184
    search for, 180–87
Thresher-class submarines, 180, 254–55
Tiernan, Michael C., 321–22, 324–31
Titanic, 287
Tomahawk Land Attack Missiles (TLAMs), 338, 340, 341
Tomlin, Edwin Ladeau, 269, 275–78, 281
Toshiba, 294–95
Tosi, Alessandro, 44
Trejo, Paul, 1–9
Trident missiles, 340
Trieste, USS, 181–83, 185–87
Trigger, USS, 184
Tsushima Strait, 200–201
Tuell, Allen, 168
Turban, Frank, 290
    on Flying Fish, 255–63
    on Seawolf, 284–87
    on Swordfish, 209–11
Turkey, Jupiter missiles in, 148, 156–57
Turks Island Passage, 97–98, 107, 118–19
TUSLOG Detachment 28, 22–23, 26–39, 49–50
Twain, Mark, 101, 141

Twenty-ninth Ballistic Missile Division, 205
Twin-screw designs, 17, 351–52n
Tyomin, Abram, 95

Underdog, 210–11
Under-hull runs, 257–58
Underwater Demolition Team (UDT) training, 234–35
Underwater photography, 217–22
Unidirectional hearing, 142
Union College, 13

V-5 program, 11
V-12 program, 11
Vadsø Station, 192–94
Varankin, Sergeant, 148–49
Velucci, Chris, 237–38
Victor-class submarines, 294–95, 296
Victor III, 294–95, 296, 301, 303–14
Vietnam War, 264
Viking 1, 218
Vindetto, Bob "Ginny," 239–40, 244–45
Vinson, Tim, 266–70, 274–81
Virginia, USS, 339
Vladivostok, 2–3, 210, 230, 276, 301, 303
Vladivostok Higher Naval School, 65
Voice of America, 99–100, 120
Von Braun, Wernher, 46
Von Hipper, Franz, 45

Wald, Bruce, 191
Walker, John, 199–200, 204–5, 214
Walnut Line, 106–7, 116–17, 125, 130, 149
Walski, Joe, 184–86, 187
Walter, Helmuth, 351n
Ward, Alfred, 147, 150
Watchstanders, 240–41
Waterman, Steve, 296–97
Water reserves, 165
Westinghouse Oceanic Division, 290–91
Whiff, 61–62
White House, 135–36
Wide Aperture Receiving System (WARS), 47
Wieghorst, Olaf, 265
Wilkinson, Eugene, 351–52n
Williams, J. D., 255–63
Will Rogers, USS, 252
Woelk, Steve, 203
Wolfe, John, 218–20
Working depth, 162, 167
World Series (1962), 100
World War I, 44–45
World War II, 9, 11, 40–41, 45–46
    Kennedy and PT 109, 137–38
Wullenweber, 41, 43–47, 188–91
Wullenweber, Jurgen, 46, 355n
Wundt, Rolf, 46

Yankee-class submarines, 257–58, 261–63
Yoshihara, Toshi, 345
Yurchenko, Vitaly S., 332–33

Zarnakov, Alexander, 206
Zellmer, Ernest "Zeke," 217, 220
Zhukov, Yuri, 75, 91–92, 95–100, 107, 113–14, 119–21, 160–61, 169–71
Zulu War, 343